Computational Intelligence and Modelling Techniques for Disease Detection in Mammogram Images

Computational Intelligence and Modelling Techniques for Disease Detection in Mammogram Images

Edited by

D. Jude Hemanth

Professor, ECE Department,
Karunya Institute of Technology and Sciences,
Coimbatore, India

ELSEVIER

ACADEMIC PRESS

An imprint of Elsevier

Academic Press is an imprint of Elsevier
125 London Wall, London EC2Y 5AS, United Kingdom
525 B Street, Suite 1650, San Diego, CA 92101, United States
50 Hampshire Street, 5th Floor, Cambridge, MA 02139, United States
The Boulevard, Langford Lane, Kidlington, Oxford OX5 1GB, United Kingdom

Notices
Knowledge and best practice in this field are constantly changing. As new research and
experience broaden our understanding, changes in research methods, professional
practices, or medical treatment may become necessary.

Practitioners and researchers must always rely on their own experience and knowledge in
evaluating and using any information, methods, compounds, or experiments described
herein. In using such information or methods they should be mindful of their own safety
and the safety of others, including parties for whom they have a professional responsibility.

To the fullest extent of the law, neither the Publisher nor the authors, contributors, or
editors, assume any liability for any injury and/or damage to persons or property as a
matter of products liability, negligence or otherwise, or from any use or operation of any
methods, products, instructions, or ideas contained in the material herein.

ISBN: 978-0-443-13999-4

For information on all Academic Press publications visit our website at
https://www.elsevier.com/books-and-journals

Publisher: Mara Conner
Acquisitions Editor: Carrie Bolger
Editorial Project Manager: Emily Thomson
Production Project Manager: Erragounta Saibabu Rao
Cover Designer: Vicky Pearson Esser

Typeset by TNQ Technologies

Contents

Karthikeyan Velayuthapandian, Gopalakrishnan Karuppiah,
Sridhar Raj Sankara Vadivel and Dani Reagan Vivek Joseph

Contributors

Prerna Ajmani
Vivekananda Institute of Professional Studies-TC, GGSIPU, New Delhi, India

M. Balamurugan
Department of Mathematics, Vel Tech Rangarajan, Dr. Sagunthala R&D Institute of Science and Technology, Chennai, Tamil Nadu, India

Ashima Bhatnagar Bhatia
Vivekananda Institute of Professional Studies, New Delhi, India

D. Bini
Robotics Engineering, Karunya Institute of Technology and Sciences, Coimbatore, Tamil Nadu, India

Daniela Danciulescu
Department of Computer Science, University of Craiova, Craiova, Romania

Dharani
Dr. N.G.P. Institute of Technology, Coimbatore, Tamil Nadu, India

Anurag Dutta
Department of Computer Science and Engineering, Government College of Engineering and Textile Technology, Serampore, Kolkata, India

V. Edward Naveen
Department of CSE, Sri Shakthi Institute of Engineering and Technology, Coimbatore, Tamil Nadu, India

S.P. Girija
Department of Electronics and Communication Engineering, Kakatiya Institute of Technology and Science, Warangal, Telangana, India

Harini
Dr. N.G.P. Institute of Technology, Coimbatore, Tamil Nadu, India

John Harshith
Department of Computer Science and Engineering, Vellore Institute of Technology, Vellore, Tamil Nadu, India

A. Jenefa
Department of CSE, Karunya Institute of Technology and Sciences, Coimbatore, Tamil Nadu, India

Dani Reagan Vivek Joseph
Mepco Schlenk Engineering College (Autonomous), Sivakasi, Tamil Nadu, India

Anitha Jude
Karunya Institute of Science and Technology, Coimbatore, Tamil Nadu, India

R. Julia
Electronics and Communication Engineering, Karunya Institute of Technology and Sciences, Coimbatore, Tamil Nadu, India

Sheetal Kalra
School of Physiotherapy, Delhi Pharmaceutical Sciences and Research University (DPSRU), New Delhi, India

Gopalakrishnan Karuppiah
Mepco Schlenk Engineering College (Autonomous), Sivakasi, Tamil Nadu, India

Shama Kouser
Department of Computer Science, Jazan University, Saudi Arabia

A. Lincy
Department of CSE, National Engineering College, Kovilpatti, Tamil Nadu, India

Vijayalakshmi G.V. Mahesh
Department of Electronics and Communication Engineering, BMS Institute of Technology and Management, Bangalore, Karnataka, India

R. Manikandan
School of Computing, SASTRA Deemed University, Thanjavur, Tamil Nadu, India

Muthu Manoj
Dr. N.G.P. Institute of Technology, Coimbatore, Tamil Nadu, India

Sindhu P. Menon
School of Computing and Information Technology, Reva University, Bengaluru, Karnataka, India

Rahul Reddy Nadikattu
University of the Cumberland, Williamsburg, KY, United States

Pramodkumar Naik
Department of Computer Science & Engineering, Dayananda Sagar University, Bengaluru, Karnataka, India

R. Deiva Nayagam
Department of Electronics and Communication Engineering, Ramco Institute of Technology, Rajapalayaam, Tamil Nadu, India

Parita Oza
Institute of Technology, Nirma University, Ahmedabad, Gujarat, India

Shajin Prince
Electronics and Communication Engineering, Karunya Institute of Technology and Sciences, Coimbatore, Tamil Nadu, India

Ganta Raghotham Reddy
Department of Electronics and Communication Engineering, Kakatiya Institute of Technology and Science, Warangal, Telangana, India

Richa Hirendra Rai
School of Physiotherapy, Delhi Pharmaceutical Sciences and Research University (DPSRU), New Delhi, India

A. Stephan Antony Raj
Department of Mathematics, THE TIPSGLOBAL INSTITUTE, Coimbatore, Tamil Nadu, India

A. Ramamoorthy
Department of Mathematics, Velammal Engineering College, Anna University, Chennai, Tamil Nadu, India

Kama Ramudu
Department of Electronics and Communication Engineering, Kakatiya Institute of Technology and Science, Warangal, Telangana, India

S.R. Reeja
VIT-AP University, Amaravati, Andhra Pradesh, India

Revathi
Dr. N.G.P. Institute of Technology, Coimbatore, Tamil Nadu, India

Sridhar Raj Sankara Vadivel
Mepco Schlenk Engineering College (Autonomous), Sivakasi, Tamil Nadu, India

B. Satheesh Kumar
Department of Computer Science and Engineering, School of Computing Science and Engineering, Galgotias University, Greater Noida, Uttar Pradesh, India

D. Selvathi
Department of Bio Medical Engineering, Mepco Schlenk Engineering College, Sivakasi, Tamil Nadu, India

Vandana Sharma
Computer Science Department, CHRIST (Deemed to be University), Delhi-NCR, India

Sharmila
Dr. N.G.P. Institute of Technology, Coimbatore, Tamil Nadu, India

Ramya Srikanteshwara
Department of Computer Science & Engineering, Nitte Meenakshi Institute of Technology, Bengaluru, Karnataka, India

Tulasi Thotakura
VIT-AP University, Amaravati, Andhra Pradesh, India

Karthikeyan Velayuthapandian
Mepco Schlenk Engineering College (Autonomous), Sivakasi, Tamil Nadu, India;
Department of ECE

J. Vijayaraj
Department of Artificial Intelligence and Data Science, Easwari Engineering
College, Chennai, Tamil Nadu, India

D. Vinodhini
Department of Mathematics, Amrita School of Agricultural Sciences, Amrita
Vishwa Vidyapeetham, Coimbatore, Tamil Nadu, India

Pawan Whig
Vivekananda Institute of Professional Studies, New Delhi, India

Ishfaq Yaseen
Prince Sattam Bin Abdul Aziz University, Al-Kharj, Saudi Arabia

Preface

Breast images are widely used to detect the different types of abnormalities associated with the breast. The medical sector is continuously exploring new ways to detect the different types of diseases with high accuracy. The improvements are visible in the scanning methods and the computational methods used for analyzing the scan images in the computer. Irrespective of the scanning methods, several efforts have been made to integrate Artificial Intelligence (AI)-based computational approaches to improve the success rate of the detection process. However, the scope for improvements still exists, which is the main motivation behind the contents of this book. This book provides novel technological concepts that can enhance the practical feasibility of the proposed subject. Apart from mammography, which is the main focus of this work, a few different modalities such as ultrasound imaging, molecular imaging, and histopathological image-based detection methods are also covered in this book. This adds more weight to the book. A wide spectrum of applications is covered in this book which will create an interest among biomedical researchers and computer scientists working in this field. This book, indeed, is a comprehensive product which will help readers to grasp the extensive point of view and the essence of the recent advances in this field. A brief introduction about each chapter follows.

Chapter 1 is an introduction chapter on the different scanning methodologies with special emphasis on mammography. It provides an overall idea on the current scenario in this field. The background information and the challenges associated in capturing high-resolution images are dealt in this chapter. The future trends of breast imaging are also discussed in this chapter. Chapter 2 deals with a brief introduction on AI approaches which are used to detect the diseases in breast images. The various difficulties associated with the current approaches are explored in detail. The future scope of AI-based disease detection in breast images is also dealt in detail. Chapter 3 deals with the Machine Learning (ML)-based approach for breast cancer detection. Random forest classifier is an ML method used in this work for disease detection. An exhaustive performance analysis is given to justify the efficiency of the proposed approach.

Chapter 4 deals with the mass detection process in mammogram images using Deep Learning (DL) approaches. YOLO is the prime DL-based method used in this work. The analysis is done in such a way to prove that DL-based approaches are efficient for the early detection of the diseases. Chapter 5 explores the possibility of application of pretrained DL models for disease detection in mammogram images. MobileNet and LSTM-based methods are used in this work for the experimental analysis. A comparative analysis is also given which validates the superior nature of the proposed approach. Breast image dimensions are usually a stumbling block to achieve results within a quick time. This problem is tackled in Chapter 6. An autoencoder-based approach is specifically framed to reduce the dimensions of the images without loss of accuracy. A basic classification approach is also implemented in this work to prove the efficiency of the autoencoder-based dimensionality reduction approach.

Chapter 7 deals with ML approaches and mastography images for the detection of cancer prognosis. Several ML approaches are used in this work to highlight the best possible method for efficient disease detection. Chapter 8 illustrates the process of microcalcification detection in breast images using ML approaches. Mammogram images along with decision trees and Support Vector Machines (SVMs) are used in this work. Convolutional Neural Network (CNN)-based implementation is also carried out for abnormality detection in breast images. Chapter 9 explores the application of pretrained models such as ResNet, AlexNet, etc. for cancer detection in mammogram images. The analysis is performed in a detailed manner to show the results of various stages of the implementation process.

An analysis of the different types of modalities for cancer detection in breast is given in Chapter 10. The pros and cons of the different modalities from the cancer detection perception are given in this chapter. The different stages of the cancer detection process are also dealt in detail. Deep Belief Network is the focus of chapter 11 in which ultrasound breast images are used for the experiments. An extensive analysis is also carried out to prove the need for DL approaches in breast cancer detection. Histopathological image-based disease detection in breast is the focal point of Chapter 12. Different transfer learning-based models are deployed to analyze the images for disease detection. ML model-based disease detection in breast images is carried out in Chapter 13. Histopathological images are used in this work. Chapter 14 deals with the fuzzy logic-based tumor detection in breast images. Mammogram images are used in this work for the experiments.

I am thankful to the contributors and reviewers for their excellent contributions to this book. My special thanks to Elsevier, especially to Ms. Carrie Bolger (Acquisition Editor), for the excellent collaboration. Finally, I would like to thank Ms. Emily Thomson who coordinated the entire proceedings. This edited book covers the fundamental concepts and application areas in detail. Being an interdisciplinary book, I hope it will be useful for both health professionals and computer scientists.

Dr. D Jude Hemanth
April, 2023

Mammogram data analysis: Trends, challenges, and future directions

1

Karthikeyan Velayuthapandian[1,2], **Gopalakrishnan Karuppiah**[1],
Sridhar Raj Sankara Vadivel[1] **and Dani Reagan Vivek Joseph**[1]

[1]*Mepco Schlenk Engineering College (Autonomous), Sivakasi, Tamil Nadu, India;* [2]*Department of ECE*

1. Introduction

Cancer is a collection of disorders in which the body's cells cultivate out of control past their normal borders and attack various bodily regions. In accordance with the International Agency for Research on Cancer, there will likely be over 1.2 billion instances of cancer by the year 2050, making it the second biggest cause of mortality globally [1]. Based on the "2020 World Cancer Assessment by the International Agency for Research on Cancer (IARC)," cancer is the leading or main mortality root for those aged 30−69 in 134 out of 183 nations. Males and females both die from lung cancer, but men are more likely to get prostate cancer, while women are more probable to develop breast cancer (BC). Agreeing to IARC, the tumor rate will increase from 14.1 million to 30.5 million between 2020 and 2050, while the amount of mortality will rise from 9.6 million to 16.4 million (Fig. 1.1). Cancer is brought on by a variety of intrinsic (like age-associated, gene vulnerability, hormones, inflammation, etc.) and extrinsic causes (like environmental influences, radioactivity, and way of life) [2]. Carcinogenesis, oncogenesis, or tumorigenesis is a multistage process that involves chronic proliferation, stimulation of oncogenes, mutation of tumor suppressor genes, tolerance to cell death, metastasis, and invasion immune cell denial, plus metabolic pathway remodeling [3,4].

Tumors can be categorized into various stages based on the anatomic severity of the condition. Tumor staging is crucial for prognosis, therapy planning, and treatment evaluation. One of the biggest obstacles to effective cancer therapy, even with the right tools for tumor categorization and staging, is early detection and tracking of the treatment's effectiveness [6]. BC ranks as one of the general prevalent and dangerous tumors affecting females overall. Conferring to Fig. 1.2, BC accounts for 14% of all malignancies worldwide, is the most common disease among women, and has a high death and morbidity rate. It affects roughly 3 million women annually

Computational Intelligence and Modelling Techniques for Disease Detection in Mammogram Images
https://doi.org/10.1016/B978-0-443-13999-4.00016-X

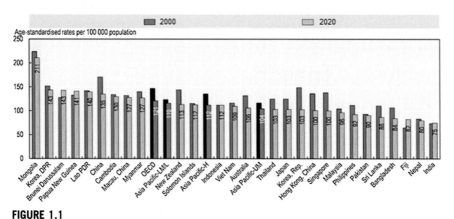

FIGURE 1.1

Cancer registration history.

Courtesy of WHO Global Burden of Disease, 2020; Department of Health, Hong Kong, China, 2020; Disease Registry, Macau, China, 2020.

Estimated New Cases				Estimated Deaths		
Females				**Females**		
	Breast	287,850	31%	Lung & bronchus	61,360	21%
	Lung & bronchus	118,830	13%	Breast	43,250	15%
	Colon & rectum	70,340	8%	Colon & rectum	24,180	8%
	Uterine corpus	65,950	7%	Pancreas	23,860	8%
	Melanoma of the skin	42,600	5%	Ovary	12,810	4%
	Non-Hodgkin lymphoma	36,350	4%	Uterine corpus	12,550	4%
	Thyroid	31,940	3%	Liver & intrahepatic bile duct	10,100	4%
	Pancreas	29,240	3%	Leukemia	9,980	3%
	Kidney & renal pelvis	28,710	3%	Non-Hodgkin lymphoma	8,550	3%
	Leukemia	24,840	3%	Brain & other nervous system	7,570	3%
	All Sites	**934,870**	**100%**	**All Sites**	**287,270**	**100%**

FIGURE 1.2

Ten common cancer classes in the United States in 2022.

Courtesy of Rebecca L, Siegel M, Kimberly D, Miller MPH, Ahmedin Jemal DVM. Cancer statistics. CA: A Cancer Journal for Clinicians 2017;67(27):7–30. https://doi.org/10.3322/caac.21708.

and raises the mortality rate for women. According to estimates, 685,000 women will pass away in BC in 2020.

The BC spread stages in four major Indian smart cities are mentioned in Table 1.1.

Changing lifestyles, postponing weddings and having children, working long hours, and using hormone replacement therapy are major contributors to the rising occurrence of BC in industrialized nations [8,9]. The fundamental causes of high BC rates and deaths in developing countries include insufficient medical arrangement,

Table 1.1 BC patient's stage-wise percentage in four Indian smart cities.

City	Stage of BC			
	I	II	III	IV
Lucknow	3.8	33.2	47.3	8.7
Chennai	1.2	29.2	52.1	23.8
Thiruvananthapuram	4.4	42.3	40.5	12.8
Mumbai	7.8	57.4	28.9	5.9

Source from Indian Medical Council for Medical Research, New Delhi.

inadequate mammogram screening, delayed identification, and scarce public-health education [10,11]. Fig. 1.3 illustrates the variety of treatments that are available to treat breast cancer [12,13].

On the contrary side, earlier tumor detection can increase the percentage of tumor patients who survive. As a result, there has been a trend toward the creation of technologies for the timely observation of tumors. The following is a comprehensive technological and theoretical background examination.

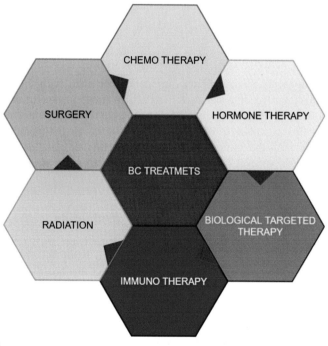

FIGURE 1.3

BC treatments.

1.1 Theoretical background

1.1.1 "Sick lobe" model

This idea of a "sick lobe." Breast carcinoma is a lobar condition that begins early in an individual's development when cells in one (or, extremely rarely, both) breast lobes gain early genetic defects that make this "sick lobe" more susceptible to the action of unpleasant stimuli. As a result, cells in the ill lobe of the mammary collect added genetic alterations than cells in the other, less vulnerable lobes of the same mammary. Over the period of a patient's lifetime, the aggregation of abnormalities might cause the tumorigenesis of these units [14−17].

1.1.2 Neoductgenesis

Despite the fact that histological analysis of most carcinoma in situ (CIS) cases indicates an approximate number of ducts and terminal duct lobular units (TDLUs), the implicated ducts are significantly dilated and deformed due to the presence of tumor tissue and its byproducts. In contrast, some forms of CIS are characterized by an abnormally high density of duct-like patterns per unit surface area. The accumulation of tumor cells "herniates" the duct's membrane to generate supplementary duct-like formations, and this phenomenon likely arises because of a blockage or disruption of the biological process of the alveolar flip, which would otherwise lead to the production of healthy TDLUs. Neoductgenesis is the term for this process, which typically includes the presence of calcifications [18−20].

1.2 Technical knowledge

Microcalcifications are so small that they cannot be seen with the human eye. Because of this, the surgeon needs radiological help at every stage of the diagnostic process for instances of calcification. To assist the physician in detaching nonpalpable tumors with acceptable restrictions, it is necessary to do preoperative radiological identification of the tumor using lead connectors or another method. The subsequent stage is specimen radiography, which directs the surgeon to take an appropriate sample from the tumor [21,22].

1.3 BC diagnosis using several imaging modalities

Screening decisions can be informed by risk assessments provided by breast cancer risk models. Multiple imaging techniques are being developed to detect this illness at its earliest stages. The identification and diagnosis of BC disorders rely heavily on medical pictures. Mammography, ultrasound, histopathology, magnetic resonance imaging (MRI), computed tomography (CT), positron emission tomography (PET), and thermography are the most common imaging tools used to diagnose BC nowadays. The various imaging techniques are all important in their own ways.

1.3.1 Mammography

When looking for BC in asymptomatic women, mammography is the bright mechanism. For initial and precise recognition of BC, nothing beats mammography. Breast X-rays are called mammograms. For breast screening and evaluation, it uses a small X-ray [23]. Mammography has done a much better job of finding microcalcifications and agglomerations of calcifications than any other method [24,25]. Screening images from mammograms have been shown to be technically superior, and hence they can be used in clinical assessment [18,26]. In addition, "Digital Tomosynthesis Mammography," a modification of the original Mammography method, now offers 3D BC images. In a new development in mammography, an iodinated substance is infused intravenously during the mammogram itself, a process known as "Contrast-Enhanced Digital Mammography" (CEDM) (Fig. 1.4) [27].

1.3.2 Ultrasound

Images obtained using ultrasound, in contrast to those obtained through other forms of computed tomography, are often black and white and have a low image quality. In most cases, ultrasound pictures of malignant regions will have an uneven shape, a blurry appearance, and unclear margins. Mammography is unable to differentiate lesions from solid masses, although ultrasonography can effectively do so [28−30].

1.3.3 MRI

MRI is the most perceptive scheme presently accessible for diagnosing BC. MRI is able to clearly show the size, form, and position of BC lesions because it supports multiplanar scanning and 3D reconstruction methods [31]. MRI scans, however, are expensive and take a long time. In addition to mammography, MRI is the principal diagnosing technique for BC.

1.3.4 Histopathology

Histopathology is the benchmark for diagnosing BC, even above [32] other forms of medical radiography. They are necessary for the identification and therapy of tumor-related illnesses because of the phenotypic data they carry. Histopathology images

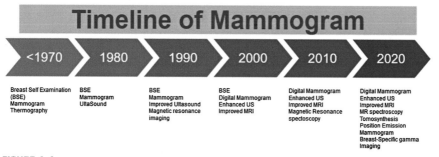

FIGURE 1.4

Mammogram timeline for BC diagnosis.

(HI) have been used for BC multiclassification, although HI has important shortcomings in this area, including strong homogeneity of malignant cells, strong intraclass variation, and poor interclass distinction [33].

1.3.5 Thermography

Another imaging method for diagnosing BC is called mammary thermography, or thermal imaging, in which aberrant thermal patterns are strong indicators of mammary anomalies because of the greater amount of heat released by tumor tissue. Unlike other BC imaging modalities, breast thermography does not cause any discomfort to the patient and poses no risk to the thermographer, making it ideal for routine exams [34]. Thermography was being used alongside mammography to recognize BC at an initial level. Mammography is advised for teenage females despite the low dissimilarity of images of thick mammaries.

1.4 Risk factors

Gender and age are the most significant determinants of risk. Once the risk of BC has progressed to that point, one's age is the key limitation. High age at birth yield, early menopause, history of BC in a parentage, the use of hormonal treatment, being tall, having a large body mass index after menopause, having dense breasts, drinking heavily, smoking tobacco, not getting enough exercise, being exposed to ionizing radiation, and having a short lactation time frame are all additional risk factors for developing BC.

1.5 Advantages in mammography

It is an original diagnostic imaging modality that utilizes a low-dose X-ray equipment to examine the mammary. Mammography is the broadly utilized BC tumor screening test. Mammograms are the outcome of the mammography technique. The mammogram is an X-ray imaging test that allows the medical professional to detect the anomaly. There are two distinct types of mammography: screening and diagnostic. Regardless of the onset of symptoms, screening mammography is used to examine the chest for any irregularities. According to medical knowledge, BC signs are only detectable after 2 years. Hence, consistent BC screening minimizes the likelihood of developing cancer. Diagnostic mammography is performed on women who have discovered anomalies such as masses, skin depressions, and nipple discharge.

In healthcare facilities, many mammograms are produced each day. As a result, radiologists must manually analyze a massive amount of images, which can take a long period to complete [35]. In reality, radiologists struggle to present a precise and consistent judgment for the vast majority of mammograms produced [36]. Additionally, their knowledge, expertise, and subjective norms all have a significant role in how accurate their assessments are. As a result, prediction accuracy tends to decline, making automatic mammography analysis and diagnosis increasingly desirable [37]. It has not been demonstrated that radiologists can effectively read mammograms visually. This happens because some physicians' interpretations have been proven to be incorrect. According to research, radiologists typically make errors between 10% and 30% of the time, according to Lehman et al. [37]. With incorrect misperceptions, the rates of false negative (FN) and false positive (FP) results rise. In the report of Blanks et al. [38], reading mammograms twice improves the accuracy of the diagnosis.

Mammary biopsy procedures are necessary in combination with mammography because it might be challenging to extricate among malignant and benign tumors. In this sense, FP and FN misinterpretations may result in many consequences: a needless biopsy that results from a false positive error results in significant expenses and may be detrimental to patients. A true tumor may go undiscovered in a false negative finding, which could result in increased costs or possibly the patient's mortality [39]. As a result, computer-assisted diagnosis (CAD) systems are crucial for assisting doctors in doing their duties with great efficiency and fewer mistakes [35]. CAD can assist in looking for worrisome symptoms and identifying tumors as benign or malignant. Microcalcifications and tiny tumors that are undetectable by self-examination are found through mammograms. The likelihood that a patient will receive effective therapy is greatly increased by early BC identification. Because mammography uses limited radiation, the risks are reduced.

The following are some of the most salient benefits of digital mammography.

- Breast mammograms improve the likelihood of detecting a cancer so small it could otherwise go undetected.
- Mammograms can find breast abnormalities before any symptoms appear.
- Because mammograms can detect BC at an initial level, without the need for invasive surgery, they are a valuable screening tool.
- Mammograms have been shown to have no adverse effects on patients.

This paper will follow the format below: BC and imaging techniques for diagnosing it are covered in the first section, followed by a brief discussion of the history of cancer screening strategies, detection schemes, and feature extraction mechanisms in the second. Sections 3–5 discuss various trends in mammogram data analysis as well as medical image analysis challenges. Following this will be a discussion of the principles and technology employed in the creation of point-of-care diagnostics as well as trends in clinical mammography practice. Future perspectives of mammography for improved clinical opportunities will be explored in the conclusion.

2. Related works

According to a review [39,40], healthcare professionals ignore 10%–40% of microcalcifications. Film and digital mammography are the two kinds of mammography, according to Gur [41] and Pisano et al. [42]. Film mammography creates the mammary using photographic film. In contrast, digital mammography digitally records the mammary, and the digitized mammogram is kept immediately in the system. According to Sampat et al. [43], irregularities in radiographs are identified by considering two indicators: mass and microcalcification. The goal of the techniques utilized for mass detection in mammograms is to locate and characterize the mass, if any, evident in the image. Form and contour are typical descriptors of the masses. Normal tissue has to be distinguished from masses using edge-based techniques. To differentiate among regular breast tissue and mass, various characteristics, including texture and shape, are retrieved from the targeted area.

2.1 Microcalcification detection

To automatically spot microcalcifications in mammograms, Oliver et al. [44] presented a boosting-based method. Employing local features collected from a filter bank [45] led to the development of the suggested methodology, which provides a description of microcalcification morphology. According to a novel quick proposed by Sankar and Thomas [46], mammograms may be modeled using a stochastic fractal coding scheme, and the presence of microcalcification, an early sign of breast cancer, can be identified. Using fractal encoding techniques [47], they contrasted the modeled mammography obtained with the baseline image, revealing the microcalcifications. Automatic detection of microcalcifications using a supervised method has been made possible due to the research of Torrent et al. [48]. The authors have detailed the suggested system's morphological categorization, which is predicated on the retrieval of feature descriptors using a set of filters.

2.2 Classification of mass

"Masses" are a symptom of breast cancer. Regions of interest (ROI) on a mammogram might include both normal tissue and tumors that can lead to false positives. To solve the problems with massive categorization, Hussain et al. [49] compared and contrasted six alternative Gabor feature extraction strategies. Using mammography data of varying densities, Anand et al. [50] provide a technique for detecting structural deformities. The techniques for feature extraction were used to compute the ROIs from the Modified National Institute of Standards and Technology (MNIST) dataset. Automatic computer-aided identification and localization of mass in mammary images were pioneered by Isa and Ting [51]. Tissue structures revealed by mammography contrast fluctuations or image restoration were fostered.

2.3 Feature-based BC detection

Garge and Bapat [52] present an inexpensive adaptive filtering image processing solution implemented with the help of MATLAB software for detecting calcification on a somewhat difficult computational architecture. Microcalcification clusters in mammograms can be difficult to spot, but D'Elia et al. [53] and another study from 2004 describes complex methods for doing just that. Classification is achieved using the support vector machine (SVM) and orientation features (0,/4, 3/4, and/2), and Gabor filter alignment is assessed by measuring the detection accuracy [54]. A mean-centered region-based subdivision scheme, presented by Zaheeruddin and Singh [55], uses the position of the initial pixel to enhance the accuracy of the segmentation method as a function of the predefined threshold. For the purpose of producing the final classification, deep learning and machine learning methods were employed [56]. The experimental results showed that the whole fine-grained groups achieved equivalent precision and that the ensemble classification algorithm significantly improved the overall categorization compared to the coarse-grained units.

2.4 Computer-aided mammography

Utilizing cutting-edge CAD systems, Mohanty et al. [57] conducted a study to evaluate preprocessing and feature extraction strategies for mammographic images. It was in 2009 that Bhattacharya et al. [58] used a precise CAD methodology to split X-ray mammograms and identify digital microcalcification patches. A neural network (NN)-based CAD system is provided by Baumgartner [59], with the ultimate classifier replaced by random forest algorithms. Specifically, an SVM classifier [60] was utilized to establish criteria for microcalcifications. To identify breast cancer, Alsaedi et al. [61] devised a method that combined microwaves as the light source with infrared thermography as the heat-imaging recorder. CNN improved recognition accuracy and successfully extracted tumor size and position. To produce the ultrasound volume deformation configuration, Hopp et al. [62] developed a 2D (mammogram picture)/3D (ultrasound image) registration approach. Some recent research has sought to use various NN to rapidly and accurately locate existing lesions using CAD models. Researchers Ribli et al. [63] created a CAD mechanism that could identify breast lesions in the INbreast dataset and classify them as malignant or benign by means of the faster R−CNN framework. Their system achieved an area under a curve of 95%. Peng et al. [64] presented a similar autonomous scheme for mass detection, which used a multiscale-feature pyramid network and a Faster R−CNN structure. Many problems in pattern recognition for medical imaging have recently been addressed by employing im2im translation due to the invention of deep CNN models. Most modern approaches depend on im2im conversion, with Pix2Pix and CycleGAN serving as the foundational frameworks, respectively, for matched and unmatched samples, for picture synthesis and reconstruction (Fig. 1.5) [65].

To increase the success rate of recognition, CAD flow analysis preprocesses the collected BC images before segmentation. Once the ROIs have been manually segmented, these methods pull relevant characteristics from them. Then, conventional Machine Learning (ML) models like "SVM," "Random Forest," "Naive Bayes," and "Deep Learning" (DL) models are used to get the final results. To classify BC images, many researchers have turned to SVM, a supervised ML methodology [66−68]. DL models are multilayered structures capable of learning subtle

FIGURE 1.5

Process flow of CAD-based BC detection.

nuances within an image [69]. In a DL model, the first layers retrieve low-level characteristics (such as edges, blobs, and circles), the second levels retrieve moderate-level characteristics (such as eyes, noses, and mouths), and the third and final layers retrieve high-level characteristics (such as objects similar to head) [70]. Some of the significant CAD analysis techniques used in the BC detection mechanism are listed in Table 1.2 as a summary.

Table 1.2 A synthesis of some of the most important works in BC detection.

Literature	Year	Inference
Xu L [27]	1993	The feature extraction process has been optimized using metrics like mean squared reconstruction error.
Smyth P [26]	1996	Feature extraction uses optimization methods like mean classification error (MCE).
Ranzato M and Szummer M [29]	2008	The authors propose a model that incorporates a sparse encoder into a CNN and makes use of deep features in an unsupervised fashion. "Sparsity regularization" has been used to limit the size of models. To that end, GAN is used as an enhancement tool.
Kallenberg et al. [28]	2016	Introduced the "convolutional sparse autoencoder" (CSAE) layout for mammography intensity ranking.
Taghanaki S.A. et al. [25]	2017	It was proposed that a novel autoencoder model be used to categorize mammographic BC subcategories. Each successive stage of this framework utilizes the sigmoid activation function, excluding the final phase, which makes use of the joint distribution.
Al-Masni M.A. et al. [24]	2018	This work presented You Only Look Once (YOLO), a regional DL-based approach for detecting and classifying BC.
Ismail N.S. and Sovuthy C [22]	2019	To categorize among benign and malignant tumors in mammograms, the researchers compared two popular CNN models. When compared to ResNet50, VGG-16 had good efficiency (94% vs. 91.7%).
Guan S and Loew M [30]	2019	Introduced a convolutional neural network classifier-enhancing technique called an autoencoder-generative adversarial network to generate a supplemental interpretation of mammographic images.
Omonigho E.L. et al. [23]	2020	Developed a novel method for classifying mammograms from the MIAS dataset into one of two groups using a modified version of the AlexNet framework

Overall, these research findings demonstrated the potential for enhanced performance across a range of tasks by combining mammographic information from recent and past assessments. To efficaciously assess breast cancer with relatively little effort, even so, a cohesive, fully automatic framework is needed.

2.5 Database for mammogram images

In this section, we list the standard databases that have pictures of mammograms that can be used to check and test how well cancer detection systems work. The majority of BC mammogram data sources are not accessible to the general public.

2.5.1 INbreast

INbreast is a digital, full-field dataset of breast cancer cases that may be used to evaluate and improve any suggested algorithms [71]. The 410 mammograms in the INBREAST database represent CC and mediolateral oblique (MLO) views from 115 different women. The INbreast contains a variety of lesions, including masses, calcifications, asymmetry, and distortion. INbreast was created in a Digital Imaging and Communications in Medicine (DICOM) file with a pixel resolution of 3328×4084 and a secondary resolution of 2560×3328. Only 116 of the photos show masses, and of those, only 41 show benign masses (18 cases), 75 show malignant masses (32 cases), and the remaining images do not show any masses at all. The INbreast features a number of annotated cases created with full-field digital mammograms. DICOM is used to store the data [71]. The Breast Imaging Reporting and Data System (BI-RADS) assessment covers 1−6 categories included in the INbreast dataset. According to Ref. [72], we classified all photos with a BI-RADS score of 1 or 2 as noncancerous; BI-RADS scores of 4−6 as cancerous; and 23 images with a BI-RADS score of 3 as outliers.

2.5.2 CBIS-DDSM

Images from the Digital Database for Screening Mammography (DDSM) have been decoded and transformed to the DICOM organization to create the dataset. Images of mass cases, calcification cases (with full mammograms available), and ROIs are all included in the Curated Breast Imaging Subset (CBIS)-DDSM. The CBIS-whole DDSM's collection of complete mammograms has been retrieved for use in training and evaluation. There are a total of 1318 photos in the mass training database and 378 images in the mass testing database, each of which has been pathologically proven to be either benign or malignant. Due to the fact that some individuals exhibit more than one anomaly, only the most severe aberration was taken into account. The training dataset included 1231 complete photos, 634 of which were benign cases and 597 were malignant instances; the testing dataset included 361 images, 218 of which were harmless cases and 143 were malicious cases [73].

2.5.3 Image retrieval in medical applications

With a focus on medico diagnostic tasks, the image retrieval in medical application (IRMA) venture plans to develop and deploy advanced computations for "content-based image retrieval." Mammogram photos are stored in IRMA, a database created by Germany's RWTH Aachens. Among the 128×128 pixel mammograms, there are 931 normal images and 584 images of malignancy [74].

2.5.4 Mammographic Image Analysis Society

There are a total of 322 images (161 pairs) in the Mammographic Image Analysis Society (MIAS) dataset, all of which are 1024×1024 pixels in size and saved in the "Portable Gray Map" (PGM) format. These images are accompanied by "truth" data that features radiologists' "truth" markings of the locations of any existent irregularities. There are 207 healthy photos, 63 benign images, and 52 cancerous images in the MIAS collection. Researchers in the United Kingdom created the MIAS Dataset [75]. There are a total of 322 digital photos in the MIAS dataset, 115 of which are abnormal while the remaining 207 are considered normal. The picture was originally 1024 pixels wide and high. Three-and-a-half percent (222 samples) of the full dataset were used for testing, while the remaining 69% (100 samples) were used for training. No data augmentation was performed on either set. The photographs' anomalies have been labeled according to their true presence in this dataset [75].

2.5.5 Breast cancer digital repository

The breast cancer digital repository (BCDR) dataset was created with assistance from the IMED research, which was directed by "INEGI, FMUP-CHS" at the University of Porto in Portugal and "CETA-CIEMAT" in Spain. There are 1734 BIRADS-classified instances in the BCDR dataset. The BDCR-FM database has 1010 pseudonymous patient cases, 1125 studies, and incidences of 3703 "mediolateral" and "craniocaudal" mammograms [76].

2.5.6 BancoWeb LAPIMO

Nearly 1400 mammographic images in TIFF format can be found in the BancoWeb Laboratório de Análise e Processamento de Imagens Médicas e Odontológicas (LAPIMO) collection. This data set contains images that have been classified as either healthy, benign, or malignant [77].

2.5.7 UCHC DigiMammo

A compilation of confidential datasets originating from the University of Connecticut Center (UCHC), which is referred to as the UCHC Digi-Mammo (UCHCDM) dataset. The database contains the screening mammography of 230 individuals. In each physician's case, a first screening was performed, which is referred to as the preceding test, and a subsequent follow-up-evaluation occurred among 1–6 years later, which is referred to as the Contemporary exam [78].

Throughout this relevant literature, we seemed to be able to validate the various methodologies being utilized to categorize tumors, the usage of various datasets, the several attributes chosen, the different methodologies employed to decrease feature dimension, and the various test strategies and performance measures used to assess and examine the findings. Due to the different methods, strategies, and methodologies, it is very hard to compare the results of one study to those of another in a quantitative way. For example, datasets, classification approaches, characteristics used, artificial intelligence techniques, etc. are all different from one study to another.

3. Current trends in mammography analysis

Mammography, the only screening tool shown to cut the death rate from BC, is one of the major successes of innovation in breast cancer care. It is the primary method radiologists use to check for breast cancer. Despite this, the industry has been attempting to address its acknowledged shortcomings ever since the technology's introduction, with the end goal of improving early invasive cancer diagnosis while decreasing false positives.

Since its introduction by German surgeon Dr. Albert Salomon in 1913, X-rays have been utilized to see breast tissue. Jacob Gershon began to push for wider X-ray screening use in the 1950s. RSNA officially introduced digital mammography in the United States in December 2005. Mammography, which was first introduced, is now the best scheme available for detecting BC. The technology has come a long way since the 1980s, and even comparisons to that era's exams show significant differences from those conducted now. Radiation exposure is minimized without sacrificing image quality in modern X-ray mammography because of the use of dedicated systems (a machine used specifically for breast X-rays). The Mammography Quality Standard Act is responsible for the strict regulation of modern mammography systems (MQSA).

3.1 Full field digital mammography

Despite the success of film-screen mammography, some estimates place the percentage of cancers missed by this method at between 20% and 30%. However, SFM suffers from a lackluster contrast resolution, which is a major downside. Contrasting densities of glandular tissue and fat make the breast a challenging organ to photograph. Women with thick breasts are shown to be 4–6 times more possible to get BC than those with less or no glandular tissue in their breasts. That is probably because the breast tissue sits above preexisting tumors, hiding them from view. Breast cancer detection with DM and screen-film mammography (SFM) yielded similar overall diagnostic accuracy; however, full-field digital mammography (FFDM) has a higher signal-to-noise ratio. The W/Rh combo is favored. Saving money and time with near-instantaneous picture acquisition (Fig. 1.6).

3.2 Digital mammography

Digital mammography (DM) captures and displays areas of varying densities with minimal loss of contrast resolution. With a spatial perseverance of 16 line combinations/millimeter, SFM can easily detect microscopic features like microcalcification. However, pixel size is a constraint on DM's spatial resolution. Despite this limitation, DM has been claimed to be as good as or better than SFM at diagnosing microcalcifications. This is because DM has higher contrast resolution, making it easier to see tiny high-contrast features like microcalcification. In comparison to SFM, DM can save the time it takes to conduct an examination and process photos by 45% (Fig. 1.7).

DM has many benefits, including improved contrast resolution, the ability to modify images for better clarity and visibility, lower false positive rates, a higher positive predictive value, and fewer tests needing to be repeated due to poor exposure. Due to this, the index of repeatability is quite high.

Recently, digital mammography has been improved through the utilization of an iodinated contrast agent injected intravenously before the scan, a process known as

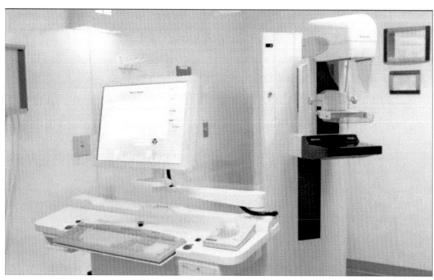

FIGURE 1.6

Full field digital mammography.

Courtesy of Feng SSJ, Sechopoulos I. Clinical digital breast tomosynthesis system: dosimetric characterization. Radiology 2012;263:35—42. https://doi.org/10.1148/radiol.11111789.

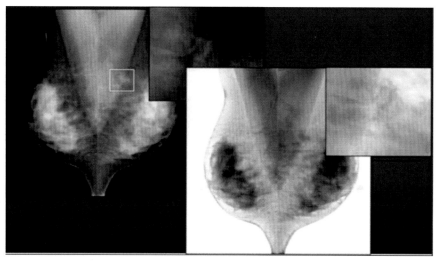

FIGURE 1.7

Inverted digital image—minute calcifications shown by square mark.

Courtesy of Chong A, Weinstein SP, McDonald ES, Conant EF. Digital breast tomosynthesis: concepts and clinical practice. Radiology 2019;292:1—14.

CEDM. The concept behind contrast-enhanced mammography is that developing cancers need more blood flow to thrive. As the contrast agent tends to collect more heavily in these regions, contrast-enhanced mammography provides a means of seeing the dissemination of contrast inside the mammary. This process can expand the detectability of mammary lesions in mammograms. The first method for conducting CEDM exams is called temporal subtraction and involves the capture of high-energy pictures both before and after the injection of contrast material. The kinetic arc of augmentation of mammary lesions may be analyzed using the temporal subtraction method. The main drawbacks of this method are that only one breast may be scanned at a time and those patients have to remain in a specific posture (often MLO) for an extended amount of time. There are additional kinetic artifacts because the breast is compressed as the contrast enters the bloodstream.

Tomosynthesis of the breast, or 3D mammography, is a relatively new concept. Similar to DM, but more advanced. Breast tomosynthesis (BT), a novel mechanism built on the gathering of 3D digital mammary information, may provide a solution to the challenge of deciphering mammographic characteristics caused by tissue overlap. A digital detector and a moving X-ray source are utilized in breast tomosynthesis. In a BT scheme, the X-ray tube travels in an area as it is exposed.

3.2.1 Computed tomography electro-optical tomographic laser mammography

Medical imaging technique incorporating computed tomography electro-optical tomographic laser mammography (CTLM) is a method for evaluating breast tissue using lasers. The angiogenesis in the breasts can be measured using this scanning method by using reasonably close laser radiation (Fig. 1.8).

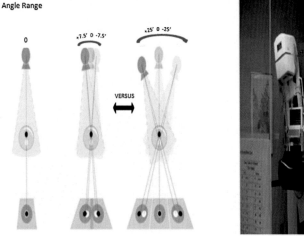
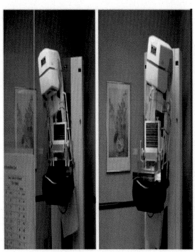

FIGURE 1.8

Angle range of CTLM.

Courtesy of Chong A, Weinstein SP, McDonald ES, Conant EF. Digital breast tomosynthesis: concepts and clinical practice. Radiology 2019;292:1—14.

3.3 Scintimammography

It is also acknowledged as nuclear medicine breast imaging, which is a test that can be performed to learn more about a breast irregularity seen on a mammogram. Breast-specific gamma imaging (BSGI) and molecular breast imaging (MBI) are other names for scintimammography. It is carried out on patients who have irregular mammograms, thick breast tissue, postoperative fatty tissue, breast implants, or who have had breast surgery. Small amounts of technetium 99 sestamibi, a radioactive material taken up by cancer cells, are injected into the patient, and gamma cameras are then used to capture photographs of the breasts. Sestamibi breast imaging is also known as a Miraluma test [80]. Smaller than 1 cm anomalies are harder to accurately assess using this method. Compared to traditional mammography, it has a better sensitivity and positive predictive value (PPV), but subjects the patient to a somewhat higher radiation dose.

3.4 Optical mammography

Diffuse optical imaging, or optical mammography, is a group of noninvasive screening methodologies that use near-infrared (NIR) light and has the potential to replace or supplement current methods. As it is well established that cancer location is significantly connected with entire hemoglobin accumulation via angiogenesis, such 3D images of hemodynamic features assist as markers of malignant levels. When doing an optical mammogram, NIR light is used to scan breast tissue, and then a CAD is used to decipher the image and data. The method can be used to detect and quantify changes in water and lipid content. The device captures dynamic photos of metabolic processes, revealing alterations in tissue oxygenation and hemoglobin concentration in real time. Compared to traditional mammography, the breasts are compressed substantially less during optical mammography. In particular, the absence of breast compression and ionizing radiation, as well as its high repetition index, contribute significantly to the technology's widespread acceptance among patients (Fig. 1.9).

3.5 Digital breast tomosynthesis

Since the utilization of mammography, digital breast tomosynthesis (DBT) has been widely regarded as the single most significant advancement in the screening process. DBT mammography is a subset of X-ray mammography in which low-dose forecast pictures of the breast are combined with tomographic image reconstruction to create a mammogram with some 3D appearance. This enables the examination of several thin breast tissue sections, which improves carcinoma diagnosis and reduces false-positive cases by eliminating the concealing influence of underlying fibroglandular breast tissue. Improved diagnostic and screening efficacy of DBT has led to its rapid adoption as the new standard in breast imaging, surpassing mammography. Tomosynthesis acquisition allows for better lesion identification, characterization, and location by minimizing the

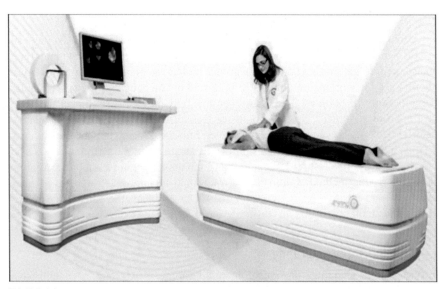

FIGURE 1.9

Computerized tomography LASER mammography.

confounding influence of overlapping tissue. The quasi-3D data from the restored DBT attribute provides a more effective imaging procedure compared to using only 2D full-field digital mammography.

When combined with traditional mammography and breast ultrasonography, positron emission mammography (PEM) has shown to be a valuable diagnostic tool, earning FDA approval and entering clinical use. For PEM, fluorine-18 fluorodeoxyglucose is administered, and then pair of gamma radiation detectors is positioned up and down of the breast, while the breast is gently compressed (18F-FDG). This method is based on the observation that cancer cells use glucose more efficiently than healthy cells. This consumption of glucose can be observed by using the radiopharmaceutical 18F-FDG, which is made by attaching the isotope fluorine-18 to the delivery molecule deoxyglucose. It is performed after a mammogram or USG if: Manifestation of a noticeable tumor or growth in a significant quantity. Invasive lobular cancer, which is more common in women who have used HRT, can be detected earlier with MRI than with mammography or ultrasound. Unique coils are placed on the surface of each breast, and they are bilateral (simultaneous examination of both breasts). MRI before surgery provides a more precise evaluation of tumor size and multisite involvement (incl. DCIS). MRI has the highest detection accuracy for ILC and the lowest FN rate for sizing. MRM's sensitivity was 71%, specificity was 85%, and accuracy was 76% when testing for extended intraductal component (EIC) (Fig. 1.10).

Since the first DBT system was approved by the US FDA a decade ago, radiologists have been able to address many of the limitations of 2-D mammography. Since then, DBT has become the bright standard in the United States for detecting BC.

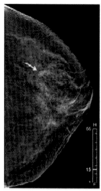

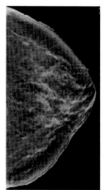

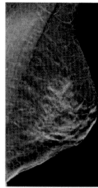

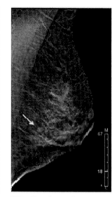

(a) A spiculated mass (arrow) that is not seen on synthetic craniocaudal specimens.

(b) Mediolateral oblique

(c) Mammograms

(d) On the evidence of its position in the craniocaudal DBT stack, the mass (arrow) can be localized on the mediolateral oblique DBT image.

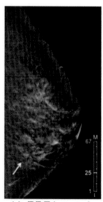

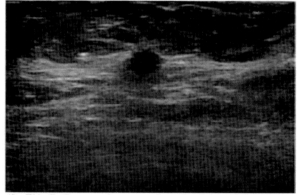

(e). DBT images in mediolateral representation show the mass (arrow) more clearly.

(f). This finding correlates to an irregular, hypoechoic mass, as determined by US inspection. The investigation of a biopsy specimen revealed invasive ductal carcinoma that was positive for the oestrogen receptor, positive for the progesterone receptor, and negative for the human epidermal growth factor.

FIGURE 1.10

Five images in 57 years old women with tumor identified merely by DBT.

Courtesy of Chong A, Weinstein SP, McDonald ES, Conant EF. Digital breast tomosynthesis: concepts and clinical practice. Radiology 2019;292:1−14.

Reductions in recall rates and increases in detection are two of the most noticeable results of DBT implementation. The 3Dimensions mammography system from Hologic, for instance, has been authorized by the FDA as superior to 2-D mammography for women with thick breasts and has been found to identify 20%−65% more invasive tumors than a 2-D mammogram alone1 [80,81]. The DBT methods have the ability to spare women the emotional toll of unneeded callbacks and to lessen the economic impact on the healthcare system because it minimizes false positive rates compared to full-field digital mammography.

3.6 Future of DBT imaging

Next-generation DBT units are now under development with a wide range of acquisition arcs in an exertion to increase spatial resolution and, by extension, lesion conspicuity in reconstructed synthetic pictures [82]. Research on contrast-enhanced DM and digital breast tomosynthesis suggests that these modalities may replace MRI as a secondary screening option for some females [83,84]. Even while DBT is linked to much longer reading durations than DM [85,86], it is anticipated that interpretation time will decrease as experience with DBT grows. An automated "slabbing" of DBT parts to form thicker overlapping portions is just one example of the new presentation modes devised with reader efficiency in mind. That expedite DBT review, researchers have developed machine learning-based detection algorithms to "flag" portions and record the fatness of a discovered lesion inside the DBT stack [87]. Lesions on SM images are more easily seen after being "computer-aided detection-enhanced," which has been proven to minimize reading interval by 29.2% in early reader studies without sacrificing accuracy [87]. Radiologists' diagnostic accuracy and efficiency can be increased by using computer-aided detection and machine learning to objectively identify breast density categories [88–90].

4. Challenges in mammogram data analysis
4.1 General challenges in BC measurement and analysis
4.1.1 Shortcomings in primary care

Patients seeking treatment for both chronic and acute conditions, as well as preventative care, should initially visit a primary healthcare facility (i.e., general pediatric and family planning hospitals in Asia). Its purpose in breast cancer detection is to help qualified women have clinical breast exams (CBEs) every other year and to educate women on how to perform CBEs on their own, that is, breast self-examination (BSE). According to nurses and physicians, a shortage of time to deliberate anticipatory screening with asymptomatic individuals is a key barrier to implementing CBE and BSE [91]. Some doctors have also noted the wide range of abilities of nurses to spot anomalies throughout a CBE, especially among more junior staff members. Due to women's discomfort with having their breasts evaluated by male employees, lengthy delay durations and a lack of women workers for CBE assessment in hospitals were noted as obstacles by focus group discussion (FGD) participants.

Female patients (aged 40) were taught BSE by physicians at prenatal and post-natal consultations, as well as in mother–child interactions. Owing to the conventional culture of the nation as well as the absence of private services in communal engagement activities, self-examination is typically trained by displaying female images of the method without any direct communication.

4.1.2 Public secondary healthcare clinic mammography concerns

This subtheme concerns the difficulties healthcare employees have with mammography screening. Similar to issues with staffing in the healthcare system, there was a shortage of personnel with expertise in BC care, notably female radiographers, who are required to do mammography examinations. Only around 30% of clinic radiologists, according to those who were interviewed, were female. Additionally, the district hospital that received subsidies could only provide a certain volume of mammograms each month, and that amount has decreased over time. As a result, many women who cannot or do not want to afford a mammogram must wait. Budget restrictions resulted in outdated or occasionally unmaintained mammography apparatus and other infrastructures, such as air conditioning, that were vulnerable to failure. Attendees in the FGD and physicians also noted that women frequently needed understanding about the value of mammograms and that the public firm's healthcare workers lacked the time to give thorough clarifications because of the excessive numbers of patients and pressure [91].

4.1.3 A gap between the BC detection strategy for primary care and secondary

According to the doctors working in hospitals, the government established key performance measurements for CBE, such as regular consultation targets for health centers to satisfy. Therefore, primary care clinicians did not endorse hospital-based BC screening programs, which led to fewer referring patients and longer waiting periods.

4.1.4 Potential risks of mammography

In addition to its questionable efficacy, mammography has been questioned on the grounds that it may be harmful. Thornton and coworkers have identified numerous types of harm, including those to one's body, emotions, relationships, finances, and future generations [92]. According to the research so far, they are moderate and transient in nature, though some works imply that they may be permanent. The cost of further examinations required to arrive at a diagnosis in the wake of abnormal mammography is a form of economic injury. Studies looking backward have found that assessing false-positive findings be able to increase up to an additional 30% of the entire charge of testing for every female [93,94]. The term "intergenerational harm" indicates the potential increase in assurance percentages for the daughters of ladies who have been identified with an illness that did not result in any negative health outcomes but was discovered during a routine screening mammogram [95].

4.1.5 Physical and mental suffering

Mammography involves the squeezing of the mammary, which can be painful for some women due to the dates of their most recent menstrual cycle or the expectation of pain [96]. There was a high degree of inconsistency in the findings of [97] studies conducted before 1999. The percentage of women who reported discomfort during mammograms ranged from 1% to up to 85%. Given the well-known difficulties associated with accurately quantifying pain, it is critical to conduct an analysis of the methodology used in these studies [98,99]. Only (15−28)% of females, depending on the study, reported experiencing moderate or intense pain. These findings come from new research that was constructed adequately.

4.1.6 Biopsies

If a mammogram comes back abnormal, further testing, such as a needle biopsy or a conventional surgical biopsy, is required to determine the cause. Unnecessary biopsies pose a serious risk to the fitness condition of women. The variation in biopsy rates for diagnosed instances of BC between nations suggests that this procedure is performed for reasons other than mammography's technical shortcomings. Besides screening, other criteria, such as practice heterogeneity and vulnerability assessments, play a role in the levels [91].

4.2 Obstacles to data analytics in BC

4.2.1 Personal encounters and obstacles to obtaining assistance

Women who participated in the FGDs spoke most about interpersonal constraints, and several healthcare experts concurred. Participants noted that several Asian women from remote regions had very little understanding of BC, were in ignorance about it (refusal), had inaccurate information about malignancy or cancer treatments, or did not believe they were at risk for BC and were not engaged in testing. Several women had the belief that conventional healers could cure cancer, either by themselves or in conjunction with modern therapy [8]. South Asian women claimed that some spouses forbade their wives from showing their bodies to other men out of worry that their partners could desert them following surgery for breast reduction. Women of all races felt anxiety when discussing breast health problems with a male doctor or exposing their bodies to him. Some women who claimed they would not have leisure for breast examinations cited child care as a competitive concern. Attendees in the FGD and interviews stated that certain women were afraid of receiving a cancer diagnosis or that the screening was uncomfortable. Additionally, primary care doctors and BC patients noted that patients are scared of having their breasts removed and of losing their hair as a result of BC therapy [8]. Many rural women, according to accounts provided by community workers, cannot read or write (e.g., brochures, etc.). If a doctor or nurse did not speak the individual's language, communication was difficult.

4.2.2 Connecting theory into practice

The difficulties in putting all of their academic knowledge into practice were cited by the medical students. A combination of fundamental physics knowledge and an in-depth understanding of exposure factors tailored to each therapeutic setting and person were recognized as necessary. It has been observed that an understanding of anatomy and pathology is essential for evaluating image quality and illumination requirements. They were exposed to the theoretical concepts of modern technology and methodologies used in mammograms as part of their training. To acquire competence in many aspects of breast imaging and mammography, it is crucial to overcome the gap between theory and practice. An understanding of equipment, procedures, and patient–radiographer interaction is crucial [42].

4.2.3 Carrying out mammograms

Medical imaging trainees cited difficulties with orientation as a major barrier to the profession. Everyone had listened to presentations on the principle of placement. The doctors also had the opportunity to get hands-on experience using radiography apparatus at educational institutes. They tended to understand orientation through hands-on practice, initially with a phantom, also with other colleagues, and lastly on actual patients under the supervision of a skilled healthcare professional. Courage is needed to go near patients/women and persistence is required to complete a thorough smear test. Mammograms vary from other imaging modalities in that they necessitate direct interaction with patients. It is a lot more personal.

4.2.4 Communication

During their time in the clinic, medical residents and undergraduates may use some guidance from an experienced doctor to better communicate with patients, especially women. The radiologists found it difficult to apply their communication abilities while still conducting mammography and caring for female patients [50].

4.3 Breast density versus mammographic sensitivity

The radiographic depth of a woman's breasts varies greatly, making it extremely difficult to detect malignancy in dense-breasted females on mammography. The amount of X-ray particles that can pass through the breast tissue and register as visible on the film determines the intensity of mammography [21]. Given that fat is radiographically transparent, it appears darker on X-ray pictures because X-rays travel through it comparatively unobstructed. The breast cells and the surrounding muscular and epithelial cells appear less dense on radiographs because fat blocks more of the radiation that is transmitted through the body. As breast tumors and microcalcifications are likely to collect additional X-ray photons, they appear on mammography as brighter spots, making them harder to identify [28].

Mammograms are often categorized on the premise of BI-RADS breast density scale established by the American College of Radiology [100] as an aspect of screening mammography assessment (Table 1.3).

Table 1.3 BC identification ratio against breast density level.

Density level	Category description	BC detection affection ratio
1	Primarily denser mass	Lower than (1/4)%
2	Sprinkled glandular mass	Between (1/4) and (1/2)%
3	Heterogeneous denser mass	Between (1/2) and (3/4)%

X-ray-free imaging methods, such as MRI and sonography, are fewer probable to be impacted by breast thickness [101].

Overweight, race or ethnicity, aging, menstrual cycle phase, and the number of children a woman has all play a role in determining her breast density (parity). In addition to age, menopausal state, obesity, and parity, these factors also contribute to 20%−30% of the total variance in dense breast tissue proportion. Because of the augmented density of their breast skin, mammography of younger women is frequently more difficult to read. Most women have substantial changes during hormonal therapy, and the degree to which breast density grows or decreases is conditional on the specific hormone regimen used [102]. In comparison to estrogen therapy alone, combined treatment with estrogen and progestin, for example, significantly improves breast density [103].

However, radiography may be more accurate if it is performed in the first 2 weeks of a woman's menstrual cycle [104] likely because women experience less agony while breast-compressing at this time. This improves the chances of getting an assessment with little patient movement, which can reduce image quality and hamper cancer detection. Ultrasound examination of all women with especially dense breasts may be one answer to the challenges created by dense breasts. This is a common practice in Korea and certain American hospitals as well, but there is no documented evidence showing that it improves outcomes or identifying which women would benefit from it [105].

4.4 False alarms

If your mammography reveals something unusual, you should seek further evaluation. Furthermore, the resulting emotional pain is typically short-lived [106], and it disappears if further testing reveals that the initial assessment was incorrect. However, for some women, adequate amounts of apprehension continue after the original mistaken result has been eliminated. One study examined the difference in worry between women who got false-positive and negative findings on their initial test. The women's anxiety levels ranged from 0 (not concerned at all) to 5 (extremely concerned). According to the measurement methodology and component of anxiousness used, there was a wide variation in baseline perceived stress among the women who received false positives, ranging from 1.5 to 2.5. Moreover, 6−8 weeks after the false positive was resolved, their stress levels continued to be "relatively nervous," not "extremely anxious" (level 5). This was despite the fact that their stress levels were regularly roughly threefold higher than those of women whose mammography had initially been negative [67]. The women's baseline anxiety levels strongly predicted their postnegative-result anxiety. The anxiety levels of roughly 800 women were assessed in this study; however, conclusions depend on a considerably minimal population of 36 females who were incorrectly identified as having anxiety. Nervousness about radiography may have been triggered by the fact that the research was done in Swiss at the start of a prototype screening program, when many of the women participating had never before had a mammogram for diagnostic purposes. Recent developments in artificial intelligence, information processing, big data, etc. must be incorporated into the screening procedure to eradicate the number of false positives [104,106].

4.5 Radiation dose and digital breast tomosynthesis

The mean glandular dose (MGD) is used to evaluate mammary radiation dose, and it is calculated using air kerma (exposure) measurements and a normalized glandular dose factor unique to the concentration of the X-ray source. Although the proportion of fibroglandular parenchyma and the X-ray spectra do perform a duty, the volume and density of breast tissue in the MLO view have a greater impact on this coefficient's departure from the focal representation than do these other factors [107]. DBT typically has a greater MGD to the breast compared to FFDM. When compared to a 2D mammography, the radiation dosage from adding two tomosynthesis views is less than half as high, at 1 mrem per view. Denser breasts result in a smaller dosage differential between DBT and FFDM [79]. To maintain higher image excellence, this calls for the adoption of scatter-reducing systems during postprocessing [108].

4.6 Artifacts caused by surgical staples

Blurring-ripple artifacts are caused by anything with a high population density that is positioned on tissue, like surgical clips or identifiers. The edges of these clips look fuzzy and broader than they really are in parts that are out of focus (blur), and the skin looks thicker than it should be. Ripples show up as the range between the real object and the recovered slice grows. This happens because of a process called "volume averaging" in CT, which happens when the percentage of forecasts taken is significantly smaller than the total count of recovered slices, which causes noise [109,110].

4.6.1 Imbalanced database

When using only one database, there is a risk that the classification will do poorly with photos from other resources, even if it does well with the dataset in question. Thus, developing a stronger, more robust model that can manage a broader range of datasets could enhance the likelihood of physicians using the tools on a regular basis, suggesting a potential discrepancy in this field of research. The development of self-improving, online-accessible systems is also an unmet need [71]. The majority of the models presented in the publications can only be trained in a group or offline setting. Building systems that can recognize and adapt to changing tendencies as new images are shown on the internet could be a huge step forward in this field. There was not a single study that used grammars, compositional methods, or graphs to construct a classifier. While these methodologies have been used by numerous academics, they have yet to be implemented in large-scale picture categorization projects. Utilizing the same dataset and identical images to construct the classifiers is essential for a valid assessment. When performing k-fold cross manipulation, for example, it is essential that all k-folds contain the same images. When developing classification techniques, the issue of imbalanced data is a significant obstacle. Those from the dominant group may skew the results. Uneven distribution occurs at both the patient and image levels in the multiclass scenario. Hence, the absence of diverse databases is another issue that needs resolving [75].

4.6.2 Insufficient standardization

While we discovered a few flaws, we also became aware of a few issues that reoccurred in a few of the works examined. The primary issue is that there is not enough standardization in relation to the measures that are utilized to assess the mechanisms. Accuracy and area under the curve (AUC) are the measures that are used the most frequently; nonetheless, we discovered a great deal of other statistics, such as sensitivity, specificity, f1-score, FP, and FN. The challenge that arises when attempting to compare two studies, one of which offers just the sensitivity and specificity and the other of which provides only the AUC, has been uncovered here. Because of these difficulties, it is also exceedingly difficult to determine the contemporary approaches in this area of research. One further widespread issue that we call attention to is the fact that, in general, researchers do not examine the findings acquired by the classification model with the information derived by physicians. For instance, an effectiveness of 90% achieved by a classification model could be acceptable or terrible based on the accuracy that might be achieved by a physician studying the same images; if this accuracy cannot be achieved, then the technique can be applied on a regular basis by physicians. Here, we can examine the rationale for not performing this assessment [39]. It is possible that the physicians are not interested in participating in this kind of research; it is also possible that the researchers do not place a high value on the opinions of the clinicians; or it is also possible that there are just not enough clinicians accessible. The second issue that we discovered was the fact that the methodology that was applied during the tests was either not described at all in the papers or was not there. Due to the wide variety of used methodologies, techniques, and procedures, quantitative comparisons between works are typically challenging [41]. This is because the results of each paper generally differ in a great number of different aspects, such as the dataset, separation methods, characteristics engaged, machine learning mechanisms, and so on.

4.7 Challenges in data analytics models

While DL models have seen a significant amount of success, there are still obstacles and constraints with DL that have to be addressed to properly identify and classify BC. We describe the main problems we found during our analysis below:

The shortage of a rich database for building DL models for computed tomography presents the biggest obstacle. Developing a large database is challenging since annotating clinical data is time-consuming, labor-intensive, and requires numerous attempts to eliminate error rates. Although supervised learning-based approaches have improved accuracy, it remains challenging to acquire medical professionals' annotations of mammary images. The majority of online mammary images lack descriptive captions. This highlights the critical necessity for unsupervised learning-based strategies for the categorization of BC data [39].

Also, most studies relied on clinical datasets (private datasets) for their analyses, which presents its own set of challenges. This makes it difficult to compare the efficacy of such systems across studies. In addition, to prevent overfitting, some research has employed data preprocessing methods rather than learning algorithms. When it comes to clinical BC imaging, pretrained systems are few. Pretrained classifiers from the "ImageNet" database can be used in a various clinical imaging practices, but they do not provide a huge boost. This highlights the importance of methods that achieve comparable or superior results without necessitating extensive data collection and analysis [111].

When it comes to training NN and DL algorithms, such as deep convolutional neural networks (DCNN), which necessitate large amounts of data, protecting the confidentiality of clinical records presents an additional significant challenge. Data sharing at a central location becomes problematic as a result of privacy rules. The training of DCNN can thus be carried out in a distributed and decentralized fashion, avoiding the requirement to share customer records. The potential of Deep Reinforcement Learning, a cutting-edge and rapidly developing methodology, for use in medical image analysis (MIA) is yet largely untapped [91].

4.8 Robustness

The robustness problem of data collection methods is another obstacle in MIA, but this can be overcome by collecting data from a variety of sources, such as multiple analyzers, multiple views and dimensions of computed tomography, multiple illumination conditions, multiple color presentations, and multiple augmentation considerations. These variations allow for the gradual introduction of additional data. However, only a small number of realistic models take into account both imaging and nonimaging factors (such as a client's family history of tumor or their genetic makeup) [41]. A plethora of such frameworks that integrate radiomic features with visual information are required. BC grows when cancerous cells move from one area of the breast to another. Therefore, the same image may be annotated with various subcategory labels because various regions of similar breast tissue may exhibit distinct phases of BC. The decision-making mechanism can become muddled if interference or biological artifacts are combined with these images [112].

4.9 Cyber security

When applied to medical research, AI has the potential to greatly hasten the pace of scientific advancement and completely revolutionize healthcare. The complex characteristics of this modern innovation include ethical, medicolegal, and compliance issues [113]. Ethical issues are exacerbated by the fact that computational methods are susceptible to problems including data bias, algorithmic prejudice, and efficiency degradation. According to a report by cyber-security firm Critical Assessments, the number of cyber-security threats in 2021 reached an all-time peak, resulting in the exposure of a record volume of patients' confidential health information (PHI). The number of people impacted by healthcare breaches increased to 4.2 billion in 2021, up from 33 million in 2020. Compared to 2020, there will be 32% more victims in 2021, which means that every year, so many records will be jeopardized. Even though the increase from 2020 to 2021 was relatively small (2.4 percentage points), the total number of breaches, at 679, was still an all-time high. Security breaches and cyber-attacks are more widespread than ever as digitalization advances exponentially, putting sensitive patient information in greater danger than ever before [114]. That's why it's so important for healthcare providers to adhere to the strictest standards of data privacy.

It is evident from the study that, in the future, the suggested data analysis approach can be utilized to provide surgeons with assistance regarding the screening of BC cases, thereby minimizing the likelihood of surgeries. In this approach, the suggested framework can assist a partner hospital's clinical procedure. However, the aforementioned obstacles must be addressed before implementing any system in clinics, as human life-related matters require greater precision and more thorough testing, but this is a positive step in the right direction.

5. Future directions of mammogram analysis

The future study is to evaluate present methods using more models. A considerable proportion of produced models are not accessible to the public. However, more work could be invested in gaining access to these models to evaluate and compare them using the same datasets. Regarding the construction of the framework, it is recommended to analyze the additional value of training with pixel-level segmentation. In the weakly supervised method, on the other hand, pixel-level classifications are only incorporated for testing and not for training. The benefit is that larger databases not tagged at the pixel level can be used for training. However, when it is determined that the impact of applying these observations throughout training is notably strong, it may be desirable to focus on gathering additional pixel-level annotations rather than constructing poorly supervised training methods.

Regarding the medical applicability of BC screening mechanisms, further study is also advised. There are numerous application choices for these devices. Moreover, there are numerous ways for a radiologist to collaborate by means of a forecast scheme. A framework could initially examine all images and then recognize those in which it feels an anomaly exists. A radiologist would then review these photos and make the final determination. Use the computer as a backup scanner, and therefore only one radiologist as a replacement [115]. A third alternative is to apply the framework to all images, display the model's conclusions alongside the original metaphors to the physicians, and continue the assessment as usual. A final alternative would be to maintain the current screening procedure but apply the model to all radiologists-classified healthy photos. Then, when humans do not detect a tumor, the model may detect it in difficult circumstances. These strategies and their consequences on clinical practice require further investigation.

All the different ways in which data analytics and deep learning might be interpreted for tumor categorization—using regional and comprehensive networks, visualization tools, and more—need to be studied in depth. This will pave the way for future research into the assessment of ML and DL techniques for cancer categorization. An additional factor that helps speed up studies is the massive amount of publicly available data. Single-cell RNA sequencing data on gene manifestation points to a promising avenue for finding biomarkers of cancer-related genes [116]. This requires additional effort, particularly while assessing with diverse databases and multiclass mammogram categories. Supervised and DL schemes are thought to be of great importance in this area. Consequently, future categorization systems for cancer will require continuous research.

The development of MR technology will reveal tiny occult lesions on mammography X-rays, ultrasound, and clinical examination. The spatial and temporal resolutions of 3D schemes with multichannel recorders will grow, and 4D sequences will give complete breast coverage. Minor scratches will be examined by means of disparity approval and structural attributes by employing unique pulse series as well as a tiny slice depth. Additional research is needed to evaluate the medical importance of these tiny invasive, specifically while they are multifocal. In research centers, BC MRI has demonstrated significant promise in evaluating mammary illness; nevertheless, its widespread adoption is hampered by the restricted availability of MR devices and lengthy wait times in many centers. Systems will be enhanced by introducing extra new options for funding [117].

BC is the main reason of mortality among women, making emerging an analytic tool crucial. Additionally, many strategies, models, and combinations can improve performance. Additionally, breast cancer can be classified into more distinct categories [118]. HIFDM (hyperbolic intuitionistic fuzzy divergence measure) techniques will enhance the existing process in numerous conducts and function on automatically detecting early-stage aberrant tissue regions, lumps, and breast masses in a real-time setting [119]. Breast tomosynthesis offers improved sensitivity, specificity, and reader confidence; consequently, most institutes and centers use this technique more frequently [120]. In addition, there is an increase in women's satisfactoriness owing to decreased false positive rates and the requirement for different views.

Several researchers have studied CAD in DBT. To establish their effectiveness, utility, and impression on BC detection, it is obvious from the survey that CAD algorithms require additional research and validation with more clinical data. Suppose CAD schemes for the diagnosis of mammary anomalies are to obtain a solid medical footstep. In that case, it is besides essential for the scientist communal to evaluate principal screenings that improve association, similarity, and understandability [121]. Despite its limits, this exhaustive analysis can assist as a comprehensive outline to the topic and guide the design and conduct of future research. Performing multiple or additional preprocessing activities and balancing the data with augmentation will provide improved data outputs [122]. Additionally, the utilization of numerous transfer learning models may enhance the metrics. In the future, ensemble learning, hybrid models, and data replication using GAN will be tested within the scope of the obtained results. Models of computer vision can aid both patients and healthcare professionals. Thus, it will improve time management and eliminate subjectivity.

Despite the ongoing development of breast cancer screening programs [123], it is anticipated that the mortality and medical expense thrust of BC will rise due to an escalation in the absolute number of BC patients [115]. In addition, females aged greater than 50 with more than 13 years of schooling and living in countryside locations exhibited decreased BC screening rates. Additional investigation is required to identify the populations most likely to be excluded from screening advantages and the underlying causes of inequalities in screening, such as sociodemographic changes and the gradients in breast cancer screening rate trends.

6. Conclusion

This article offered a comprehensive review of the application of data analysis tools and data modeling approaches to the detection of malignancies in mammograms. We reckon there is a considerable amount of interest in this field of study, as each year several analyses employing various methodologies and characteristics are published. Also, a lot of the papers that were evaluated had a rate of recognition of over 90%, which suggests that this field may have reached its peak.

Studies indicate that the artificial NN, SVM, and k-nearest neighbors rule this field in terms of data analysis methodologies. In a few studies, classic methods such as Naive Bayes, regression models, and random forests were employed, but infrequently.

The absence of syntactic approaches in the reviewed literature stands out as a void in this field of study. The fact that most of the investigations relied on a centralized database also leaves room for improvement. Classifiers can be made extremely robust by using images from many resources. One final potential weakness in this field of study is the absence of classifiers capable of online learning.

Additionally, we indicated obstacles that must be addressed. We could indeed cite the insufficiency of standardization of the performance measures used to analyze the study results, the exclusion of a standardizations dataset or framework to enhance the comparative evaluation of classifiers, the fact that the majority of research teams do not correlate their outcomes with those of healthcare professionals, the fact that the training and testing procedures are frequently unclear, and the large number of insufficient studies published.

Finally, we intend to emphasize the significance of this study topic to society in general. BC is one of the extreme health issues confronting us today and merits the attention of corporate entities and government research organizations. With earlier and more accurate diagnoses, a greater number of women will overcome this struggle.

References

[1] Ferlay J, et al. Global cancer observatory: cancer tomorrow. Lyon, France: International Agency for Research on Cancer; 2020. https://gco.iarc.fr/tomorrow. [Accessed January 2021].

[2] Wu S, et al. Evaluating intrinsic and non-intrinsic cancer risk factors. Nature Communications 2018;9(3490). https://doi.org/10.1038/s41467-018-05467-z.

[3] Hanahan D, Weinberg RA. Hallmarks of cancer: the next generation. Cell 2011; 144(5):646−74. https://doi.org/10.1016/j.cell.2011.02.013.

[4] Gonzalez DM, Medici D. Signaling mechanisms of the epithelial-mesenchymal transition. Science Signaling 2014;7(344). https://doi.org/10.1126/scisignal.2005189. re8.

[5] WHO Global Burden of Disease, 2020; Department of Health, Hong Kong, China, 2020; Disease Registry, Macau, China, 2020.

[6] Pita-Fernández S, et al. Effect of diagnostic delay on survival in patients with colorectal cancer: a retrospective cohort study. BMC Cancer 2016;16(1):1−11. https://doi.org/10.1186/s12885-016-2717-z.

[7] Rebecca L, Siegel MPH, Kimberly D, Miller MPH, Hannah E, Fuchs BS, Ahmedin Jemal DVM. Cancer statistics. CA: A Cancer Journal for Clinicians 2017;67(27): 7−30. https://doi.org/10.3322/caac.21708.

[8] Lehmann BD, Pietenpol JA, Tan AR. Triple-negative breast cancer: molecular subtypes and new targets for therapy. American Society of Clinical Oncology Educational Book 2015;35(35):31−9.

[9] Tfayli A, Temraz S, Abou Mrad R, Shamseddine A. Breast cancer in low- and middle-income countries: an emerging and challenging epidemic. Journal of Oncology 2010; 2010.

[10] Da Costa Vieira RA, Biller G, Uemura G, Ruiz CA, Curado MP. Breast cancer screening in developing countries. Clinic 2017;72.

[11] Shulman LN, Willett W, Sievers A, Knaul FM. Breast cancer in developing countries: opportunities for improved survival. Journal of Oncology 2010;2010.

[12] Nounou MI, Elamrawy F, Ahmed N, Abdelraouf K, Goda S, Syed-Sha-Qhattal H. Breast cancer: conventional diagnosis and treatment modalities and recent patents and technologies supplementary issue: targeted therapies in breast cancer treatment. Breast Cancer: Basic and Clinical Research 2015;9(Suppl. 2):17−34.

[13] Sharma GN, Dave R, Sanadya J, Sharma P, Sharma KK. Various types and management of breast cancer: an overview. Journal of Advanced Pharmaceutical Technology & Research 2010;1(2):109.

[14] Tot T. Correlating the ground truth of mammographic histology with the success or failure of imaging. Technology in Cancer Research and Treatment 2005;4(1):23—8. https://doi.org/10.1177/153303460500400104.

[15] Tot T. DCIS, cytokeratins and the theory of the sick lobe. Virchows Archiv 2005;447: 1—8. https://doi.org/10.1007/s00428-005-1274-7.

[16] Tot T. The theory of the sick lobe and the possible consequences. International Journal of Surgical Pathology 2007;15(4):369—75. https://doi.org/10.1177/1066896907302225.

[17] Tot T. The theory of the sick lobe. In: Tot T, editor. Breast cancer — a lobar disease. London: Springer; 2011. p. 1—17. https://doi.org/10.1007/978-1-84996-314-5_1.

[18] Tabár L, Vitak B, Chen H-HT, Yen M-F, Duffy SW, Smith RA. Beyond randomized controlled trials: organized mammographic screening substantially reduces breast carcinoma mortality. Cancer 2001;91(9):1724—31.

[19] Zhou W, Sollie T, Tot T, et al. Breast cancer with neoductgenesis: histopathological criteria and its correlation with mammographic and tumour features. International Journal of Breast Cancer 2014;2014:581706. https://doi.org/10.1155/2014/581706.

[20] Holland R, Hendriks JHCL. Microcalcifications associated with ductal carcinoma in situ: mammographic-pathologic correlation. Seminars in Diagnostic Pathology 1994;11(3):181—92.

[21] Tot T, Gere M. Radiological-pathological correlation in diagnosing breast carcinoma: the role of pathology in the multimodality era. Pathology and Oncology Research 2008;14(2):173—8.

[22] Tot T, Tabár L. Mammographic — pathologic correlation of ductal carcinoma in situ of the breast using two- and three-dimensional large histological sections. Seminars in Breast Disease 2005;8:144—51.

[23] Tang J, Rangayyan RM, Xu J, El Naqa I, Yang Y. Computer-aided detection and diagnosis of breast cancer with mammography: recent advances. IEEE Transactions on Information Technology in Biomedicine 2009;13(2):236—51.

[24] Horsch K, Giger ML, Vyborny CJ, Lan L, Mendelson EB, Hendrick RE. Classification of breast lesions with multimodality computer-aided diagnosis: observer study results on an independent clinical data set. Radiology 2006;240(2):357—68.

[25] Malur S, Wurdinger S, Moritz A, Michels W, Schneider A. Comparison of written reports of mammography, sonography and magnetic resonance mammography for preoperative evaluation of breast lesions, with special emphasis on magnetic resonance mammography. Breast Cancer Research 2000;3(1):1—6.

[26] Kopans DB. Sonography should not be used for breast cancer screening until its efficacy has been proven scientifically. American Journal of Roentgenology 2004;182(2): 489—91.

[27] Dromain C, Balleyguier C. Contrast-enhanced digital mammography. In: Digital mammography. Springer; 2010. p. 187—98.

[28] Kelly KM, Dean J, Comulada WS, Lee S-J. Breast cancer detection using automated whole breast ultrasound and mammography in radiographically dense breasts. European Radiology 2010;20(3):734—42.

[29] Corsetti V, Houssami N, Ghirardi M, Ferrari A, Speziani M, Bellarosa S, et al. Evidence of the effect of adjunct ultrasound screening in women with mammography-negative

dense breasts: interval breast cancers at 1 year follow-up. European Journal of Cancer 2011;47(7):1021−6.

[30] Truhn D, Schrading S, Haarburger C, Schneider H, Merhof D, Kuhl C. Radiomic versus convolutional neural networks analysis for classification of contrast-enhancing lesions at multiparametric breast MRI. Radiology 2019;290(2):290−7.

[31] Honda E, Nakayama R, Koyama H, Yamashita A. Computer-aided diagnosis scheme for distinguishing between benign and malignant masses in breast DCE-MRI. Journal of Digital Imaging 2016;29(3):388−93.

[32] Yang X, Wu L, Ye W, Zhao K, Wang Y, Liu W, et al. Deep learning signature based on staging CT for preoperative prediction of sentinel lymph node metastasis in breast cancer. Academic Radiology 2020;27(9):1226−33.

[33] Han Z, Wei B, Zheng Y, Yin Y, Li K, Li S. Breast cancer multi-classification from histopathological images with structured deep learning model. Scientific Reports 2017; 7(1):1−10.

[34] Frize M, Herry C, Roberge R. Processing of thermal images to detect breast cancer: comparison with previous work. In: Proceedings of the second joint 24th annual conference and the annual fall meeting of the biomedical engineering society, engineering in medicine and biology, vol 2. IEEE; 2002. p. 1159−60.

[35] Ahirwar A, Jadon RS. Characterization of tumor region using SOM and neuro fuzzy techniques in digital mammography. International Journal of Computer Science and Information Technology 2011;3(1):199−211.

[36] Amjath AJ, Janet J. Discrete Shearlet transform based classification of microcalcification in digital mammograms. Journal of Computer Applications 2013;6(1):19−21.

[37] Lehman CD, Arao RF, Sprague BL, Lee JM, Buist DS, Kerlikowske K, et al. National performance benchmarks for modern screening digital mammography: update from the Breast Cancer Surveillance Consortium. Radiology 2016;283(1):49−58.

[38] Blanks RG, Wallis MG, Moss SM. A comparison of cancer detection rates achieved by breast cancer screening programmes by number of readers, for one and two view mammography: results from the UK National Health Service breast screening programme. Journal of Medical Screening 1998;5(4):195−201.

[39] Agarwal G, Ramakant P. Breast cancer care in India: the current scenario and the challenges for the future. Breast Care 2008;3(1):21−7.

[40] Olson JS. Bathsheba's breast: women, cancer, and history. JHU Press; 2002.

[41] Gur D. Digital mammography: do we need to convert now? Radiology 2007;245(1): 10−1.

[42] Pisano ED, et al. Should breast imaging practices convert to digital mammography? A response from members of the DMIST executive committee. Radiology 2007;245(1): 12−3.

[43] Sampat MP, Markey MK, Bovik AC. Computer-aided detection and diagnosis in mammography. Handbook of Image and Video Processing 2005;2(1):1195−217.

[44] Oliver A, Torrent A, Lladó X, Tortajada M, Tortajada L, Sentís M, et al. Automatic microcalcification and cluster detection for digital and digitised mammograms. Knowledge-Based Systems 2012;28:68−75.

[45] Nakayama R, Uchiyama Y, Yamamoto K, Watanabe R, Namba K. Computer-aided diagnosis scheme using a filter bank for detection of microcalcification clusters in mammograms. IEEE Transactions on Biomedical Engineering 2006;53(2):273−83.

[46] Sankar D, Thomas T. A new fast fractal modeling approach for the detection of microcalcifications in mammograms. Journal of Digital Imaging 2010;23(5):538−46.

[47] Qi D, Lei Y. Multifractal spectrum theory used to medical image from CT testing. 2008 IEEE International Conference on Advanced Intelligent Mechatronics 2008;1(1): 68–73.

[48] Torrent A, Oliver A, Lladó X, Marti R, Freixenet J. A supervised micro-calcification detection approach in digitized mammograms. IEEE seventeenth International Conference on Image Processing (ICIP) 2010:4345–8.

[49] Hussain M, Khan S, Muhammad G, Bebis G. A comparison of different gabor features for mass classification in mammography. In: IEEE eighth international conference on signal image technology and internet based systems (SITIS); 2012. p. 142–8.

[50] Anand S, Aynesh Vijaya Rathna R. Detection of architectural distortion in mammogram images using contourlet transform. In: IEEE international conference on emerging trends in computing, communication and nanotechnology; 2013.

[51] Isa NAM, Ting SS. Automatic segmentation and detection of mass in digital mammograms. In: Proceedings of the eleventh international conference on telecommunications and informatics (TELE-INFO 12) & proceedings of the eleventh international conference on signal processing (SIP 12); 2012. p. 143–6.

[52] Garge DM, Bapat VN. A low cost wavelet based mammogram image processing for early detection of breast cancer. Indian Journal of Science and Technology 2009; 2(9):63–5.

[53] D'Elia C, Marrocco C, Molinara M, Tortorella F. Detection of clusters of microcalcifications in mammograms: a multi classifier approach. In: IEEE TwentyFirst international symposium on computer based medical systems CBMS'08; 2008. p. 572–7.

[54] Jasmine JS, Baskaran S, Govardhan A. Non subsampled contourlet transform based classification of microcalcification in digital mammograms. Procedia Engineering 2012;38:622–31.

[55] Zaheeruddin ZA, Singh L. Detection and shape feature extraction of breast tumor in mammograms. In: Proceedings of the world congress on engineering; 2012. p. 719–24.

[56] Wei L, Peng D. A new contour detection in mammogram using sequential edge linking. In: IEEE second international symposium on intelligent information technology application' 08; 2008. p. 197–200.

[57] Mohanty AK, Beberta S, Lenka SK. Classifying benign and malignant mass using GLCM and GLRLM based texture features from mammogram. International Journal of Engineering Research in Africa 2011;1(3):687–93.

[58] Bhattacharya M, Das A. Fuzzy logic based segmentation of microcalcification in breast using digital mammograms considering multiresolution. In: Proceedings of the international conference in machine vision and image processing IMVIP 2007; 2007. p. 98–105.

[59] Baumgartner C. Computer-aided detection of malignant breast masses in mammography using random forests. Semester thesis. Swiss Federal Institute of Technology; 2010.

[60] Dheeba J, Selvi ST. Classification of malignant and benign microcalcification using SVM classifier. In: International conference in emerging trends in electrical and computer technology, Nagercoil, India; 2011. p. 686–90.

[61] Alsaedi D, Melnikov A, Muzaffar K, Mandelis A, Ramahi OM. A microwave-thermography hybrid technique for breast cancer detection. In: IEEE Journal of Electromagnetics, RF and Microwaves in Medicine and Biology; 2021.

[62] Hopp T, Duric N, Ruiter NV. Image fusion of Ultrasound Computer Tomography volumes with X-ray mammograms using a biomechanical model based 2D/3D registration. Computerized Medical Imaging and Graphics 2015;40:170−81.

[63] Ribli D, Horváth A, Unger Z, Pollner P, Csabai I. Detecting and classifying lesions in mammograms with deep learning. Scientific Reports 2018;8(1):1−7.

[64] Peng J, Bao C, Hu C, Wang X, Jian W, Liu W. Automated mammographic mass detection using deformable convolution and multiscale features. Medical & Biological Engineering & Computing 2020;58(7):1405−17.

[65] Kaji S, Kida S. Overview of image-to-image translation by use of deep neural networks: denoising, super-resolution, modality conversion, and reconstruction in medical imaging. Radiological Physics and Technology 2019;12(3):235−48.

[66] de Nazaré Silva J, de Carvalho Filho AO, Corrêa Silva A, Cardoso de Paiva A, Gattass M. Automatic detection of masses in mammograms using quality threshold clustering, correlogram function, and SVM. Journal of Digital Imaging 2015;28(3):323−37.

[67] Sanae B, Mounir AK, Youssef F, et al. Statistical block-based DWT features for digital mammograms classification. In: 2014 9th international conference on intelligent systems: theories and applications (SITA-14), IEEE; 2014. p. 1−7.

[68] de Oliveira FSS, de Carvalho Filho AO, Silva AC, de Paiva AC, Gattass M. Classification of breast regions as mass and non-mass based on digital mammograms using taxonomic indexes and SVM. Computers in Biology and Medicine 2015;57:42−53.

[69] LeCun Y, Bengio Y, Hinton G. Deep learning. Nature 2015;521(7553):436−44.

[70] Jiao Z, Gao X, Wang Y, Li J. A deep feature based framework for breast masses classification. Neurocomputing 2016;197:221−31.

[71] Moreira IC, Amaral I, Domingues I, Cardoso A, Cardoso MJ, Cardoso JS. Inbreast: toward a full-field digital mammographic database. Academic Radiology 2012;19(2):236−48.

[72] Shen L, Margolies LR, Rothstein JH, Fluder E, McBride R, Sieh W. Deep learning to improve breast cancer detection on screening mammography. Scientific Reports 2019;9(1):1−12.

[73] Lee RS, Gimenez F, Hoogi A, Miyake KK, Gorovoy M, Rubin DL. A curated mammography data set for use in computer-aided detection and diagnosis research. Scientific Data 2017;4(1):1−9.

[74] Oliveira JE, Gueld MO, Araújo ADA, Ott B, Deserno TM. Toward a standard reference database for computer-aided mammography. In: Medical imaging 2008: computer-aided diagnosis, vol 6915. SPIE; 2008. p. 606−14.

[75] Fernandes F, Bonifacio R, Brasil L, Guadagnin R, Lamas J, Uchiyama N. MIDAS−mammographic image database for automated analysis. In: Mammography-recent advances; 2012. p. 243−60.

[76] Lopez MG, Posada N, Moura DC, Pollán RR, Valiente JMF, Ortega CS, et al. BCDR: a breast cancer digital repository. 15th International Conference on Experimental Mechanics 2012;1215:113−20.

[77] Matheus BRN, Schiabel H. Online mammographic images database for development and comparison of CAD schemes. Journal of Digital Imaging 2011;24(3):500−6.

[78] Zheng Y, Yang C, Merkulov A, Bandari M. Early breast cancer detection with digital mammograms using Haar-like features and AdaBoost algorithm. In: SPIE proceedings, 9871, sensing and analysis technologies for biomedical and cognitive applications 2016; 2016. 98710D.

[79] Feng SSJ, Sechopoulos I. Clinical digital breast tomosynthesis system: dosimetric characterization. Radiology 2012;263:35—42. https://doi.org/10.1148/radiol.11111789.

[80] Chong A, Weinstein SP, McDonald ES, Conant EF. Digital breast tomosynthesis: concepts and clinical practice. Radiology 2019;292:1—14.

[81] ORSI MA, Cellina M, Rosti C, Gibelli D, Belloni E, Oliva G. Digital breast tomosynthesis: a state-of-the-art review. Nuclear Medicine and Biomedical Imaging 2018;3(4): 1—9.

[82] Calliste J, Wu G, Laganis PE, et al. Second generation stationary digital breast tomosynthesis system with faster scan time and wider angular span. Medical Physics 2017; 44(9):4482—95.

[83] Jochelson MS, Pinker K, Dershaw DD, et al. Comparison of screening CEDM and MRI for women at increased risk for breast cancer: a pilot study. European Journal of Radiology 2017;97:37—43.

[84] Chou CP, Lewin JM, Chiang CL, et al. Clinical evaluation of contrast-enhanced digital mammography and contrast enhanced tomosynthesis: comparison to contrast-enhanced breast MRI. European Journal of Radiology 2015;84(12):2501—8.

[85] Dang PA, Freer PE, Humphrey KL, Halpern EF, Rafferty EA. Addition of tomosynthesis to conventional digital mammography: effect on image interpretation time of screening examinations. Radiology 2014;270(1):49—56.

[86] Skaane P, Bandos AI, Gullien R, et al. Comparison of digital mammography alone and digital mammography plus tomosynthesis in a population-based screening program. Radiology 2013;267(1):47—56.

[87] Benedikt RA, Boatsman JE, Swann CA, Kirkpatrick AD, Toledano AY. Concurrent computer-aided detection improves reading time of digital breast tomosynthesis and maintains interpretation performance in a multireader multicase study. American Journal of Roentgenology 2018;210(3):685—94.

[88] Mohamed AA, Berg WA, Peng H, Luo Y, Jankowitz RC, Wu S. A deep learning method for classifying mammographic breast density categories. Medical Physics 2018;45(1):314—21.

[89] Chougrad H, Zouaki H, Alheyane O. Deep convolutional neural networks for breast cancer screening. Computer Methods and Programs in Biomedicine 2018;157:19—30.

[90] Becker AS, Marcon M, Ghafoor S, Wurnig MC, Frauenfelder T, Boss A. Deep learning in mammography: diagnostic accuracy of a multipurpose image analysis software in the detection of breast cancer. Investigative Radiology 2017;52(7):434—40.

[91] Schliemann D, Hoe WMK, Mohan D, Allotey P, Reidpath DD, Tan MM, et al. Challenges and opportunities for breast cancer early detection among rural dwelling women in Segamat District, Malaysia: a qualitative study. PLoS One May 20, 2022; 17(5):e0267308. https://doi.org/10.1371/journal.pone.0267308.

[92] Thornton H, Edwards A, Baum M. Women need better information about routine mammography. BMJ 2003;327(7406):101—3.

[93] Elmore JG, Barton MB, Moceri VM, Polk S, Arena PJ, Fletcher SW. Ten-year risk of false positive screening mammograms and clinical breast examinations. New England Journal of Medicine 1998;338(16):1089—96.

[94] Lidbrink E, Elfving J, Frisell J, Jonsson E. Neglected aspects of false positive findings of mammography in breast cancer screening: analysis of false positive cases from the Stockholm trial. BMJ 1996;312(7026):273—6.

[95] Davey C, White V, Ward JE. Insurance repercussions of mammographic screening: what do women think? Medical Science Monitor 2002;8(12):LE54—5.

[96] Drossaert CH, Boer H, Seydel ER. Does mammographic screening and a negative result affect attitudes towards future breast screening? Journal of Medical Screening 2001;8(4):204−12.

[97] Drossaert CH, Boer H, Seydel ER. Monitoring women's experiences during three rounds of breast cancer screening: results from a longitudinal study. Journal of Medical Screening 2002;9(4):168−75.

[98] Andrews FJ. Pain during mammography: implications for breast screening programmes. Australasian Radiology 2001;45(2):113−7.

[99] Sapir R, Patlas M, Strano SD, Hadas-Halpern I, Cherny NI. Does mammography hurt? Journal of Pain and Symptom Management 2003;25(1):53−63.

[100] Fletcher SW, Elmore JG. Clinical practice. Mammographic screening for breast cancer. New England Journal of Medicine 2003;348(17):1672−80.

[101] Kolb TM, Lichy J, Newhouse JH. Comparison of the performance of screening mammography, physical examination, and breast US and evaluation of factors that influence them: an analysis of 27,825 patient evaluations. Radiology 2002;225(1): 165−75.

[102] Slanetz PJ. Hormone replacement therapy and breast tissue density on mammography. Menopause 2002;9(2):82−3.

[103] Dixon JM. Hormone replacement therapy and the breast. Surgical Oncology 2003; 12(4):251−63.

[104] White E, Velentgas P, Mandelson MT, Lehman CD, Elmore JG, Porter P, et al. Variation in mammographic breast density by time in menstrual cycle among women aged 40−49 years. Journal of the National Cancer Institute 1998;90(12):906−10.

[105] Harvey J, Bovbjerg VE. Quantitative assessment of mammographic breast density: relationship with breast cancer risk. Radiology 2004;230:29−41.

[106] Sandin B, Chorot P, Valiente RM, Lostao L, Santed MA. Adverse psychological effects in women attending a second-stage breast cancer screening. Journal of Psychosomatic Research 2002;52(5):303−9.

[107] Sechopoulos I, Suryanarayanan S, Vedantham S, D'Orsi C, Karellas A. Computation of the glandular radiation dose in digital tomosynthesis of the breast: computation of dose in digital breast tomosynthesis. Medical Physics 2006;34:221−32. https://doi.org/10.1118/1.2400836.

[108] Sechopoulos I, Suryanarayanan S, Vedantham S, D'Orsi CJ, Karellas A. Scatter radiation in digital tomosynthesis of the breast: scatter radiation in breast tomosynthesis. Medical Physics 2007;34:564−76.

[109] Hu Y-H, Zhao B, Zhao W. Image artifacts in digital breast tomosynthesis: investigation of the effects of system geometry and reconstruction parameters using a linear system approach: image artifacts in digital breast tomosynthesis. Medical Physics 2008;35: 5242−52. https://doi.org/10.1118/1.2996110.

[110] Machida H, Yuhara T, Mori T, Ueno E, Moribe Y, Sabol JM. Optimizing parameters for flat-panel detector digital tomosynthesis. RadioGraphics 2010;30:549−62. https://doi.org/10.1148/rg.302095097.

[111] Dar RA, Rasool M, Assad A. Breast cancer detection using deep learning: datasets, methods, and challenges ahead. Computers in Biology and Medicine 2022:106073.

[112] Becker AS, Jendele L, Skopek O, Berger N, Ghafoor S, Marcon M, et al. Injecting and removing suspicious features in breast imaging with CycleGAN: a pilot study of automated adversarial attacks using neural networks on small images. European Journal of Radiology 2019;120:108649.

[113] Pangalos G, Gritzalis D, Khair M, Bozios L. Improving the security of medical database systems53$. In: Eloff JHP, von Solms SH, editors. Information security — the next decade. IFIP advances in information and communication technology. Boston, MA: Springer; 1995.

[114] Patel V. A framework for secure and decentralized sharing of medical imaging data via blockchain consensus. Health Informatics Journal December 2019;25(4):1398—411.

[115] Smilde E. ILC detection Applying image processing and deep learning to improve the detection of Invasive Lobular Carcinoma using mammography Jan 2022.

[116] Mazlan AU, Sahabudin NA, Remli MA, Ismail NSN, Mohamad MS, Nies HW, et al. A review on recent progress in machine learning and deep learning methods for cancer classification on gene expression data. Processes 2021;9(8):1466. https://doi.org/10.3390/pr9081466.

[117] Kneeshaw PJ, Turnbull LW, Drew PJ. Current applications and future direction of MR mammography. British Journal of Cancer January 13, 2003;88(1):4—10. https://doi.org/10.1038/sj.bjc.6600713.

[118] El Houby EM, Yassin IR. Malignant and nonmalignant classification of breast lesions in mammograms using convolutional neural networks. Biomedical Signal Processing and Control September 2021;70:102954. https://doi.org/10.1016/j.bspc.2021.102954.

[119] KrGhosh S, Ghosh A. A novel hyperbolic intuitionistic fuzzy divergence measure based mammogram enhancement for visual elucidation of breast lesions. Biomedical Signal Processing and Control May 2022;75:103586. https://doi.org/10.1016/j.bspc.2022.103586.

[120] Dhamija E, Gulati M, Deo SVS, Gogia A, Hari S. Digital breast tomosynthesis: an overview. Indian Journal of Surgical Oncology 2021;12:315—29.

[121] Loizidou K, Elia R, Pitris C. Computer-aided breast cancer detection and classification in mammography: a comprehensive review. Computers in Biology and Medicine February 2023;153:106554. https://doi.org/10.1016/j.compbiomed.2023.106554.

[122] Jabeen K, Attique Khan M, Alhaisoni M, Tariq U, Zhang Y-D, Hamza A, Mickus A, Damaševičius R. Breast cancer classification from ultrasound images using probability-based optimal deep learning feature fusion. Journal of Sensors 2022;22(3).

[123] Song SY, Lee YY, Shin HY, Park B, Suh M, Choi KS, et al. Trends in breast cancer screening rates among Korean women: results from the Korean national cancer screening survey (KNCSS), 2005—2020. Epidemiology and Health November 24, 2022:e2022111. https://doi.org/10.4178/epih.e2022111.

AI in breast imaging: Applications, challenges, and future research

2

Parita Oza

Institute of Technology, Nirma University, Ahmedabad, Gujarat, India

1. Introduction

1.1 Breast cancer: Statistics

Breast cancer is the most common malignancy among women all over the world. With an estimated 2.3 million new cases, which is around 11.7% of all cancer cases, this disease has surpassed lung cancer, which was the top cause of cancer incidence globally in 2020 [1]. As per the study presented by Globocon 2020, every 4 min, a woman is diagnosed with breast cancer in India. Breast cancer is now the most prevalent cancer among Indian women, with over 178,000 new cases being identified each year. What is even more concerning is that, as compared to the West, in India, this illness is being progressively diagnosed at an early age (almost a decade earlier). Tragically, breast cancer claims the lives of 90,000 people a year. Additionally, a woman loses her life every 8 min due to this disease in India [2]. According to a study by FICCI and EY titled "Call for Action: Making Quality Cancer Care More Accessible and Affordable in India," the reported cancer incidence in India in 2022 is predicted to be between 19 and 20 lacks, whereas the actual incidence is likely to be 1.5−3 times higher than the reported cases [3]. Additionally, The younger age group also experiences it more frequently. The age range of 25−50 accounts for about 50% of breast cancer cases. Additionally, poor survival and high mortality were seen in more than 70% of the patients in the advanced stage [4]. Early identification and prompt treatment are the most effective interventions for breast cancer management, according to the World Cancer Report 2020 [5]. Additionally, earlier diagnosis of this disease can also result in less expensive treatment [6]. Fig. 2.1 presents some alarming facts about breast cancer in India.

1.2 Breast imaging techniques and common breast abnormalities

A wide variety of tools and technologies are used by radiologists for the purpose of screening and also for the detection and diagnosis of breast cancer. Once cancer is detected, with the help of these diagnostics tests, doctors can determine its type and its exact location and stage. Various imaging modalities are continuously being developed for disease diagnosis to improve the accuracy of results. Some of these

Computational Intelligence and Modelling Techniques for Disease Detection in Mammogram Images
https://doi.org/10.1016/B978-0-443-13999-4.00005-5

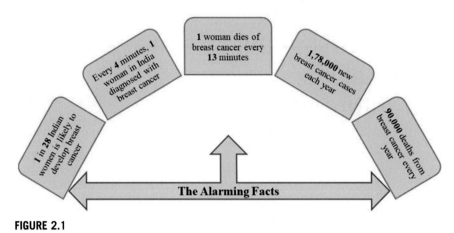

FIGURE 2.1

Alarming facts about breast cancer in India.

imaging modalities are used only for screening processes, some for diagnosis purposes, and very few for supportive evaluation. There is overlap among these technologies, and the purpose of their usage can vary. Modalities used for screening are expected to be cost-effective and efficient in reaching the masses. Once cancer is detected, other diagnostic modalities can be used for detailed evaluation. Supportive imaging modalities are helpful to the practitioner for providing an added reliance for the primary diagnosis and providing the clinicians with additional confidence in their initial diagnosis. Imaging modalities used nowadays include digital mammography, breast ultrasound, breast magnetic resonance imaging (MRI), positron emission tomography and computed tomography (PET CT), and thermal imaging [7]. Digital mammography is the most widely used method of breast imaging. It is an X-ray picture of the women's breast, which uses very low-dose X-rays for the examination. This technique is used to check for breast cancer in women having no symptoms or signs of the disease. Abnormal masses and tiny calcium spots called microcalcifications are seen better on the mammogram. This method has become the gold standard for the early detection of breast cancer before the masses become clinically intense [8]. Breast cancer datasets of imaging modalities like Breast MRI, Breast Ultrasound, and Breast thermogram are also available. Even yet, mammogram-based datasets are primarily utilized for research as mammograms are regarded as the gold standard for the detection of breast cancer.

There are mainly two types of findings seen in mammograms: masses and calcifications. A mass in the breast can either be benign or malignant. Benign masses do not have malignancy, grow exceptionally slowly, and do not spread to nearby tissues and organs. Where malignant masses are cancerous, develop much more quickly than benign masses, have the propensity to spread, and can affect nearby breast tissues and organs [9]. Both these masses can further be identified by their shape, margins, and density. The shape of the mass can be round, oval, or irregular. The malignancy probability increases as the mass's shape become even more irregular.

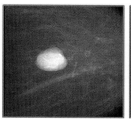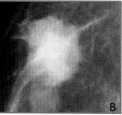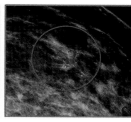

FIGURE 2.2

(A) Benign mass, (B) malignant mass, and (C) microcalcification.

The margins can be indistinct, microlobulated, spiculated, circumscribed, or obscured. Spiculated masses are linked to a high probability of malignancy. Mass with high density is also likely to be malignant mass. Calcifications can be classified as microcalcifications or macrocalcifications. Microcalcifications are tiny calcium spots and are thought to be an early indicator of breast cancer if they appear in specific patterns [10]. We show some examples of these abnormalities in Fig. 2.2 [11].

1.3 Mammogram datasets

There is a need for a proper mammographic database to train, validate, and test deep learning-based approaches. In addition, the deep learning method needs a substantial amount of data compared to conventional machine learning approaches. Abnormalities available in mammogram datasets are mainly calcifications, masses, architectural distortion, and asymmetries. Many public mammogram repositories exist and are widely adopted by researchers in the domain that includes DDSM [11], CBIS-DDSM [12], MIAS [13], INBreast [14], BCDR [15], and IRMA [16], and SuReMaPP [17]. There are also other recently published mammograms such as CSAW [18], CSAW-S [19], and KAU-BCMD [20]. Annotations are available either at pixel-level boundaries of abnormalities or the center of the radius of suspicious regions or supported with a file containing a class of abnormality and other essential details. Digital mammogram images are initially available in DICOM format with some metadata that is not useful for deep learning methods; hence, we need to extract the image matrix out of each file in the set. Some datasets have different formats, like PGM (potable gray map) and LJPEG (lossless JPEG). Image size and resolution, as well as abnormality distribution, may vary in these datasets (Table 2.1).

2. Toward AI for breast cancer diagnosis

Artificial intelligence (AI) is considered a system with the capacity to interpret external input appropriately, to learn from this input, and to apply those learnings to fulfill specified objectives and tasks through flexible adaptation. Applications of AI have expanded into new fields over the past many years as a result of the

Table 2.1 Some popular mammogram repositories available in the literature.

Dataset	No of images	Type of image	Multiview	BIRADS	Year
DDSM	10,000	Ljpeg	Yes	Yes	1998
MIAS	322	Pgm	No	No	2003
IRMA	10,509	Several	Yes	Yes	2008
INBreast	422	Dcm	Yes	Yes	2011
BCDR	7315	Tiff	Yes	Unknown	2012
CBIS-DDSM	10,239	Dcm	Yes	Yes	2017
SureMaPP	343	Dcm	No	No	2020
CSAW	1,000,000	Dcm	Yes	Unknown	2020
CSAW-S	Unknown	Dcm	Yes	No	2020
KAU-BCMD	Unknown	dcm/jpg	Yes	Yes	2021

explosive rise of computer functions connected to significant data intrusion. Currently, AI is used in driverless cars, face recognition, speech recognition, and other innovative technologies. In addition, the use of AI in medical imaging has progressively grown in importance as a study area. AI-based computer-aided diagnosis (CAD) systems have been widely employed in diagnostic imaging for the last many years, including breast cancer diagnosis [9]. The radiologist can use these tools as a second reader tool. Due to their higher accuracy rates, AI approaches like machine learning (ML) and deep learning (DL) have gradually replaced the basic image processing-based techniques such as low-level image descriptors, including texture, local key points, and boundaries.

Machine learning connects the challenge of deducing information from samples of input data to the fundamental principles of inference. The algorithm may derive information from training data using this method's analytic, statistical, and mathematical methodologies without direct programming. For developing computer-aided breast cancer diagnosis systems, some machine learning algorithms [21−23], including support vector machine (SVM), nave Bayes, and artificial neural network (ANN) have become highly popular. The extraction of visual characteristics is typically the initial stage in machine learning approaches. Therefore, the appropriate feature selection is crucial to the overall training accuracy. Moreover, accurate feature selection is a complex and time-consuming task and requires expert knowledge. These difficulties led to the development of a novel deep learning strategy, which improved the conventional machine learning paradigm.

Deep learning delivers higher levels of generalization than traditional machine learning because it focuses on knowledge inference methods from data. Convolutional neural network (CNN) is one of the most persuasive DL networks characterized by several convolutional layers. Due to the large number of inner layers that partially conduct feature extraction along the way using layer-embedded operators, DL methods are independent of feature extraction procedures. These algorithms are not taught to recognize breast masses by providing details on the shape, patterns, and other aspects of the mass; instead, the algorithm picks up on this information on its own while being trained [10]. We discuss applications of these methods for breast cancer diagnosis domain in the corresponding sections of the paper.

2.1 AI applications for mammogram-based breast cancer analysis

Digital mammography is the most commonly used imaging modality for breast cancer diagnosis and screening. This technique successfully identifies benign and malignant masses and can easily detect microcalcifications. AI has been used to analyze mammography images called mammograms for a certain application; these applications are discussed in this section. Fig. 2.3 presents a pictorial view of AI applications for breast mammogram analysis. A lot of work has been done in this area; in this section, we discuss a few of those methods; we additionally summarize a few recent state-of-the-art applications of mammogram-based breast cancer diagnosis in Table 2.2.

2.1.1 Breast abnormality identification and categorization

The most prevalent study areas include mammography image analysis for breast abnormality (mass or calcifications) identification and categorization. Various

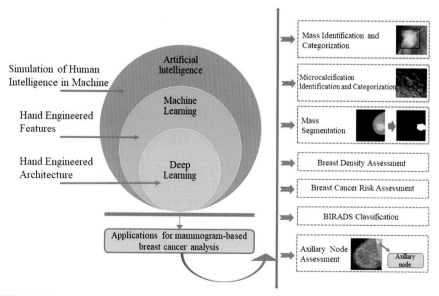

FIGURE 2.3

AI applications for mammogram-based breast cancer analysis.

Table 2.2 Summary of recent state-of-the-arts in mammogram based breast cancer diagnosis.

Ref.	Application	Type of abnormality	Dataset	Model
[24]	Classification	Mass	MIAS	CNN
[25]	Classification	Mass	CBIS-DDSM	CNN
[26]	Classification	Mass	MIAS, CBIS-DDSM, private	Ensemble CNN
[27]	Classification	Microcalcification	Private	GCN
[28]	Detection	Microcalcification	Private	CNN
[29]	Detection, segmentation, classification	Mass	INbreast	YOLO, FrCNN
[30]	Segmentation	Mass	DDSM	Level sets
[31]	Detection, classification	Mass	DDSM	DBN
[32]	Segmentation	Mass	INbreast, CBIS-DDSM	CNN
[33]	Segmentation	Mass	INbreast, CBIS-DDSM, private	UNet
[34]	Segmentation, classification	Mass	MIAS, DDSM, CBIS-DDSM	CNN
[35]	Breast density assessment	—	Private	CNN
[36]	Breast density classification	—	Private	CNN
[37]	Breast density classification	—	TCIA	CNN
[38]	Breast density assessment	—	Private	CNN
[39]	Risk assessment	—	Private	CNN
[40]	Risk assessment	—	Private	CNN, NLP
[41]	Risk assessment	—	FFDM	DNN
[42]	BI-RADS classification	—	Private	DNN
[43]	BI-RADS classification	—	Inbreast	CNN
[44]	BI-RADS classification	—	Inbreast	CNN
[45]	Axillary node metastasis	—	Private	ANN
[46]	Axillary node detection	—	Private	DCNN

DBN, *Deep Belief Network;* FrCNN, *Full Resolution Convolutional Network;* GCN, *Graph Convolutional Network;* NLP, *Natural Language Processing.*

algorithms have been developed for automated breast mass identification and categorization, and this topic is continuously being researched. Mammogram mass identification is an important computer vision task; the CAD system further examines these detected abnormalities. Masses that have been found abnormal can be classified as benign or malignant. These tasks are exceedingly difficult due to anatomical variances in masses' sizes, shapes, and placement in mammograms [47]. Transfer learning is an excellent method for dealing with the small training set, enabling us to utilize deep learning's extendibility fully. In one of our previous works [24], we could increase training accuracy (96.14%) compared to the model created from scratch by using a pretrained network on the preprocessed MIAS dataset. In another work, we trained and tested various variants of pretrained CNN models to classify breast mass into benign or malignant class [25]. An ensemble model is utilized in the study [26] to propose a strategy for clinical decision support systems that can precisely diagnose benign and malignant masses from breast X-ray pictures. The work used an average, weighted average, and concatenation mechanism that relies on transfer learning to combine pretrained CNN.

Breast calcifications, which manifest as little white spots on mammography, are localized deposits of calcium spots in the breast tissue. Calcifications come in two main varieties: macrocalcifications and microcalcifications. Large, granular macrocalcifications are mostly harmless and age-related [10]. Microcalcifications range in size from 0.1 to 1 mm. They may be considered an early indicator of breast cancer; hence, if detected in advance, proper treatment planning is possible. They could be regarded as early breast cancer warning signs, so if they are identified early, appropriate treatment can be planned. As microcalcifications are so small, even an experienced radiologist has a significant probability of misreading them. To prevent such overlook mistakes, the radiologist can undoubtedly benefit from using CAD technologies to identify microcalcifications. To classify clustered microcalcifications, in the work [27], the discriminant model based on image convolution is utilized to learn the image characteristics. Another study [28] presented a DL-based CAD system for microcalcification detection. The authors extracted fixed sizes of patches from the mammograms and developed a patch classifier. Pretrained CNN model was trained to classify patches into normal and calcifications groups. The proposed classifier could achieve a classification accuracy of 96.7%.

2.1.2 Breast mass segmentation

Another crucial application that extracts distinguishing characteristics of specific lesion sites is breast mass segmentation [29]. Breast mass segmentation or segmentation of any anatomical structure is the most basic image processing operation. However, there is a strong correlation between breast masses and their erratic, low contrast, and hazy borders, making mass segmentation difficult. Clinicians can segment lesions, but it is a highly time-consuming process. Many techniques have been utilized in the literature for mass segmentation, including region growth, active contour, and Chan-Vese approach [30]. However, because of handcrafted features, these approaches could not understand complicated changes in the shape [29]. By automatically extracting high-level characteristics for the segmentation process, a few research based on deep learning models have demonstrated improved results when compared to standard segmentation approaches [31−34].

2.1.3 Breast density assessment

Breast density is often assessed by two-dimensional (2D) mammography and is a significant risk factor for breast cancer. As per ACR-BIRADS [48], breast density can be classified into four categories: ACR A (entirely fatty), ACR B (dense fibroglandular tissue), ACR C (glandular and connective tissue), and ACR D (extremely dense) (Fig. 2.4). An abnormal breast condition is not thought to exist in those with dense breasts, but the risk of breast cancer is increased in dense breasts. In other words, women with dense breasts are more likely to get breast cancer than those with fatty breasts. The impact of dense breast tissue on a mammography reader's skill is a different concern from this one [49]. Various research has demonstrated that AI technology can support the assessment of mammographic breast density. In the work proposed in Ref. [35], authors have trained and tested DCNN to assess the ACRBIRADS. Around 41,479 mammograms were used in the study. The clinical implementation sets and binary classifications of dense and nondense breasts showed high agreement between the DL model and radiologists. Another work [36] proposed CNN-based model to classify breast density into "scattered density" and "heterogeneously dense." In this research, the authors utilized 22,000 mammograms. The accuracy of the classifier was evaluated using the area under the curve (AUC) and receiver operating characteristic (ROC) curves. Authors of [37] created a model to categorize breast density automatically. The authors' utilized wavelet transform and modified CNN models. The AlexNet [50] model was used, with custom layers added and the last FC (fully connected) layer deleted. According to experimental data, the suggested technique encourages classification performance in separating scattered density and heterogeneously dense. One more study [38] suggested using the recently developed deep learning technique to examine how radiologists may have used the MLO and CC view pictures from a mammography examination to determine a BI-RADS density category. Using around 15,415

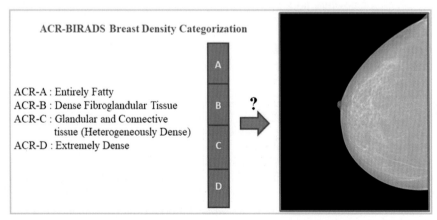

FIGURE 2.4

ACR-BIRADS breast density categorization.

mammography images, authors developed a deep learning model based on CNNs to identify the different breast density categories. Their findings demonstrated that the area under the receiver operating characteristic curve for classifying density categories using MLO view images is much larger than that for the CC view.

2.1.4 Breast cancer risk assessment

Breast cancer occurs as a result of several circumstances called risk factors. Some of them cannot be changed, while others can. Age, genetic mutations, personal and family history, reproductive history, and radiation therapy are risk factors that cannot be controlled. Inactivity or insufficient physical activity, being overweight, using hormone replacement medication, and drinking are additional risk factors [9]. Raising awareness of breast cancer risk makes it possible to encourage early identification and prevention of the disease [51]. According to the literature, there also has been a lot of AI research on predicting the risk of breast cancer. The authors assessed the DL model for evaluating breast cancer risk in Ref. [39] using full-field digital mammography (FFDM) images with three distinct instances: high-risk gene mutation carriers, high-risk unilateral cancer patients, and low-risk dataset. To directly extract parenchymal properties from the images, CNN with a transfer learning technique is applied. In Ref. [40], authors introduced a tool that employs a multidisciplinary artificial intelligence approach that combines natural language processing, image processing, a deep learning-based analytic model, and thousands of BI-RADS-4 patients details to provide accurate breast cancer biopsy risk assessment and decision assistance. The authors proved that this technique retains excellent accuracy by comparing the suggested model with the findings of a manual review. Using deep neural networks to analyze breast cancer risk shortly is a novel approach proposed in Ref. [41]. The authors used 420 instances with two consecutive mammography exams; half of these cases received a positive diagnosis during the second screening, while the other half remained negative. A deep neural network was created using four pairs of convolution neural networks and one fully connected layer in place of human-specified features. Each breast image was split into 100 ROIs with a size of 52×52 pixels, and each ROI underwent deep neural network training. The final predictions for each instance were based on the sum of the risk scores from all 100 ROIs. In Ref. [52], authors reviewed papers that employed machine learning algorithms and textural data taken from mammograms to determine the likelihood of developing breast cancer. The same-goal-focused deep learning approaches were also examined in this review article.

2.1.5 BI-RADS classification

The American College of Radiology created a well-defined instrument for risk assessment and quality control called BI-RADS (Breast Imaging-Reporting and Data System) [53]. This reporting method assigned a BIRADS score to a breast image using descriptors like shape and margin and their morphological properties. BIRADS evaluation is divided into a total of seven categories shown below [10]. In addition, BIRADS 4A (low degree of suspicion for malignancy), BIRADS 4B (moderate level of suspicion for malignancy), and BIRADS 4C (high level of suspicion for malignancy) are further classifications for BI-RADS 4 category.

- BI-RADS 0: Incomplete Assessment and Need further assistance.
- BI-RADS 1: Normal, No evidence of lesion.
- BI-RADS 2: Benign lesion
- BI-RADS 3: Probably benign lesion
- BI-RADS 4: Suspicious lesion
- BI-RADS 5: High probability of malignancy
- BI-RADS 6: Biopsy-proven malignancy

AI approaches have also progressed in automated BI-RADS classification from mammograms for many years. The research work in Ref. [42] provides a deep neural network-based model as an accurate tool to aid radiologists in mammogram interpretation. The proposed model was trained using a private mammography dataset and block-based image segmentation technique. A block-based image was fed into the model as an input, and as an output, a BI-RADS category was predicted. The overall accuracy of 94.22% demonstrated the effectiveness of this work. One more research work [43] proposed an approach employing a ResNet-based customized neural network (RN-BCNN) for the six categories of BI-RADS classification, as opposed to the conventional ConvNet model, using the data augmentation and pyramid of scales methodologies on an unbalanced dataset. The proposed model and elastic deformation played significant roles in the effectiveness of the suggested technique, which further enhanced accuracy by up to 85.9. In the study [44], we categorize mammography images using transfer learning on NASNet Mobile and finetuning on VGG16 and VGG19 on the INbreast dataset. Our suggested approach outperformed some of the comparable works reported in the literature study, achieving an accuracy (ACC) of 90.9% and a macroaveraged area under the receiver operating characteristic curve (AUC) of 99.0%.

2.1.6 Axillary node assessment

The most significant predictor of overall survival and recurrence in patients with breast cancer is axillary lymph node (LN) metastases, and proper evaluation of axillary LN involvement is a crucial part of staging breast cancer. Breast radiologists and surgeons must collaborate closely as techniques for axillary nodal assessment advance if they are to fully realize the potential of imaging and offer patients the best possible care [54]. AI has excelled in tasks like the identification, segmentation, and classification of breast masses for breast cancer diagnosis. These methods have recently begun to function well on tasks like axillary node evaluation. Authors of [45] offer a novel deep-learning method for predicting axillary lymph node metastasis in Patients. The Erasmus Medical Center provided the dataset of axillary lymph nodes from breast cancer patients. Pathologic investigation verified the presence of axillary lymph nodes. ANNs are utilized for training 84% of the Erasmus Medical Center dataset and evaluating the remaining 16% of the independent dataset. These ANNs include feed forward, radial basis function, and Kohonen self-organizing.

According to the study's findings, the suggested mechanism outperforms the radiologists' models in terms of sensitivity, specificity, and accuracy, achieving 95%, 96%, and 98%, respectively. One more recent work [46] used a deep convolutional network to detect abnormal axillary lymph nodes on mammograms. A private dataset was used in the work, labeled into three classes; breast tissue, benign lymph nodes, and suspicious lymph nodes. After preprocessing the data, a model was trained and tested on 5385 breast images. The performance of the trained model was then evaluated against that of human readers using a real-world dataset. For the training and validation set, the accuracy was 98% and 99%, respectively. The accuracy for breast tissue, benign lymph nodes, and suspicious lymph nodes in confusion matrices of the "real-world" dataset using radiological reports as ground truth was 98.51%, 98.63%, and 95.96%, respectively.

2.2 Challenges and future research

Although AI has advanced quickly in the medical profession over the past few years, it is still a long way from its ultimate objective of being fully integrated into the work of physicians and having widespread global applicability. In addition, several scientific problems still require further examination to produce a viable CAD for breast cancer diagnosis. We highlight such issues below.

- There is a requirement for a large number of breast images with expert annotation to train AI models. The most common problem is a *lack of significant public databases* with skilled annotation. Additionally, using fewer samples for training results in two new issues: *overfitting* and *model generalization problem.*
- Another obstacle that makes training of AI models even more challenging is *class imbalance problem.* There may not be an equal number of diseases and nondiseases samples in the data collection. This skewed data distribution largely occurs in many medical applications, including breast imaging, when multiple positive samples occur with lower frequency than negative samples [47]. This problem can further be solved by incorporating methods like image augmentation, and sampling [23].
- The majority of ML or DL models are developed as binary classifiers; however, it is sometimes necessary to create a model that is a *multiclass classifier.*
- Majority of the models can either detect and/or classify breast mass or microcalcification. Therefore, there is a need for a CAD system to integrate both of these abnormalities.
- Building *multimodal CAD system* for breast cancer diagnosis is also very challenging. Such a system should integrate data from multiple imaging modalities to confirm the diagnosis. For example, a mammogram reading can be combined with a sonogram's readings to diagnose cancer accurately.

Along with its use in conventional imaging modalities, CAD systems based on AI models are advancing rapidly in many other areas also, which includes contrast-enhanced mammography [55], breast ultrasound, breast MRI, and digital breast tomosynthesis [56], and ultrasound elastography [57]. Additionally, AI in breast imaging is not only utilized for the detection, classification, segmentation, and prediction of breast disorders but also for further classifying particular breast diseases such as breast fibroplasia and for predicting lymph node metastasis [58]. Moreover, it can also be utilized for survival prediction [59], disease progression, and disease recurrence [60]. Technological advancements in AI can help radiologists diagnose diseases more accurately and lead to an early diagnosis of the disease, which will benefit a huge population of cancer patients.

3. Conclusion

The use of AI techniques, particularly DL models, is growing in the field of medical imaging, and these models excel in tasks requiring medical image processing or analysis, such as breast imaging. These techniques can benefit physicians by minimizing their burden and oversight errors and giving them unbiased and practical information. This study presents various mammogram-based breast cancer research applications of AI models. However, there are also several uses of AI in other imaging modalities, including histopathology, breast MRI, breast ultrasound, and others. Due to their quick computation times, strong repeatability, lack of tiredness, and increased accuracy, AI-based CAD systems have consistently demonstrated their value to both patients and medical practitioners.

Acknowledgments

The author would like to thank Dr. Rajiv Oza (Consultant Radiologist) for advising on breast cancer procedures from a medical perspective.

References

[1] Sung H, Ferlay J, Siegel RL, Laversanne M, Soerjomataram I, Jemal A, et al. Global cancer statistics 2020: globocan estimates of incidence and mortality worldwide for 36 cancers in 185 countries. CA: A Cancer Journal for Clinicians 2021;71(3):209−49.

[2] Vijya Mary SB. Breast cancer awareness month: stay vigilant, stay healthy. 2022. https://www.thehindu.com/life-and-style/october-marks-breast-cancer-awareness-monthheres-what-we-need-to-know-and-do-to-keep-the-disease-away/article66030432.ece.

[3] Cancer cases reported in India in 2022. 2022. https://www.dailypioneer.com/2022/india/19-20-lakhcancer-cases-reported-in-india-in-2022.html.

[4] Statistics of breast cancer in India. 2022. https://cytecare.com/blog/breast-cancer/statistics-of-breastcancer.

[5] International agency for research on cancer. World cancer report. 2020. https://www.iarc.who.int/cards_page/world-cancer-report.

[6] Sun L, Legood R, dos Santos-Silva I, Gaiha SM, Sadique Z. Global treatment costs of breast cancer by stage: a systematic review. PLoS One 2018;13(11):e0207993.

[7] Sree SV, Ng EY-K, Acharya RU, Faust O. Breast imaging: a survey. World Journal of Clinical Oncology 2011;2(4):171.

[8] Islam MS, Kaabouch N, Hu WC. A survey of medical imaging techniques used for breast cancer detection. In: IEEE international conference on electroInformation technology, EIT 2013. IEEE; 2013. p. 1−5.

[9] Oza P, Sharma P, Patel S. A drive through computer-aided diagnosis of breast cancer: a comprehensive study of clinical and technical aspects. Recent Innovations in Computing 2022:233−49.

[10] Oza P, Sharma P, Patel S, Bruno A. A bottom-up review of image analysis methods for suspicious region detection in mammograms. Journal of Imaging 2021;7(9):190.

[11] Heath M, Bowyer K, Kopans D, Kegelmeyer P, Moore R, Chang K, et al. Current status of the digital database for screening mammography. In: Digital mammography. Springer; 1998. p. 457−60.

[12] Lee RS, Gimenez F, Hoogi A, Miyake KK, Gorovoy M, Rubin DL. A curated mammography data set for use in computer-aided detection and diagnosis research. Scientific Data 2017;4(1):1−9.

[13] Suckling J, Parker J, Dance D, Astley S, Hutt I, Boggis C, et al. Mammographic image analysis society (mias) database v1. 21. 2015.

[14] Moreira IC, Amaral I, Domingues I, Cardoso A, Cardoso MJ, Cardoso JS. Inbreast: toward a full-field digital mammographic database. Academic Radiology 2012;19(2): 236−48.

[15] Lopez MG, Posada N, Moura DC, Ramos Pollán R, Valiente JMF, Ortega CS, et al. BCDR: a breast cancer digital repository. In: 15th international conference on experimental mechanics, vol. 1215; 2012.

[16] Oliveira JEE, Gueld MO, Araújo Ade A, Ott B, Deserno TM. Toward a standard reference database for computer-aided mammography. In: Medical imaging 2008: computer-aided diagnosis, vol. 6915. SPIE; 2008. p. 606−14.

[17] Bruno A, Ardizzone E, Vitabile S, Midiri M. A novel solution based on scale invariant feature transform descriptors and deep learning for the detection of suspicious regions in mammogram images. Journal of Medical Signals and Sensors 2020;10(3):158.

[18] Dembrower K, Lindholm P, Strand F. A multi-million mammography image dataset and population-based screening cohort for the training and evaluation of deep neural network—the cohort of screen-aged women (CSAW). Journal of Digital Imaging 2020;33(2):408−13.

[19] Matsoukas C, Hernandez AB, Liu Y, Dembrower K, Miranda G, Konuk E, et al. Adding seemingly uninformative labels helps in low data regimes. In: International conference on machine learning. PMLR; 2020. p. 6775−84.

[20] Alsolami AS, Shalash W, Alsaggaf W, Ashoor S, Refaat H, Elmogy M. King Abdulaziz University breast cancer mammogram dataset (KAU-BCMD). Data 2021;6(11):111.

[21] Oza P, Sharma P, Patel S. Machine learning applications for computeraided medical diagnostics. In: Proceedings of second international conference on computing, communications, and cyber-security. Springer; 2021. p. 377−92.

[22] Oza P, Shah Y, Vegda M. A comprehensive study of mammogram classification techniques. In: Tracking and preventing diseases with artificial intelligence. Springer; 2022. p. 217−38.

[23] Oza P, Sharma P, Patel S, Kumar P. Computer-aided breast cancer diagnosis: a comparative analysis of breast imaging modalities and mammogram repositories. Current Medical Imaging 2022;19:456—68.

[24] Oza P, Sharma P, Patel S. Transfer learning assisted classification of artefacts removed and contrast improved digital mammograms. Scalable Computing: Practice and Experience 2022;23(3):115—27.

[25] Oza P, Sharma P, Patel S. A transfer representation learning approach for breast cancer diagnosis from mammograms using efficientnet models. Scalable Computing: Practice and Experience 2022;23(2):51—8.

[26] Oza P, Sharma P, Patel S. Deep ensemble transfer learning-based framework for mammographic image classification. The Journal of Supercomputing 2022:1—22.

[27] Zhang Y, Han J, Chen B, Chang L, Song T, Cai G. Classification of microcalcification clusters using bilateral features based on graph convolutional network. Frontiers in Oncology 2022;12.

[28] Kahnouei MS, Giti M, Akhaee MA, Ameri A. Microcalcification detection in mammograms using deep learning. Iranian Journal of Radiology 2022;19(1).

[29] Al-Antari MA, Al-Masni MA, Choi M-T, Han S-M, Kim TS. A fully integrated computer-aided diagnosis system for digital x-ray mammograms via deep learning detection, segmentation, and classification. International Journal of Medical Informatics 2018;117:44—54.

[30] Rahmati P, Adler A, Hamarneh G. Mammography segmentation with maximum likelihood active contours. Medical Image Analysis 2012;16(6):1167—86.

[31] Al-Antari MA, Al-Masni MA, Park S-U, Park JH, Metwally MK, Kadah YM, et al. An automatic computeraided diagnosis system for breast cancer in digital mammograms via deep belief network. Journal of Medical and Biological Engineering 2018;38(3): 443—56.

[32] Alkhaleefah M, Tan T-H, Chang C-H, Wang T-C, Ma S-C, Chang L, et al. Connected-segnets: a deep learning model for breast tumor segmentation from x-ray images. Cancers 2022;14(16):4030.

[33] Baccouche A, Garcia-Zapirain B, Olea CC, Elmaghraby AS. Connected-unets: a deep learning architecture for breast mass segmentation. NPJ Breast Cancer 2021;7(1):1—12.

[34] Salama WM, Aly MH. Deep learning in mammography images segmentation and classification: automated CNN approach. Alexandria Engineering Journal 2021;60(5): 4701—9.

[35] Lehman CD, Yala A, Schuster T, Dontchos B, Bahl M, Swanson K, et al. Mammographic breast density assessment using deep learning: clinical implementation. Radiology 2019;290(1):52—8.

[36] Mohamed AA, Berg WA, Peng H, Luo Y, Jankowitz RC, Wu S. A deep learning method for classifying mammographic breast density categories. Medical Physics 2018;45(1): 314—21.

[37] Matsuyama E, Takehara M, Tsai D-Y, et al. Using a wavelet-based and fine-tuned convolutional neural network for classification of breast density in mammographic images. Open Journal of Medical Imaging 2020;10(01):17.

[38] Mohamed AA, Luo Y, Peng H, Jankowitz RC, Wu S. Understanding clinical mammographic breast density assessment: a deep learning perspective. Journal of Digital Imaging 2018;31(4):387—92.

[39] Li H, Giger ML, Huynh BQ, Antropova NO. Deep learning in breast cancer risk assessment: evaluation of convolutional neural networks on a clinical dataset of full-field digital mammograms. Journal of Medical Imaging 2017;4(4):041304.

[40] He T, Puppala M, Ezeana CF, Huang Y-S, Chou P-H, Yu X, et al. A deep learning–based decision support tool for precision risk assessment of breast cancer. JCO Clinical Cancer Informatics 2019;3:1–12.

[41] Sun W, Tseng T-LB, Zheng B, Qian W. A preliminary study on breast cancer risk analysis using deep neural network. In: International workshop on breast imaging. Springer; 2016. p. 385–91.

[42] Tsai K-J, Chou M-C, Li H-M, Liu S-T, Hsu J-H, Yeh W-C, et al. A high-performance deep neural network model for bi-rads classification of screening mammography. Sensors 2022;22(3):1160.

[43] Siddeeq S, Li J, Bhatti HMA, Manzoor A, Malhi US. Deep learning RN-BCNN model for breast cancer BI-RADS classification. In: 2021 the 4th international conference on image and graphics processing; 2021. p. 219–25.

[44] Falconí L, Pérez M, Aguilar W, Conci A. Transfer learning and fine tuning in mammogram bi-rads classification. In: 2020 IEEE 33rd international symposium on computer-based medical systems (CBMS). IEEE; 2020. p. 475–80.

[45] Ashokkumar N, Meera S, Anandan P, Murthy MYB, Kalaivani KS, Alahmadi TA, et al. Deep learning mechanism for predicting the axillary lymph node metastasis in patients with primary breast cancer. BioMed Research International 2022:2022.

[46] Abel F, Landsmann A, Hejduk P, Ruppert C, Borkowski K, Ciritsis A, et al. Detecting abnormal axillary lymph nodes on mammograms using a deep convolutional neural network. Diagnostics 2022;12(6):1347.

[47] Oza P, Sharma P, Patel S, Kumar P. Deep convolutional neural networks for computer-aided breast cancer diagnostic: a survey. Neural Computing & Applications 2022:1–22.

[48] Sickles EA, ZOrsi CJDA', Bassett LW, Appleton CM, Berg WA', Burnside ES, et al. ACR BI-RADS R mammography. ACR BI-RADS R atlas, breast imaging reporting and data system. 5; 2013. p. 2013.

[49] Fowler EEE, Smallwood AM, Khan NZ, Kilpatrick K, Sellers TA, Heine J. Technical challenges in generalizing calibration techniques for breast density measurements. Medical Physics 2019;46(2):679–88.

[50] Krizhevsky A, Sutskever I, Hinton GE. Imagenet classification with deep convolutional neural networks. In: Pereira F, Burges CJ, Bottou L, Weinberger KQ, editors. Advances in neural information processing systems, vol. 25. Curran Associates, Inc.; 2012.

[51] Lei Y-M, Yin M, Mei-Hui Y, Jing Y, Zeng S-E, Lv W-Z, et al. Artificial intelligence in medical imaging of the breast. Frontiers in Oncology 2021:2892.

[52] Mendes J, Matela N. Breast cancer risk assessment: a review on mammography based approaches. Journal of Imaging 2021;7(6):98.

[53] Sickles EA, ZOrsi CJDA', Bassett LW, Appleton CM, Berg WA', Burnside ES, et al. Anleitung/hilfestellung. In: ACR BI-RADS R-atlas der mammadiagnostik. Springer; 2016. p. 131–54.

[54] Chang JM, Leung JWT, Moy L, Ha SM, Moon WK. Axillary nodal evaluation in breast cancer: state of the art. Radiology 2020;295(3):500–15.

[55] Khaled R, Helal M, Alfarghaly O, Mokhtar O, Elkorany A, El Kassas H, et al. Categorized contrast enhanced mammography dataset for diagnostic and artificial intelligence research. Scientific Data 2022;9(1):1–10.

[56] Dahlblom V, Dustler M, Tingberg A, Zackrisson S. Breast cancer screening with digital breast tomosynthesis: comparison of different reading strategies implementing artificial intelligence. European Radiology 2022:1−12.

[57] Mao Y-J, Lim H-J, Ni M, Yan W-H, Wong DW-C, Cheung JCW. Breast tumour classification using ultrasound elastography with machine learning: a systematic scoping review. Cancers 2022;14(2):367.

[58] Zhou LQ, Wu XL, Huang SY, Wu GG, Ye HR, Wei Q, et al. Lymph node metastasis prediction from primary breast cancer US images using deep learning. Radiology 2020;294(1):19.

[59] Jang W, Jeong C, Kwon KA, In Yoon T, Yi O, Kim KW, et al. Artificial intelligence for predicting five-year survival in stage iv metastatic breast cancer patients: a focus on sarcopenia and other host factors. Frontiers in Physiology 2022;13:2062.

[60] Chan H-P, Samala RK, Hadjiiski LM. CAD and AI for breast cancer—recent development and challenges. British Journal of Radiology 2019;93(1108).

Prediction of breast cancer diagnosis using random forest classifier

3

Pawan Whig[1], Shama Kouser[2], Ashima Bhatnagar Bhatia[1], Rahul Reddy Nadikattu[3]

[1]*Vivekananda Institute of Professional Studies, New Delhi, India;* [2]*Department of Computer Science, Jazan University, Saudi Arabia;* [3]*University of the Cumberland, Williamsburg, KY, United States*

1. Introduction

The primary cause of mortality for many women throughout the world is increasingly breast cancer. Examined as the primary factor in women's fatalities is the manner in which the sickness is transmitted. We still fall short in accurately identifying this deadly illness when it is still in its early stages, despite the fact that innovation has had a tremendous impact on how we live. For a certain age range of anxious women, the rate of mammography has grown as the problem is not usually detected in its early stages [1]. Breast cancer is curable, and there is a potential that it might even save a life. It would start by examining. Numerous elements, such as hormone problems, familial heritage, obesity, radiation therapy, and others, have been speculated as probable causes of this horrible disease. The diagnosis of this illness was performed using in-depth learning and artificial intelligence (AI) calculations [2].

An initial analysis of breast cancer (BC) can significantly recover the forecast and existence chances for patients because it can stimulate immediate clinical therapy [3]. A more accurate classification of benign tumors might save patients from obtaining unnecessary medical attention.

The correct diagnosis of BC and the categorization of people into benign or malignant categories have thus received a great deal of research. Due to its clear benefits in crucial feature identification from challenging BC datasets, machine learning (ML) is widely accepted as the preferable technique in BC design categorization and prediction demonstration [4]. Fig. 3.1 depicts the process flow for the machine learning-based therapy.

The three fundamental phases of the numerous data mining and machine learning techniques created over the past few decades for the detection and classification of breast cancer are preprocessing, feature extraction, and classification [5]. Preprocessing mammography films makes it simpler to comprehend and analyze results by improving the visibility of outskirt regions and intensity distribution. It has been observed that this operation can be aided by a number of strategies.

Computational Intelligence and Modelling Techniques for Disease Detection in Mammogram Images
https://doi.org/10.1016/B978-0-443-13999-4.00011-0

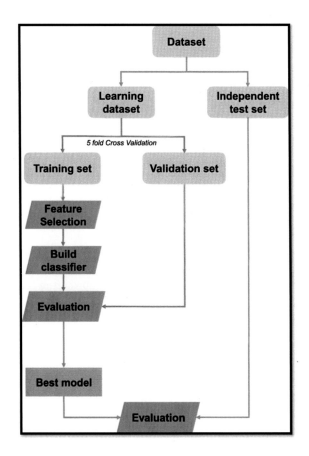

FIGURE 3.1

Flow of steps for the treatment using machine learning.

Several alter founded texture examination methods are used to transform the picture into a new form by means of the three-dimensional incidence possessions of the pixel intensity differences. Typical methods include the Wavelet transform, Fast Fourier Transform (FFT), and singular value decomposition (SVD). The dimensionality of the eye picture may be decreased using principal component analysis (PCA). Several efforts have focused on automating the diagnosis of breast cancer using machine-learning algorithms [6].

Despite significant efforts, methods for diagnosing breast cancer reported in the literature can be considered to be semiautomatic, as illustrated in Fig. 3.2. According to Kuhn and Johnson, the hyperparameters are those parameters that are difficult to deduce from the data. In most cases, specific model parameters must be changed for an algorithm to perform as predicted [7]. For example, the education rate for

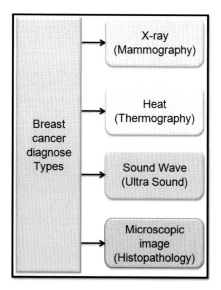

FIGURE 3.2

Methods for diagnosing breast cancer.

exercising neuronal net parameters of SVMs is physical because there is no analytical method to determine the optimal value. Thus, the ultimate tuning parameters for any given model are still up for discussion [8].

Machine learning is in increasing demand and will soon be offered as a service. Unfortunately, the science of machine learning still faces major barriers to entrance and specialized expertise [9]. A real ML perfect that incorporates the preprocessing phases needs a certain set of services and know how an example of a machine learning model or data processing pipeline. Every stage of the pipeline has a number of choices. The preprocessing phase and classification stage's techniques and parameters are automatically defined by the specified pipeline. A skilled ML chooses the optimum strategy for the present problematic area [10].

2. Data set used

The University of California, Irvine (UCI) Mechanism Knowledge Repository was used to obtain the Wisconsin Breast Cancer dataset that was used as a backup in this training [11]. Fig. 3.3 displays the identical dataset that Bennett employs to discriminate between cancerous and noncancerous tumors. Digital photographs of a fine-needle aspirate of a breast tumor were used to extract the attributes that best define the nucleus of the present image. The Wisconsin Diagnostic Breas (WDBC) database, which discovered 357 benign cases and 212 malignant instances, had an impact on 569 patients at Wisconsin hospitals. A measurement from an fine needle

No. Instances	WDBC		ST-ONCODIAG	
	Benign	Malignant	Not-Breast Cancer	Breast Cancer
100	35	65	51	49
200	96	104	102	98
300	154	146	142	158
400	227	173	179	221
500	305	195	218	282
559	353	206	245	314

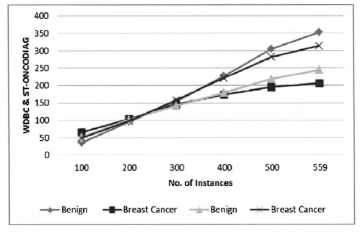

FIGURE 3.3

Wisconsin breast cancer dataset used.

aspiration (FNA) test is included in each observation. The first two traits in this collection stand in for the identification number and the diagnostic status. The 30 real traits, along with the mean, standard deviation, and the bottom 10 percentiles, make up the final values.

In machine learning, the process of choosing a subset of relevant features from a set of alternative subsets is known as feature selection, and it is a crucial step in creating a model. The features of a successful prediction model must be carefully chosen. The use of feature selection techniques offers a number of benefits, including (1) faster and more effective training of machine learning algorithms, (2) simpler and easier to comprehend models, (3) higher model accuracy with the right subset, and (4) reduced overfitting [12].

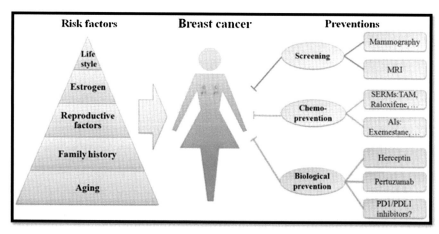

FIGURE 3.4

Hazard issues for breast cancer.

3. Several breast cancer risk factors

As seen in Fig. 3.4, approximately of the recognized risk issues for breast growth are mentioned below. Though, the majority of breast cancer occurrences cannot be related to a single reason. With your doctor, go through your exact medical history, current symptoms, and any concerns you may have to ensure a comprehensive evaluation and appropriate medical guidance [13].

Age. As females become older, their probabilities of developing breast cancer rise. Women over the age of 50 are the ones who develop breast cancer at a rate of about 80%.

Personal experience. A female who has already had chest cancer in one chest is more likely to have it in the other breast.

History of breast cancer in the family. If a woman's mother, sister, or daughter had breast cancer, particularly when she was young, her chance of developing breast cancer is increased (before 40). Another risk factor is having relatives who have breast cancer.

Genetic influences. A woman's lifelong chance of getting breast cancer increases if she has specific genetic alterations, including changes to the BRCA1 and BRCA2 genes. Other gene alterations may also increase the risk of breast cancer.

History of menstruation and childbearing. A woman's risk of breast cancer increases with her age at the time of her first childbirth. Also more at risk are

- Women who begin menstruating at a young age
- Those who experience climacteric later
- Females who have not ever given birth

---

4. Various machine learning algorithms

Machine learning has increased significantly in popularity over the past few years as a result of the massive demand and technical advancements. Machine learning's ability to derive value from data has attracted businesses from a variety of industries. Most ML solutions are intended to leverage ML algorithms, usually with approximate adjustments, to optimize performance and address complex problems in various domains [14−16].

4.1 Linear regression

The supervised learning method of lined reversion makes an effort to define the association between an incessant board parameter and sovereign variable stars by appropriate linear reckoning to the data [17].

Only two of the many methods available to examine the relationship between variables are scattered plots and correlation matrices. The scatter figure below, for example, shows a strong correlation between a self-governing mutable (x-axis) and dependency on a mutable (y-axis). The second one rises in tandem with the first [18,19].

A linear regression model seeks to identify the regression line that most closely approximates the connections or correlations between the data points. The most used technique is ordinary-least squares (OLE). This method allows for the identification of the ideal reversion line by reducing the number of quadrangles of the coldness amid the information opinions and the degradation stroke. OLE-obtained regression line looks to be as depicted in Fig. 3.5 for the aforementioned data points.

4.2 SVM

Although it may also be used for regression tasks, the supervised learning method SVM is primarily employed for classification issues.

As seen in Fig. 3.6, SVM develops a decision boundary to distinguish between classes. The process of drawing or selecting the decision boundary is the most crucial element of SVM algorithms. Before the construction of the decision border, every comment is presented in an n-dimensional interplanetary. There are "n" features being used.

One may consider the kernel function to be a similarity metric. Similarity in this context denotes a level of closeness. It costs money to really convert information points into a height dimensional ear interplanetary. The method does not create a completely new, high-dimensional feature space from the input points [20−24].

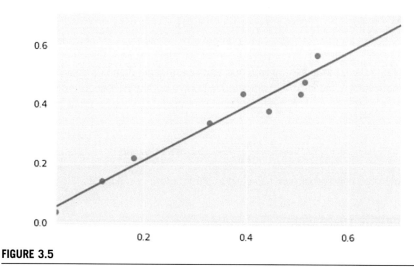

FIGURE 3.5

OLE-obtained regression line.

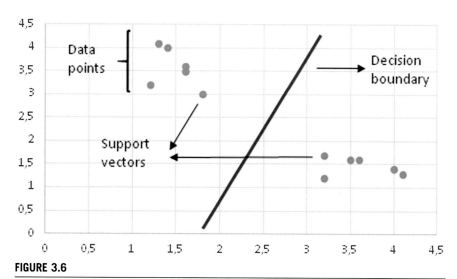

FIGURE 3.6

Decision boundary to discriminate between classes.

SVM is particularly useful when the number of dimensions exceeds the number of samples. SVM employs a little amount of training data to determine the decision boundary, making it memory efficient. Large datasets, on the other hand, lengthen training time, which has a negative effect on performance [25–29].

4.3 Naive Bayes

Naive Bayes is an oversaw knowledge approach secondhand for categorization issues. For this reason, it is often referred to as the Naive Bayes classifier. The naïve Bayes method makes the assumption that traits are independent and do not correlate. This is not actually the case. This method is known as "naive" because it falsely assumes that traits do not correlate with one another. The naive Bayes method is founded on Bayes' Theorem.

$$p(A|B) = \frac{\mathbf{p(A) \cdot p(B/A)}}{\mathbf{p(B)}} \text{ (\textbf{Bayes' Theorem})} \tag{3.1}$$

More speed can sometimes be selected above more precision.

4.4 Logistic regression

As seen in Fig. 3.7, scenarios requiring binary classification usually call for the employment of the supervised learning method known as logistic regression. Despite the fact that the phrases "regression" and "classification" are mutually incompatible, "logistic" refers to the logistic function, which is what this approach employs to do the classification task. As logistic regression is an easy-to-use but incredibly effective classification approach, it is a popular choice for many binary classification tasks. Among the issues that logistic regression successfully fixes are predictions of website or ad clicks, spam emails, and client attrition.

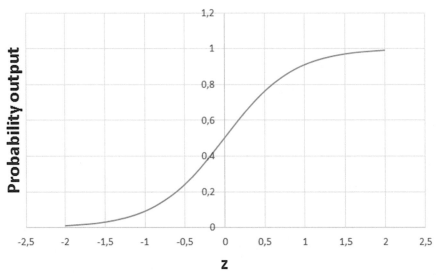

FIGURE 3.7

Supervised learning approach.

It is not continuously recommended to select the optimistic class for all likelihood standards larger than 60%. Before classifying a communication as spam, we must be almost positive that it is. We need to pay much closer attention when classifying a health-related concern. We have a remote hunch that a certain cell could be malignant. As a result, the issue determines the worth that delineates the positive and negative classes.

4.5 *k*-Nearest neighbors

Organization and reversion issues can be solved using the supervised learning method known as *k*-nearest neighbors (kNN). The core idea behind kNN is that neighboring data points determine a data point's value or class.

The kNN classifier uses the majority voting principle to identify the class of a data item. For instance, if *k* is set to 5 as in Fig. 3.8, the classes of the five closest points are investigated. Based on the dominating class, predictions are formed. In kNN reversion, the nasty worth of the five neighboring opinions is also utilized. Examine one example. Consider the four distinct data categories in the examples below.

The selection of an optimum *k* value is very important. If *k* is chosen also little, the perfect is very specific and difficult to generalize. On the Pullman set, the ideal is highly accurate, but determination does badly when used as a forecaster for new, unanticipated data items. As a result, we are probably going to end up with an overfit model.

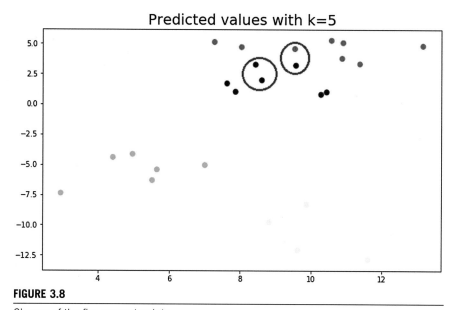

FIGURE 3.8

Classes of the five nearest points.

4.6 Decision trees

By repeatedly classifying data into categories using questions, a decision tree emerges.

Fig. 3.9 depicts a decision tree for predicting client turnover. The initial split is determined by the total monthly fees. Further queries are then made by the system to generate distinct class tags.

The DT method aims to upsurge predictability by continually learning more. Typically, splitting the topographies arbitrarily does not provide us with information that is helpful. Higher node purity separations offer more information. A node's purity is negatively correlated with the distribution of different classes within it. The questions were chosen to encourage purity or reduce impurity.

4.7 RF algorithm

A random forest (RF) is a collection of many decision trees, as seen in Fig. 3.10. The bagging approach is used to build random forests, while DT are utilized as similar estimators. When used to address an organizational problem, the consequence is decided by the effectiveness of the chosen solution and the actions taken to implement it. The prediction in a reversal is the nasty worth of the board standards in a leaf node. RF reversion accounts for the nasty worth of the decision tree consequences.

Compared to a single decision tree, random forests are significantly more accurate and provide less of an overfitting risk. To avoid time becoming a bottleneck, decision trees in a random forest also run simultaneously.

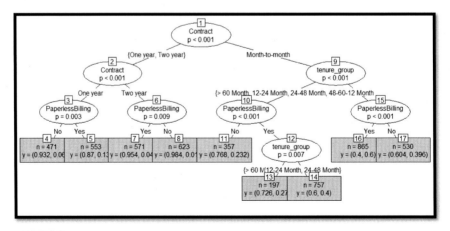

FIGURE 3.9

Decision tree to forecast client turnover.

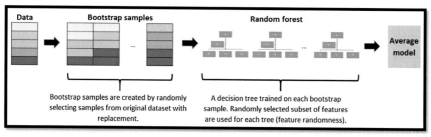

FIGURE 3.10

Random forest algorithm flow.

4.8 Boosted gradient decision trees

As seen in Fig. 3.11, gradient-boosted decision trees (GBDT) is an ensemble technique that combines several decision trees using the boosting method. In GBDT, decision trees act as feeble beginners.

Every tree tries to lessen the mistakes of the tree before it. Although increasing trees are not good learners, by putting additional trees in a row and having each one focus on the errors produced by the one before it, boosting transforms into a very accurate and effective model. In contrast to bagging, boosting does not employ bootstrap sampling.

The model's pace of information acquisition is referred to as its "learning rate." Each additional tree changes the overall model. The degree of the change depends on the rate of learning. The n estimator shows how many trees were used in the model. But when deciding how many trees to plant, we must be very picky. Overfitting is more likely to occur when there are too many trees being used.

GBDT provides more accurate predictions and performs both classification and regression tasks better than random forests. It does not require any preprocessing and can handle features of mixed types.

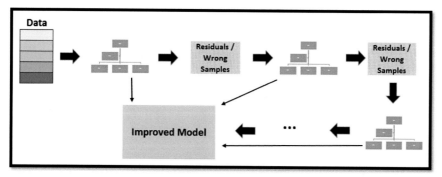

FIGURE 3.11

Boosting is the process of successively combining learning algorithms.

4.9 Clustering with *k*-means

Clustering is a method for organizing a set of data opinions so that similar data opinions are grouped together. As a result, bunch processes look for overlaps or discrepancies across data viewpoints. As bunch is an unproven approach for gaining information, data points have no labels. Numerous methods seek for the underlying structure of the data.

To make the clusters more closely spaced and the individual data points more scattered, information is separated into *k* bunches using *k*-means bunch, as illustrated in Fig. 3.12. As a result, it is a divider-founded clustering method. The resemblance of two places is strongminded by their coldness from one another.

k-Means bunch seeks to maximize detachments across clusters while minimizing distances within each cluster. The number of clusters cannot be determined using the *k*-means method. When building the *k*-means object, we must define it, which may be difficult.

4.10 Analysis by principal components

With as much data as possible retained, a dimensionality reduction method known as PCA generates new features from the ones that currently exist. As illustrated in Fig. 3.13, PCA is an unverified knowledge technique that is frequently used as a preprocessing stage for oversaw knowledge systems. PCA may extract additional topographies by determining how the characteristics in a dataset are related to one another.

The objective of PCA is to utilize as few features as possible to explain the variance in the original dataset (or columns). The newly determined qualities are the

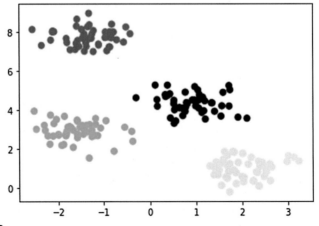

FIGURE 3.12

k-Means clustering.

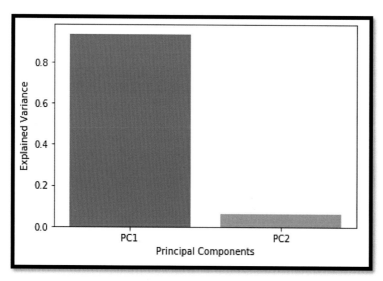

FIGURE 3.13

Analysis by principal components.

principal components. Each principle component is ranked according to how much variance in the original dataset it explains.

5. Case study

The objective of this study is to discover general patterns that may aid in the selection of the appropriate model and hyperparameters, as well as to evaluate which traits are most effective in predicting whether a tumor is malignant or benign. Finding out if breast cancer is aggressive or benign is the main objective. To do this, we fitted a functional that can anticipate the categorical class of fresh data using algorithmic classification techniques.

5.1 Various machine learning libraries used

An ML library is often a collection of readily usable functions and procedures. A developer's armory must have a strong collection of libraries to do research and create sophisticated applications without having to write a lot of code.

Libraries save programmers from repeatedly creating unnecessary code. Additionally, there are other libraries for handling various issues. For instance, we provide libraries for text processing, graphics, data manipulation, and scientific calculation. The top five headers of the dataset are shown in Fig. 3.14.

	id	radius_mean	...	fractal_dimension_worst	diagnosis
0	842302	17.99	...	0.11890	M
1	842517	20.57	...	0.08902	M
2	84300903	19.69	...	0.08758	M
3	84348301	11.42	...	0.17300	M
4	84358402	20.29	...	0.07678	M

FIGURE 3.14

Top five headers of the dataset.

Using the Dimension command, we can see the dimension of the dataset. The data set has 569 rows and 32 columns, as can be seen in Fig. 3.15. The column labeled "Diagnosis" will tell us if the cancer is M = malignant or B = benign. Cancer is indicated by a 1 and benign cancer by a 0. We can see that 357 of the 569 people have the label "B" (benign), while 212 have the label "M" (malignant).

5.1.1 Handling missing values

There are no missing values; this can be checked using

dataset.isnull().sum()

dataset.isna().sum()

There is no missing value as shown in Fig. 3.16.

Variables with categorical data have label values rather than numerical values. Frequently, just a fixed set of values are available.

Users are frequently characterized in terms of their place of origin, gender, age group, etc.

To label the categorical data, we will utilize Label Encoder. Our predictive models can better interpret categorical input when it is converted into numbers using

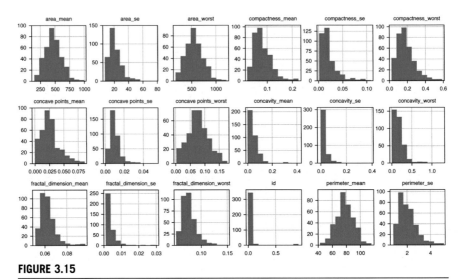

FIGURE 3.15

Visualization of the dataset.

```
id                          0
radius_mean                 0
texture_mean                0
perimeter_mean              0
area_mean                   0
smoothness_mean             0
compactness_mean            0
concavity_mean              0
concave points_mean         0
symmetry_mean               0
fractal_dimension_mean      0
radius_se                   0
texture_se                  0
perimeter_se                0
area_se                     0
smoothness_se               0
compactness_se              0
concavity_se                0
concave points_se           0
symmetry_se                 0
fractal_dimension_se        0
radius_worst                0
texture_worst               0
perimeter_worst             0
area_worst                  0
smoothness_worst            0
compactness_worst           0
concavity_worst             0
concave points_worst        0
symmetry_worst              0
fractal_dimension_worst     0
diagnosis                   0
dtype: int64
```

FIGURE 3.16

Missing values checking.

the Label Encoder function of the SciKit Learn module in Python as shown in Fig. 3.17.

6. Experimental results and discussions

The tests that were run using various methodologies on the dataset mentioned earlier are presented in Table 3.1. Based on training and testing accuracy and the amount of time spent on the dataset, several strategies are being compared as shown in Table 3.1. As it provides 99% accuracy in less time, the extreme learning machine (ELM) is the best among the competition, according to the data shown in Fig. 3.18.

A bar chart comparison of all the models is shown below, broken down by accuracy and time.

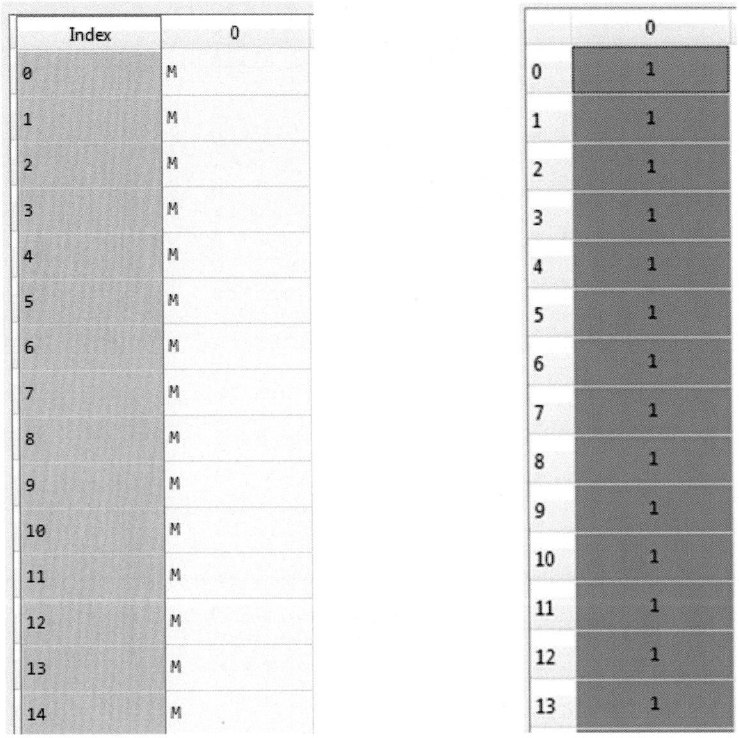

Index	0
0	M
1	M
2	M
3	M
4	M
5	M
6	M
7	M
8	M
9	M
10	M
11	M
12	M
13	M
14	M

	0
0	1
1	1
2	1
3	1
4	1
5	1
6	1
7	1
8	1
9	1
10	1
11	1
12	1
13	1

FIGURE 3.17

Diagnosis data with and without encoding.

Table 3.1 Comparative analysis.

	Accuracy		Time	
Model	**Training (%)**	**Testing (%)**	**Training (ms)**	**Testing (ms)**
Decision tree (DT)	84	89	0.04	0.01
k-Nearest neighbor (kNN)	89	90	0.35	0.32
Support vector machine (SVM)	91	91	0.06	0.01
Random forest (RF)	98	98	0.15	0.14
Extreme learning machine (ELM)	95	99	0.04	0.01

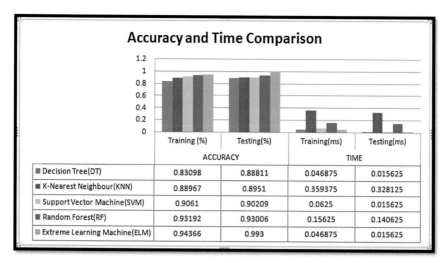

FIGURE 3.18

Accuracy and time comparison.

7. Conclusion

Machine learning has effectively detected breast cancer with an accuracy of up to 97.4%. By utilizing machine learning, the result is more efficient, quicker, and less complicated. To create this highly accurate and effective model, a mix of classifiers and algorithms, including the decision tree algorithm, the random algorithm, and logistic regression, was utilized. The classification of patients is based on the predictions made by decision tree classifiers and their actual presentation of one as malignant (cancerous) and zero as benign (noncancerous). More accurate predictions can be made by this model than by incorrect ones. Early breast cancer detection makes it possible to treat the disease and save patients from grueling operations. After 50 years, Breast cancer will be predicted with an estimated 94% accuracy rate using ELM. This precision is provided using PCA and ELM as the primary tools. This component can be applied to mammography techniques in the future to discriminate between benign and hazardous cells in the nascent stages. Every day presents a new chance to improve. This study offers assistance to colleagues conducting related research.

References

[1] Alkali Y, Routray I, Whig P. Strategy for reliable, efficient, and secure IoT using artificial intelligence. IUP Journal of Computer Sciences 2022;16(2).

[2] Anand M, Velu A, Whig P. Prediction of loan behaviour with machine learning models for secure banking. Journal of Computer Science and Engineering (JCSE) 2022;3(1): 1–13.

[3] Arun Velu PW. Impact of covid vaccination on the globe using data analytics. International Journal of Sustainable Development in Computing Science 2021a;3(2).

[4] Jupalle H, Kouser S, Bhatia AB, Alam N, Nadikattu RR, Whig P. Automation of human behaviors and their prediction using machine learning. Microsystem Technologies 2022:1−9.

[5] Khera Y, Whig P, Velu A. Efficient effective and secured electronic billing system using AI. Vivekananda Journal of Research 2021a;10:53−60.

[6] Nadikattu RR, Mohammad SM, Whig P. Novel economical social distancing smart device for covid-19. International Journal of Electrical Engineering and Technology (IJEET) 2020:14.

[7] Sharma A, Kumar A, Whig P. On the performance of CDTA-based novel analog inverse low pass filter using 0.35 μm CMOS parameter. International Journal of Services Technology and Management 2015;4(1):594−601.

[8] Tomar U, Chakroborty N, Sharma H, Whig P. AI-Based smart agriculture system. Transactions on Latest Trends in Artificial Intelligence 2021;2(2).

[9] Velu A, Whig P. Studying the impact of the COVID vaccination on the world using data analytics. Vivekananda Journal of Research 2022;10(1):147−60.

[10] Whig P. A novel multi-center and threshold ternary pattern. International Journal of Machine Learning for Sustainable Development 2019;1(2):1−10.

[11] Ferlay J, Soerjomataram I, Dikshit R, et al. Cancer incidence and mortality worldwide: sources, methods and major patterns in GLOBOCAN 2012. International Journal of Cancer 2014;136(5):359−89.

[12] Elmore JG, Wells CK, Lee CH, Howard DH, Feinstein AR. Variability in radiologists' interpretations of mammograms. New England Journal of Medicine 1994;331(22): 1493−9.

[13] Mao N, Yin P, Wang Q, et al. Added value of radionics on mammography for breast cancer diagnosis: a feasibility study. Journal of the American College of Radiology 2019;16(4):485−91.

[14] Wang H, Feng J, Bu Q, et al. Breast mass detection in digital mammogram based on gestalt psychology. Journal of Healthcare Engineering 2018;2018. Article ID 4015613, 13 pages.

[15] Whig P, Ahmad SN. A CMOS-integrated CC-ISFET device for water quality monitoring. International Journal of Computer Science Issues 2012;9(4):1694−814.

[16] Whig P, Kouser S, Velu A, Nadikattu RR. Fog-IoT-assisted-based smart agriculture application. In: Demystifying federated learning for blockchain and industrial internet of things. IGI Global; 2022. p. 74−93.

[17] Valvano G, Santini G, Martini N, et al. Convolutional neural networks for the segmentation of microcalcification in mammography imaging. Journal of Healthcare Engineering 2019;2019. Article ID 9360941, 9 pages.

[18] Whig P, Nadikattu RR, Velu A. COVID-19 pandemic analysis using the application of AI. Healthcare Monitoring and Data Analysis Using IoT: Technologies and Applications 2022:1.

[19] Akay MF. Support vector machines combined with feature selection for breast cancer diagnosis. Expert Systems with Applications 2009;36(2):3240−7.

[20] Whig P, Velu A, Bhatia AB. Protect nature and reduce the carbon footprint with an application of blockchain for IoT. In: Demystifying federated learning for blockchain and industrial internet of things. IGI Global; 2022. p. 123−42.

[21] Whig P, Velu A, Naddikatu RR. The economic impact of AI-enabled blockchain in 6G-based industry. In: AI and blockchain technology in 6G wireless network. Singapore: Springer; 2022. p. 205–24.

[22] Whig P, Velu A, Nadikattu RR. Blockchain platform to resolve security issues in IoT and smart networks. In: AI-enabled agile internet of things for sustainable FinTech ecosystems. IGI Global; 2022. p. 46–65.

[23] Narain Ponraj D, Evangelin Jenifer M, Poongodi P, Samuel Manoharan J. A survey of the preprocessing techniques of mammogram for the detection of breast cancer. Journal of Emerging Trends in Computing and Information Sciences 2011;2(12):656–64.

[24] Charite P, Jamie SB. The preprocessing methods of mammogram images for breast cancer detection. International Journal on Recent and Innovation Trends in Computing and Communication 2017;5(1):261–4.

[25] Salembier P, Garrido L. Binary partition tree as an efficient representation for image processing, segmentation, and information retrieval. IEEE Transactions on Image Processing 2000;9(4):561–76.

[26] Whig P, Velu A, Ready R. Demystifying federated learning in artificial intelligence with human-computer interaction. In: Demystifying federated learning for blockchain and industrial internet of things. IGI Global; 2022. p. 94–122.

[27] Antonini M, Barlaud M, Mathieu P, Daubechies I. Image coding using wavelet transform. IEEE Transactions on Image Processing 1992;1(2):205–20.

[28] Carter G, Knapp C, Nuttall A. Estimation of the magnitude-squared coherence function via overlapped fast Fourier transform processing. IEEE Transactions on Audio and Electroacoustics 1973;21(4):337–44.

[29] Whig P, Velu A, Sharma P. Demystifying federated learning for blockchain: a case study. In: Demystifying federated learning for blockchain and industrial internet of things. IGI Global; 2022. p. 143–65.

Further reading

[1] Cruz J, Wishart DS. Applications of machine learning in cancer prediction and prognosis. Cancer Informatics 2006;2:59–77.

Medical image analysis of masses in mammography using deep learning model for early diagnosis of cancer tissues

R. Julia[1], Shajin Prince[1], D. Bini[2]

[1]*Electronics and Communication Engineering, Karunya Institute of Technology and Sciences, Coimbatore, Tamil Nadu, India;* [2]*Robotics Engineering, Karunya Institute of Technology and Sciences, Coimbatore, Tamil Nadu, India*

1. Introduction

In recent decades, there has been a steady rise in the prevalence of breast cancer (BC). It is the second most common reason for mortality in females. There is an effective chance of recovery when the condition is identified early. Although mammography has been shown to be a good screening method for detecting breast tumors, it is still very difficult to detect and classify breast tumors in mammograms. Most cases of BC develop in the fascicles as well as ducts. Although BC is diagnosed earlier, individuals can be given the appropriate treatment, improving their likelihood of surviving. Globally, 8.3 million people die from BC each year following acquiring an estimated 14.2 million diagnoses. In developing countries, 70% of newly diagnosed cancer patients are diagnosed, and therefore by 2025, it is predicted that there would be about 18.3 million invasive BC cases annually [1]. In contrast to previous BC testing methods, diverse clinical imaging is frequently employed to diagnose BC effectively. For early diagnosis of BC, a variety of medical imaging diagnoses are frequently used, including magnetic resonance imaging (MRI), ultrasound sonograms, histopathological imaging, breast X-ray images, and mammograms [2]. Depending on its nature, a tumor may be either benign or malignant (cancerous or noncancerous). The malignancy that forms from mammary cells is termed as BC. BC can slowly invade the nearby healthy cells as well as tissues and penetrate the lymphatic system underneath the arms, where it can then spread to other parts of the human body. The BC stage identifies the degree to which cancerous cells have spread mostly from the main tumor. According to research, BC is set on by the damage or mutation to the DNA within the cells. The three major distinctions of BC are "invasive carcinoma," "invasive lobular carcinoma," and "ductal carcinoma." One of the most common forms of BC is "invasive

Computational Intelligence and Modelling Techniques for Disease Detection in Mammogram Images
https://doi.org/10.1016/B978-0-443-13999-4.00006-7

carcinoma," which starts in the milk-producing ducts and spreads to the nearby tissues. An instance of invasive cancer that appears in glands includes "invasive lobular carcinoma" [3].

In recent decades, many researchers have devised a variety of methods for the automatic detection and classification of cells in BC diagnosis [4]. The benign as well as malignant subtypes of cells in BC can be distinguished by their nucleus characteristics. However, because of the intricacy of standard machine learning techniques like preprocessing, annotation, segmentation, and feature extraction, as well as others, the system's effectiveness and precision are reduced. The best strategy for identifying "ductal carcinoma in situ" (DCIS) is mammography. Employing MR mammography and mammogram, respectively, 77.9% and 69.4% of patients were found to have "carcinoma in situ" [5]. Mammogram imaging may be used for regular screening due to the fact that these images are scientifically more appropriate for screening [6]. Early detection of substantial alterations in cancer allows for the prevention of more rigorous treatments and an enhancement in the likelihood that a BC patient will survive. Malignant regions generally have irregularly shaped, blurry, and hazy border ultrasound pictures. Mammography cannot reliably distinguish cysts with solid masses but ultrasonography can [7].

Due to its provision for multiplanar screening as well as 3D reconstruction methods, MRI effectively depicts the shape, size, as well as localization of breast lesions [8]. Compared to the other two BC imaging technologies, breast thermography is painless, noninvasive, safe, and noncontact for both the patients and the physician, which makes it suitable for use during annual physicals [9]. A thermal infrared sensor that converts infrared radiation in and out as electrical impulses and displays it on a thermogram is used in thermography, an efficient screening procedure, to diagnose BC by identifying body areas that exhibit an unusual temperature shift [10]. Histopathology is recognized as a key element for diagnosing BC when compared [11] to other types of medical imaging. They include phenotypic data and are vital for the treatment and diagnosis of cancer illnesses. Over the past few decades, scientific studies have centered on machine learning techniques for employing thermography to diagnose BC; some of these studies have focused on detecting the size and position of tumors, whereas others have emphasized acquisition processes and breast quadrants. Recent history has seen scientists use CNNs to diagnose BC with encouraging results.

In the recent past, CNNs were not frequently utilized for the diagnosis of BC using thermal pictures. This may have been due to the effectiveness of CNNs in comparison to textural or analytical features, or it may have been due to the high computing burden [12]. Early diagnosis of cancerous regions significantly enhances the probability of producing optimal treatment decisions. Medical imaging analysis can improve from the deployment of computer-aided diagnostic (CAD) systems. Implementation of CAD systems enhances the diagnosis and identification of BC. Machine learning and deep learning models are used in CAD systems to help in visual interpretation. Radiologists or pathologists must carefully examine medical images to diagnose BC. This delicate process takes a lot of time but also expertise. The healthcare industry has many potentials uses for CAD systems that are helped by artificial intelligence (AI). Automated intelligent feature extraction from a training dataset is possible with deep learning [13].

Employing CNNs for the diagnosis of BC, researchers have made significant progress in recent years. Future investigations into the creation of machine vision-based BC detection systems are anticipated to benefit from the findings of this work. In this article, we provide a deep-learning model that has been trained to classify aberrant breast tissues. Rescaling, area segmentation, with deep learning models for detection and classification make up the three primary parts of the suggested methodology. The purpose of the resizing phase is to speed up calculation by downsizing the photos. The primary objective of this article includes the classification of healthy and unhealthy breast tissues as well as the localization of the cancerous region using two-stage deep learning models such as YOLO object detectors. The following is the layout of the article. Section 2 defines the literature review, whereas Section 3 describes the suggested strategy. The evaluation results can be found in Section 4. Sections 5 and 6 explore the paper's conclusion and provide a wrap-up.

2. Related work

Researchers face difficulties in making a diagnosis of BC illness. Machine learning, deep learning, and transfer learning are only a few of the models and methods that are employed to address the issue of BC. In this article, a framework was proposed for the CAD-based segmentation and classification of tumors. This study makes use of the MIAS datasets and the DDSM datasets. This system employs preprocessing, annotation, augmentation, segmentation, classification, and detection of each component. A CNN model with eight convolution layers, max-pooling, and completely connected layers is included in the CAD system. Many researchers have been using the popular dataset DMR-IR, which incorporates infrared images, for BC detection. These investigations, however, are premised on machine learning algorithms. Recent years have seen a rise in the use of machine learning (ML) in research, featuring applications as diverse as text mining, spam filtering, video recommendation, and image segmentation and classification.

DL is the technology that is most frequently employed in these implementations among the many ML methods. In contrast to earlier machine learning techniques, DL learns features entirely on its own [14]. Previously, handmade feature extraction methods (including "wavelet," "texture," "geometric," and "morphology") were employed to identify and extract features. Only a small number of studies used deep learning as its BC classifier when using thermal imaging and CNN model. Authors of Refs. [15,16] demonstrated the network could quickly distinguish between a stable and sick breast using the DMR-IR dataset and an InceptionV3-KNN ensemble model. The results indicate that thermography can serve as a feasible alternative to traditional BC detection methods because it improves the localization of cancerous and tumor cells, even while these cells go through angiogenesis. Mambou and

Maresova [16] proposed a transfer learning-based CNN (FTM-LL) that used thermal breast pictures to classify breasts as either healthy or unwell and produced a strong binary classification. Researchers did not choose images of young women in the dataset the researchers used, which instead had adults aged 29–85.

Furthermore, due primarily to their capacity for detecting small lesions as well as real-time detection time, the majority of the suggested models are unable to match the healthcare needs for inadequacies. It is necessary to figure out how to swiftly locate lesions and appropriately categorize them in such clinical images. Application research, meanwhile, is few and insufficiently thorough in the medical field. In this investigation, the series of YOLOv3, YOLOv5, and YOLOv7's capability to identify and categorize malignant lesions on CBIS-DDSM was evaluated and analyzed with the findings to the academic research. The results demonstrate that improving the performance of the model does not depend on how sophisticated the submodel is. This work offers a fresh perspective on how to identify and characterize breast lesions using CBIS-DDSM. Table 4.1 lists the various deep learning and machine learning classifiers for cancer cell classification and detection.

3. Proposed methodology

3.1 Datasets

For BC detection and classification, a benchmark dataset and dataset preprocessing methods are to be employed. An extensive analysis of the public databases utilized in various studies for BC categorization is presented in this section. Additionally, this section gives an overview of the typical preprocessing techniques employed in earlier investigations. For the purpose of diagnosing BC classification, numerous datasets have really been presented. Clinical data were indeed occasionally employed in studies; instead, a lot of them employed datasets of medical images. Massive datasets are necessary for the training of deep learning models. Employing deep learning models to medical diagnosis has a number of challenges due to a shortage of data. For BC classification, several researchers worked with their own data that they had obtained; therefore, termed these datasets to be private datasets (PD), which are then annotated and labeled. The fact that researchers only used a tiny dataset to test and confirm the suggested methodology is the gravest criticism leveled against our study. To address this problem, the majority of researchers employed transfer learning. Nevertheless, many studies exploited openly available databases.

The dataset that is utilized the most widely is the Digital Database for Screening Mammography (DDSM). This is the biggest publicly accessible dataset, with 2620 samples comprising two breast images per each breast, the Medio Lateral Oblique (MLO) and the CranioCaudal (CC) perspectives, and an average of 10,480 image features all types of observations, from routine sample images to samples containing both benign as well as malignant lesions. Although this is no longer recommended.

Table 4.1 Summary of the related literature work.

Reference no.	Dataset	Classifier	Application	Performance metrics
[17]	DDSM and MIAS	SVM, kNN, Naïve Bayes, Random Forest	Classify benign or malignant	Sensitivity = 96.87%, Specificity = 95.94%, Accuracy = 96.47%
[18]	DDSM	CNN, GAN	Detect normal ROIs or abnormal ROIs	Skewness; mean of GAN ROIs
[19]	RIDER	SVM, DWT, PCA	MRI breast tumor detection	ACC: 98.03%, Sensitivity = 0.964; specificity = 0.96.
[20]	DDSM	YOLO	Mass detection	Accuracy = 97%
[21]	Private datasets	CNN and SVM	Malignant or benign	Accuracy = 92%; specificity = 86%
[22]	IRMA	VGG16, Inception v3, ResNet50	Malignant or benign	Accuracy = 94% and 91.7%, respectively

CBIS-DDSM, an improved and streamlined version of DDSM database for the evaluation of CAD methods in mammography, was released by Clark et al. [27], as shown in Fig. 4.1. A readily useable dataset is also included, as well as better ROI-segmented imageries. There are 753 patients with microcalcification and 891 mass cases in the sample, respectively. The Mammographic Image Analysis Society

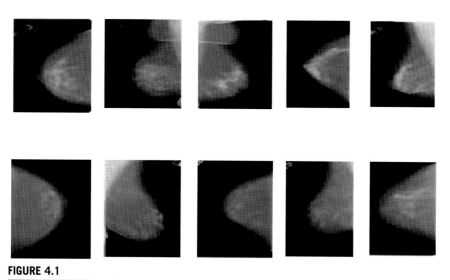

FIGURE 4.1

Sample images from CBIS-DDSM dataset.

Digital Mammogram Database (MIAS), which was the first database ever created and is still in use today, has been extensively used in studies. A total of 161 samples with 322 digitized MLO scans with a range of results, encompassing benign and malignant lesions standard images, constitute the MIAS database. To establish a new database dubbed mini-MIAS, the resolution of this one was reduced. The imaging techniques, class, with imaging structure of a dataset can affect how those three factors differ. Unique image formats are present in the datasets that are available. Images in the TIFF format, DICOM format, or PNG format can all be accessible in a few datasets.

3.2 Preprocessing

The numerous preprocessing methods employed for medical image analysis in the diagnosis of BC are covered in this section. As there were few research that employed patient statistical data and most studies included imaging techniques, we focused on image preprocessing. The image preprocessing processes for BC diagnosis typically involve annotation, augmentation, labeling, scaling, ROI extraction, image normalization, and enhancements to get rid of features and artifacts. The utilization of unprocessed images in machine learning typically distorts categorization results and can occasionally produce subpar output.

3.3 Image augmentation

Researchers artificially increase the amount of images by utilizing image augmentation. As there are not many annotated medical images readily accessible, image augmentation is being used. Furthermore, we discovered that the ROI was recovered from BC pictures in 30 out of 80 investigations. Neural networks always had the ability to learn representations associated with normal and abnormal areas rather than using the entire image, which typically contains extraneous information. Merely 20 out of the 80 trials reduced the proportion of the images before loading them to the deep neural networks (DNN). Rescaling is required when the images are provided straight to the DNNs, as CNNs do.

Before BC classification, however, only a smaller number of research (25 out of 80) used image-normalization and enhancement techniques. This process equalizes all pictures, minimizes high- and low-intensity distortion, and aids DNNs in producing correct features for pathological and normal BC tissue components. However, only six out of the 80 investigations eliminate various visual imperfections such as opacity, labels, markings, wedges, and pectoralis major muscles. This technique eliminates nonbreast regions from the images before doing a BC diagnosis. Only certain image types, like mammograms, ultrasounds, and MRIs, require artifact elimination from images, which is why so very few studies actually do it. In certain research, stain normalization preprocessing procedures were used (12 out of 80).

3.4 Deep convolutional neural network

Machine learning (ML) is being employed in the medical field to detect BC. ML is regarded as a subfield of artificial intelligence that employs logical, analytical, and mathematical procedures to enable a computer that learn from experience without programming by integrating the learning challenges from dataset to the broad framework of inference [23]. In contrast to conventional machine learning feature extraction techniques, deep learning is a subclass of machine learning and artificial intelligence that integrates a complicated hierarchy underlying image properties. To make significant advancements in the technique computers retrieve information from images, deep learning models are applied to various recently utilized computer models. These models have been employed to complete tasks in many different medical fields, particularly radiology and pathology, but in certain instances, the efficiency has been on pace with that of human professionals.

Multilevel neural networks have made deep learning (DL), which utilizes raw input pixel images to derive a hierarchical system of features. The technological advancement of efficient DL algorithms that can train with thousands of images and therefore are insensitive toward image differences has been made possible by the rapid improvement in graphic processing power. Due to its most recent successes, especially in applications for image segmentation, detection, and classification, DL has gained more recognition. For a wide range of applications, such as the segmentation, identification and classification of images containing objects voice recognition, the identification of the genotypes and the phenotypes, and medical disease diagnosis, numerous ranges of DL algorithms have been proposed. Stack autoencoders, deep neural networks, Deep Boltzmann machines, and Deep Convolution Neural Networks (CNN) are a few prominent DL-algorithms [2]. The proposed methodology and evaluation of the CAD system have been represented in Fig. 4.2.

3.4.1 Training and testing

The datasets are annotated, augmented, and are split into train, test, and validation datasets. Table 4.2 shows the training parameters employed for training the neural network.

3.4.2 YOLO series object detectors

A head and a backbone are often the two fundamental components of deep learning-based object detectors. Although the head is utilized to detect class labels and bounding boxes, its backbone pulls features using high-dimensional inputs and therefore is frequently pretrained using ImageNet data. With regard to the head, the object detectors can be divided into two-stage and one-stage categories according to whether a separate algorithm is used to create region recommendations. The one-stage heads of all YOLO object detectors employ dense sampling to classify and localize semantic objects with one shot. An object detector's neck is made up of layers that are frequently put between the backbone and the head to collect feature mAPs from various stages. Enhanced feature integration models, improved

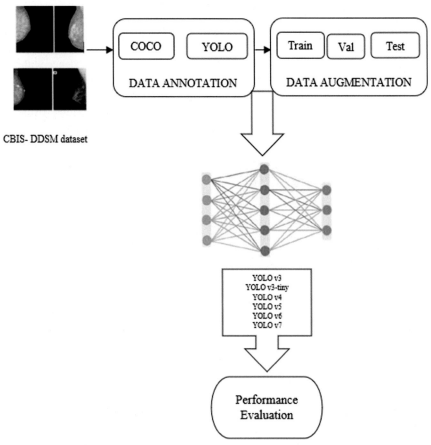

CBIS- DDSM dataset

FIGURE 4.2

Proposed workflow of cancer cell classification and detection.

Table 4.2 Training parameters for the deep learning network model.

Network model	Pixels	Epoch
YOLOv3	640	500
YOLOv3-tiny	640	500
YOLOv4	640	500
YOLOv4 CSP	640	500
YOLOv5	1280	500
YOLOv5s	1280	500
YOLOvR	1280	500
YOLOv6	1280	500
YOLOv7	1280	500

backbone models, and a variety of training methods have been the main areas of focus in object detection.

In comparison to its predecessors, YOLOv3 [24] offers the ideal balance among speed, precision, and accuracy for real-time object recognition. For feature extraction for input images, YOLOv3 employs the Darknet-53 backbone with feature extraction from input images, whereas YOLOv3-tiny, its compressed counterpart, is built on the Darknet-19. A variety of characteristics set YOLOv4 apart from YOLOv3. Spatial Pyramid Pooling with Path Aggregation Network block necks is used, along with the CSPDarknet-53 backbone. YOLOv4 [25] uses a variety of training techniques to improve detection accuracy, which includes "bag of freebies" (techniques that really lengthen training time) like Mosaic data augmentation, Cut-Mix, and DropBlock regularization, and "bag of specials" (techniques that only slightly lengthen inference time but can sharply increase detection accuracy) like mish activation, cross-stage partial interconnection, and DIoU-NMS. Diverse computing devices, such as generic, low-end, and high-end GPUs, are indeed utilized to focus on the cross-stage partial network enabling model scaling. The scaled models with the highest state-of-the-art detection accuracy on the COCO dataset were YOLOv4-large scaled for cloud GPUs. Among the most recent YOLO object detectors, along with its contemporaries YOLOF, YOLOX, and PPYOLOv2, is YOLOR. A broad, unified representation for multiple tasking multilabel image classification is learned by YOLOR using the YOLOv4-CSP. When put through object detection challenges, YOLOR produced results that were equivalent to Scaled-YOLOv4 in terms of accuracy but significantly faster in terms of inference speed.

In terms of real-time classification and object detection, the two most current YOLO versions, YOLOv6 and YOLOv7, both exhibit greater effectiveness. The proposed framework, labeling, data augmentation, loss function, and normalization are among the revised designs in YOLOv6 that are featured for industrial implementation. To improve network training and detection accuracy without raising inference costs, the designers of YOLOv4 or even Scaled-YOLOv4 have developed YOLOv7, which employs a prolonged efficient layer aggregation network in addition to a number of trainable bag of freebies methodologies (like planned reparameterization and coarse-to-fine lead guided label assignment). YOLOv7 outperformed earlier object detectors such as YOLOX, Scaled-YOLOv4, YOLOv5, YOLOR, and others in terms of both speed and accuracy after being trained from the base on the COCO dataset.

A short time after the release of YOLOv4, Ultralytics LLC launched YOLOv5, which made the claim that it performed better than all of its YOLO predecessors. Although YOLOv5 was not disclosed in a formal peer-reviewed paper like prior YOLO detectors and because it shares structural similarities with YOLOv4, there has been some debate in the computer vision community about whether their proposed term is appropriate. Apparently, it implies that the implementation of the anchor box selection mechanism in the model represented the only substantial change between YOLOv5 and YOLOv4.

However, YOLOv5 has grown significantly in popularity as a quick and effective object detector and demonstrated improved detection accuracy in several situations. A variety of devices can be utilized to run the YOLOv5 versions in the PyTorch framework, which is actively updating with the most recent version of YOLOv5-v6. Furthermore, it offers variable model size selection.

The most recent version of the YOLO model detector, entitled YOLOv7, is an object detection network with features including real-time detection, high accuracy, and flexibility in training and deployment. The network outperforms already-known classifier model object detectors with speed and precision in the range between 5 and 160 frames per second (FPS). The YOLO network model with thick blocks of multi-layer architecture is represented in Fig. 4.3.

Although earlier studies evaluated a few of these object detectors, researchers only analyzed one or two YOLO models for just a few classes. There has not been any investigation into a thorough assessment of a variety of YOLO detectors for multiclass detection. It is crucial to note that the developers of the chosen YOLO detectors have provided open-sourced software packages, which were modified in this research to train detection models.

4. Performance analysis

The performance measurements that are employed to evaluate CAD systems have been explained in this section. If BC has been diagnosed properly or accurately, it might be classified as true-positive (TP) and true-negative (TN), whereas if it is detected wrongly, it can be classified as false-positive (FP) and false-negative (FN). The most prevalent evaluation criteria used for BC classification include accuracy, sensitivity, precision, FMeasure, AUC (area under the curve), and volume under the ROC surface [26].

In a broader sense, these measures are as follows.

4.1 Accuracy

The number of instances categorized absolutely correct is measured by this procedure.

Accuracy is represented by Eq. (4.1)

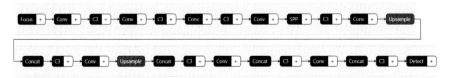

FIGURE 4.3

YOLOv5 network model.

$$Acccuracy = \frac{(TP + TN)}{(TP + TN + FP + FN)} \tag{4.1}$$

4.2 Sensitivity

The outcomes of the measure certainly show how many of the total positive samples are actually just right estimates. In general, it shows the proportion of atypical patients with BC whose numbers are appropriately predicted. This can be shown by Eq. (4.2)

$$Sensitivity = \frac{(TP)}{(TP + FN)} \tag{4.2}$$

4.3 Specificity

Specificity depicts how accurate the general optimistic predictions really are as well as shows how much of what is often expected is correct. This can be denoted by Eq. (4.3)

$$Specificity = \frac{(TN)}{(TN + FP)} \tag{4.3}$$

4.4 Precision

This basically indicates how accurate the diagnosis of malignant BC was to a certain extent. To avoid misdiagnosing cancer patients, sensitivity and precision values for medical image diagnostics should be excellent. This can be expressed by Eq. (4.4)

$$Precision = \frac{(TP)}{(TP + FP)} \tag{4.4}$$

The precision measures how well the model can distinguish between negative samples. The neural network model's capacity to discriminate negative random samples becomes stronger with increasing precision. The model's capacity to recognize positive samples is reflected by recall. The model's capacity to recognize positive samples is enhanced by higher recall. The sum of both is indeed the F1 score. The model is more stable the higher the F1 score.

5. Experimental results and discussions

The training of the datasets for detection was employed on an NVIDIA GEFORCE GTX GPU with windows 10 environment. The input images were resized to 640×640 input pixels. Over a series of research experiments, hyperparameters

Table 4.3 Comparison of YOLO as object detectors on a test dataset.

Network model	Precision %	Recall %	Specificity %	Sensitivity %
YOLOv3	92	93	89	90
YOLOv3-tiny	89	86	90.5	91
YOLOv4	91	95	90.8	90
YOLOv4 CSP	94	94	92	94
YOLOv5	93	90	96	95
YOLOv5s	96	92	96.5	95
YOLOvR	94	91	92	93.5
YOLOv6	93	93	95	94
YOLOv7	94	95	94	96

have been specifically selected; nonetheless, each hyperparameter must be determined in accordance with prior knowledge. To enhance the quantity of the training dataset, the data augmentation technique was implemented. The raw images were augmented by flipping, cropping, rotating, and shifting, which preserved their original medical features. Table 4.3 represents the performance metrics of the various YOLO network models. The test images are predicted with diverse YOLO models and the result samples of prediction with bounding boxes are depicted in Fig. 4.4.

6. Conclusion

The earlier diagnosis of malignant or benign cancer cells can enhance the life expectancy of the patients. However, improvements in medical image analysis have enabled the feasibility to identify BC in its initial stage. With retrieving the pertinent features from massive datasets, AI-based medical image analysis (MIA) approaches have facilitated the ability to explore and exploit massive data for enhanced BC diagnosis and treatment plan. The YOLO series object detector network models achieved excellent results, with YOLOv3-tiny scoring 89% and YOLOv5 scoring 96%. Significantly, the YOLOv7-based models outperformed YOLOv3 and YOLOv5, whereas YOLOvR and YOLOv5s showed benefits in quick inference speeds while attaining comparable detection accuracies. The efficiency of YOLO detectors, specifically YOLOv3-tiny, may be enhanced by data augmentation. This performance benchmark aims to provide an unbiased source for evaluating YOLO object detectors for cancer cell classification and detection.

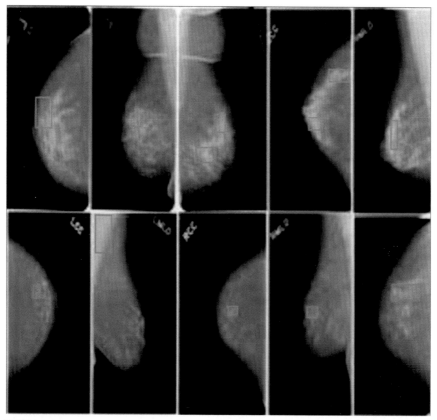

FIGURE 4.4

Result samples of images with bounding boxes predicted by YOLOv5.

References

[1] Sung H, et al. Global cancer statistics 2020: GLOBOCAN estimates of incidence and mortality worldwide for 36 cancers in 185 countries. CA Cancer Journal Clinicians 2021;71(3):209–49. https://doi.org/10.3322/caac.21660.

[2] Yap MH, et al. Automated breast ultrasound lesions detection using convolutional neural networks. IEEE Journal of Biomedical and Health Informatics 2018;22(4):1218–26. https://doi.org/10.1109/JBHI.2017.2731873.

[3] Dar RA, Rasool M, Assad A. Breast cancer detection using deep learning: datasets, methods, and challenges ahead. Computers in Biology and Medicine 2022; 149(August):106073. https://doi.org/10.1016/j.compbiomed.2022.106073.

[4] Jim Y. Applied sciences deep-learning-based computer-aided systems for breast cancer imaging: a critical review. Applied Sciences 2020;10:8298.

[5] Malur S, Wurdinger S, Moritz A, Michels W, Schneider A. Comparison of written reports of mammography, sonography and magnetic resonance mammography for preoperative evaluation of breast lesions, with special emphasis on magnetic resonance mammography. Breast Cancer Research 2001;3(1):55−60. https://doi.org/10.1186/bcr271.

[6] Kopans DB. Sonography should not be used for breast cancer screening until its efficacy has been proven scientifically. American Journal of Roentgenology 2004;182(2):489−91. https://doi.org/10.2214/ajr.182.2.1820489.

[7] Kelly KM, Dean J, Comulada WS, Lee SJ. Breast cancer detection using automated whole breast ultrasound and mammography in radiographically dense breasts. European Radiology 2010;20(3):734−42. https://doi.org/10.1007/s00330-009-1588-y.

[8] Hizukuri A, Nakayama R, Nara M, Suzuki M, Namba K. Computer-aided diagnosis scheme for distinguishing between benign and malignant masses on breast DCE-MRI images using deep convolutional neural network with Bayesian optimization. Journal of Digital Imaging 2021;34(1):116−23. https://doi.org/10.1007/s10278-020-00394-2.

[9] Frize M, Herry C, Roberge R. Processing of thermal images to detect breast cancer: comparison with previous work. Proceedings of the Second Joint 24th Annual Conference and the Annual Fall Meeting of the Biomedical Engineering Society 2002;2:1159−60. https://doi.org/10.1109/iembs.2002.1106325.

[10] Hossam A, Harb HM, Abd El Kader HM. Automatic image segmentation method for breast cancer analysis using thermography. Journal of Engineering Sciences 2018;46(1):12−32. https://doi.org/10.21608/jesaun.2017.114377.

[11] Yang X, et al. Deep learning signature based on staging CT for preoperative prediction of sentinel lymph node metastasis in breast cancer. Academic Radiology 2020;27(9):1226−33. https://doi.org/10.1016/j.acra.2019.11.007.

[12] Zuluaga-Gomez J, Al Masry Z, Benaggoune K, Meraghni S, Zerhouni N. A CNN-based methodology for breast cancer diagnosis using thermal images. Computer Methods in Biomechanics and Biomedical Engineering: Imaging & Visualization 2021;9(2):131−45. https://doi.org/10.1080/21681163.2020.1824685.

[13] Hassanien AE, Ali JMH. Feature extraction and rule classification algorithm of digital mammography based on rough set theory. pp. 1−9.

[14] Darwin B, Dharmaraj P, Prince S, Popescu DE, Hemanth DJ. Recognition of bloom/yield in crop images using deep learning models for smart agriculture: a review. Agronomy 2021;11(4):1−20. https://doi.org/10.3390/agronomy11040646.

[15] Ekici S, Jawzal H. Breast cancer diagnosis using thermography and convolutional neural networks. Medical Hypotheses 2020;137(December 2019):109542. https://doi.org/10.1016/j.mehy.2019.109542.

[16] Mambou SJ, Maresova P. Imaging and a deep learning model. Sensors 2018;18:2799. https://doi.org/10.3390/s18092799.

[17] Rouhi R, Jafari M, Kasaei S, Keshavarzian P. Benign and malignant breast tumors classification based on region growing and CNN segmentation. Expert Systems with Applications 2014;42:990−1002. https://doi.org/10.1016/j.eswa.2014.09.020.

[18] Guan S, Loew M. Breast cancer detection using synthetic mammograms from generative adversarial networks in convolutional neural networks. Journal of Medical Imaging 2019;6(3). https://doi.org/10.1117/1.JMI.6.3.031411.

[19] Ibraheem AM. Automatic MRI breast tumor detection using discrete wavelet transform and support vector machines. Novel Intelligent and Leading Emerging Sciences Conference (NILES) 2019;1:88−91.

[20] Al-masni MA, et al. Computer Methods and Programs in Biomedicine Simultaneous detection and classification of breast masses in digital mammograms via a deep learning YOLO-based CAD system. Computer Methods and Programs in Biomedicine 2018; 157:85—94. https://doi.org/10.1016/j.cmpb.2018.01.017.

[21] Alkhaleefah M, Wu C. A hybrid CNN and RBF-based SVM approach for breast cancer classification in mammograms. In: 2018 IEEE International Conference on Systems, Man, and Cybernetics; 2018. p. 894—9. https://doi.org/10.1109/SMC.2018.00159.

[22] Rodrigues MF. Breast cancer detection using deep learning techniques. 2020.

[23] Montazeri M, Montazeri M, Montazeri M, Beigzadeh A. Machine learning models in breast cancer survival prediction. Technology and Health Care 2016;24:31—42. https://doi.org/10.3233/THC-151071.

[24] Redmon J. YOLOv3: an incremental improvement. 2018.

[25] Wang C, Liao HM. YOLOv4: optimal speed and accuracy of object detection. 2020.

[26] Murtaza G, et al. Deep learning-based breast cancer classification through medical imaging modalities: state of the art and research challenges. Artificial Intelligence Review 2019;53:1655—720. https://doi.org/10.1007/s10462-019-09716-5.

[27] Clark K, Vendt B, Smith K, Freymann J, Kirby J, Koppel P, et al. The Cancer Imaging Archive (TCIA): maintaining and operating a public information repository. Journal of Digital Imaging 2013;26(6):1045—57.

A framework for breast cancer diagnostics based on MobileNetV2 and LSTM-based deep learning

A. Jenefa[1], A. Lincy[2] and V. Edward Naveen[3]

[1]*Department of CSE, Karunya Institute of Technology and Sciences, Coimbatore, Tamil Nadu, India;* [2]*Department of CSE, National Engineering College, Kovilpatti, Tamil Nadu, India;* [3]*Department of CSE, Sri Shakthi Institute of Engineering and Technology, Coimbatore, Tamil Nadu, India*

1. Introduction

Breast cancer is a frequent disease among women globally, claiming the lives of a large number of women and ranking as one of the leading causes of female mortality. The American Cancer Society reports that the number of men and women who have recently passed away from breast cancer is staggering, with over 500 men and 41,760 women losing their lives to the disease [1]. Breast cancer originates in the breast tissue, specifically in the lobules and ducts and can spread to other parts of the body, such as lymph nodes and healthy tissue. Symptoms may include tenderness, discharge, changes in size, inverted nipples, blisters under the arms, and tenderness. Certain risk factors such as age, obesity, inactivity, excessive alcohol, dense breast tissue, genetics, early onset of menstruation, and advanced maternal age may increase the chance of developing breast cancer.

Diagnosing breast cancer manually is difficult and time consuming, leading to the need for automatic diagnosis. Although existing healthcare systems can be helpful, they are prone to errors. Therefore, the use of CAD through the classification of medical images has emerged as a useful tool to assist medical professionals in identifying diseases earlier, resulting in faster diagnoses and treatments. Machine learning and deep learning techniques have been developed to diagnose breast cancer more accurately and earlier, reducing the need for readmissions in clinics and hospitals. The application of AI techniques in healthcare can improve diagnostic accuracy and lower healthcare costs associated with incorrect diagnoses.

Deep learning has recently been applied to the field of breast cancer diagnosis, with promising results [2]. A survey conducted on a large dataset of mammograms revealed that deep learning algorithms were able to accurately identify malignant

Computational Intelligence and Modelling Techniques for Disease Detection in Mammogram Images
https://doi.org/10.1016/B978-0-443-13999-4.00013-4

breast tumors with high sensitivity and specificity. In a study referenced in Zhang et al. [3], a CNN was trained using a dataset consisting of over 25,000 mammograms. The results showed that the CNN outperformed radiologists, with an area under the curve (AUC) of 0.96. Another study by Barnett et al. [6] utilized a deep learning algorithm that incorporated both mammographic images and patient demographic information, resulting in an AUC of 0.98. One of the advantages of using deep learning for breast cancer diagnosis is its ability to learn and identify subtle features in mammograms that may be missed by human radiologists. This can lead to earlier detection of breast cancer and ultimately improved patient outcomes. Additionally, deep learning algorithms have the potential to assist radiologists in making more accurate and consistent diagnoses, reducing the number of false negatives and false positives. It is important to note that the use of deep learning in breast cancer diagnosis is still in its early stages and further research is necessary to improve the performance of these algorithms and ensure their ability to accurately diagnose across different populations and imaging techniques.

The proposed deep learning framework combines the strengths of MobileNetV2, a lightweight and efficient CNN architecture, and long short-term memory (LSTM), a type of RNN that is able to capture temporal dependencies in sequential data. The MobileNetV2 architecture is used to extract high-level features from breast mammogram images, which are then fed into the LSTM layer for further analysis. The LSTM layer is able to analyze the temporal relationships between different features extracted by the MobileNetV2 and make a diagnosis based on these relationships. The deep learning framework being proposed utilizes a large dataset of mammogram images and their diagnoses to train the model, leading to a high level of accuracy in breast cancer diagnosis. In addition, the lightweight and efficient nature of MobileNetV2 allows for the framework to be easily deployed on mobile devices, making it a practical solution for breast cancer diagnosis in resource-limited settings.

MobileNetV2 is a deep learning model that has already been trained on a huge image dataset. The model can be adapted to the specific task of classifying breast cancer in mammograms by fine-tuning on a dataset of mammogram images. The fine-tuning process involves adjusting the model's parameters to adapt to the new dataset and task. The model can take as input a single mammogram and output a label indicating whether the mammogram is normal or abnormal. LSTM is a type of RNN utilized frequently for natural language recognition and time-series prediction tasks. In this case, the LSTM model takes as input a sequence of mammograms for the same patient over time. The LSTM model can extract spatiotemporal features from the sequence of images that are indicative of breast cancer. By analyzing the changes in mammograms over time, the output of the LSTM model can indicate whether a patient has breast tumor or not.

The paper's remaining sections are structured as follows: Section 2 highlights the findings pertinent to the most current breast cancer detection advancements. Section 3 addresses the suggested classification scheme for breast diseases utilizing MobileNetV2 and LSTM architecture. Section 4 analyses the findings and results and follows a conclusion and suggestions for additional research in Section 5.

2. Related work

Current CNN models enable radiologists to detect even the smallest breast tumors in their earliest stages [7,8]. Due to the availability of vast data sources and the capabilities of parallel and distributed computing, the adoption of DL techniques has led to improvements in pattern detection and classification challenges. Both [9] and [10] in the paper utilize the AlexNet model and transfer learning (TL) to analyze mammograms. They used a pretrained version of AlexNet for group diagnosis without any modifications. They evaluated the classification performance using support vector machine (SVM) and features from multiple intermediate network layers. They compared the outcomes of a classifier based on manually produced features with a soft voting ensemble that included both. The author of [10] presented the design for a pretrained CNN that would be stored on the Digital Database for Dcreening Mammography (DDSM) repository. Using deep learning medical imaging technology, the author of the paper [11] develops a breast tumor classification method from the ground up. This method increases the ability to discern between normal and cancerous breast tissue and helps classify breast tumors [12]. The researcher innovated a new approach to deep learning with his method. During the classification process, various algorithms such as CNN, RNN, LSTM, and ANN were employed. The authors utilized a pretrained AlexNet model that was slightly modified and fine-tuned for use on the CBIS-DDSM dataset. They selected the three methods that performed the best and combined the predictions from those models during the inference process. InceptionV3 and a reconfigurable convolutional net were the machine learning tools that the author used to analyze a private mammographic dataset [13]. This approach is regarded as a computationally complex one because of the fact that TL is carried out utilizing two distinct models. In Albashish et al. [14], the author searched for a comprehensive design by focusing on the overall mammography image models. Recent research has centered on utilizing pretrained networks as opposed to training networks from scratch [15−17]. The creation, training, and testing of pretrained CNN architectures, on the other hand, takes place on large datasets that are more varied and expansive than the mammographic datasets already in existence [18]. The complexity and power of these networks may become excessive when trained on large datasets, leading to negative consequences. Additionally, the limitations of these networks in mobile devices further limit their applicability [19−22]. Using such a large number of connections on mobile devices limits memory and processing power, hence decreasing performance [23−25].

This study presents a transfer learning strategy based on simple MobileNetV2 and LSTM models to circumvent this limitation. Consequently, they require less memory space and give better and faster results than alternative inception techniques. Table 5.1 shows the background work of deep learning framework for breast cancer diagnosis.

Table 5.1 Background work of deep learning framework for breast cancer diagnosis.

Related work	Algorithm	Dataset	Pros	Cons	Findings	Accuracy
[4]	CNN	Various	Can learn complex patterns in data	Can require a large amount of data and computation power	Used for image classification and feature extraction	Varies depending on the specific architecture and dataset
[5]	MobileNet	ImageNet	Efficient and lightweight	May not be as accurate as larger models	Used for image classification	Varies depending on the specific architecture and dataset
[7]	LSTM	Various	Can process long sequences of data	Can be computationally intensive	Used for time series analysis and natural language processing	Varies depending on the specific architecture and dataset
[8]	ANN	Various	Can learn complex patterns in data	Can require a large amount of data and computation power	Utilized for regression and classification	Varies depending on the specific architecture and dataset
[15]	RNN	Various	Can process long sequences of data	Can be difficult to train and may suffer from vanishing gradients	Used for time series analysis and natural language processing	Varies depending on the specific architecture and dataset

3. Deep learning framework for breast cancer diagnosis

MobileNetV2 is a pretrained deep learning model that has been trained using a vast dataset of images. It is a CNN architecture optimized for real-time object detection and classification tasks, while being lightweight and efficient. The depthwise separable convolutions, a unique feature of MobileNetV2, reduce both the number of parameters and computational requirements, making it ideal for use on mobile and embedded devices. The model can be fine-tuned on a specific dataset of mammograms to learn to classify breast cancer. Fine-tuning is a process where a pretrained model is further trained on a new dataset for a specific task. The fine-tuning process involves adjusting the model's parameters to adapt to the new dataset and task. The model can take as input a single mammogram and output a label indicating whether the mammogram is normal or abnormal. This can be used as a diagnostic tool for breast cancer. LSTM is a popular type of RNN often utilized in natural language processing and time-series prediction. LSTMs excel in capturing long-term relationships in sequential data, making them appropriate for tasks where the sequence of data matters. In the case of analyzing mammograms, LSTM takes as input a sequence of mammograms for the same patient over time. The LSTM model can extract spatiotemporal features from the sequence of images that are indicative of breast cancer. The LSTM model can provide a more complete assessment of a patient's mammograms by analyzing changes in mammograms over time. The model can then produce a label that determines whether the patient has a breast tumor or not instead of simply evaluating a single mammogram.

Algorithm 1: A deep learning system for diagnosing breast cancer utilizing MobileNetV2 and LSTM.

Input: A sequence of mammograms for the same patient over time.
   Output: A label indicating whether the mammogram is normal or abnormal.
   Step 1: Load a pretrained MobileNetV2 model.
   Step 2: Fine-tune the model on a specific dataset of mammograms using the model.fit() method.
   Step 3: Use the fine-tuned MobileNetV2 model for feature extraction by inputting a single mammogram into the model.predict() method.
   Step 4: Extract deep features from the last convolutional layer of the MobileNetV2 using $f(x) = model.predict(x)$
   Step 5: Load a pretrained LSTM model.
   Step 6: Fine-tune the LSTM model on a specific dataset of sequences of mammograms using the model.fit() method.
   Step 7: Use the fine-tuned LSTM model for feature extraction by inputting a sequence of mammograms into the model.predict() method.
   Step 8: Extract spatiotemporal features from the last hidden state of the LSTM using $g(x) = model.predict(x)$
   Step 9: Concatenate the deep features $f(x)$ and spatiotemporal features $g(x)$ obtained from MobileNetV2 and LSTM using $h(x) = concatenate(f(x), g(x))$
   Step 10: Pass the concatenated features $h(x)$ through a fully connected layer for classification using $y = softmax(W * h(x) + b)$
   Step 11: Output a label identifying whether or not the patient has tumor using.

Continued

Algorithm 1: A deep learning system for diagnosing breast cancer utilizing MobileNetV2 and LSTM.—cont'd

---

if y > threshold:
output="abnormal" (patient has breast cancer)
else:
output="normal" (patient does not have breast cancer)
Step 12: Use the final prediction to make a diagnosis and provide recommendations for further treatment or monitoring.
Step 13: Continuously monitor and evaluate the performance of the combined model using appropriate metrics and make adjustments as necessary.

The deep learning framework uses a pretrained MobileNetV2 model, optimized for real-time object detection and classification, which can be fine-tuned for breast cancer diagnosis using mammograms. Additionally, the framework employs the use of LSTM, a type of recurrent neural network, for analyzing mammograms over time and predicting the presence of breast cancer. It can be used to analyze a sequence of mammograms for the same patient over time to extract spatiotemporal features that are indicative of breast cancer and provide a more comprehensive analysis. Both models are capable of diagnosing breast cancer, with MobileNetV2 analyzing a single mammogram and LSTM analyzing a sequence of mammograms over time.

The pipeline described in the pseudocode is a general representation of how a combination of a pretrained MobileNetV2 and LSTM model can be utilized for breast tumor diagnosis and classification. The input to the deep learning framework is a series of mammograms taken from the same patient over a period of time, and the output is a label identifying the presence or absence of breast cancer. The first step is to load a pretrained MobileNetV2 model and fine-tune it on a specific dataset of mammograms using the model.fit() method. Then the fine-tuned MobileNetV2 model is used for feature extraction by inputting a single mammogram into the model.predict() method. The MobileNetV2 model is applied to collect deep information from a sequence of mammograms by inputting the images into the last convolutional layer. Then, a pretrained LSTM model is fine-tuned on the specific dataset of mammogram sequences through the model.fit() method. Finally, the fine-tuned LSTM model is utilized for feature extraction by passing the sequence of mammograms through the model.predict() method. The combination of features from MobileNetV2 and LSTM is formed into a single feature vector by concatenating deep features extracted from the last convolutional layer of the MobileNetV2 and spatiotemporal features obtained from the last hidden state of the LSTM. The resulting features are then fed into a fully connected layer for classification. The model outputs a diagnosis, indicating if the patient has breast cancer or not.

3.1 MobileNet

MobileNet equation for convolutional layer:

The MobileNet architecture utilizes depthwise separable convolutions, which can be represented by the following equation:

$$Output[i,j,k] = \sum Input[i,p,q] * Filter[k,p,q]$$

where Output is the output feature map produced by the convolutional layer. Input is the input feature map to the convolutional layer. Filter is the collection of filters applied to the input feature map i, j, and k are the indices for the output feature map, with i and j representing the spatial dimensions and k representing the channel dimension. p, q are the indices for the input feature map, with p and q representing the spatial dimensions. This equation represents a standard convolution operation, where the input feature map is convolved with a set of filters to produce the output feature map. In the MobileNet architecture, this operation is divided into two separate steps: depthwise convolution and pointwise convolution. Depthwise convolution convolves each input channel independently with its own set of filters, while pointwise convolution combines the output channels produced by depthwise convolution into a single output feature map using a 1×1 convolution. This permits the MobileNet framework to decrease computational complexity and enhance network efficiency.

$$convolutional\_layer = input\_data * weights + biases$$

This equation represents the standard operation of a convolutional layer in a deep learning network. The input data (input_data) is multiplied by a set of weights (weights) and then added to a set of biases (biases). The result of this operation is the output of the convolutional layer (convolutional_layer). The equation signifies a solitary convolutional layer in the neural network; however, in actuality, a deep learning network will have several convolutional layers arranged one above the other. The result of one layer acts as the input for the subsequent layer, and the parameters (weights and biases) of each layer are learned through the training process. The MobileNet architecture, in particular, makes use of multiple convolutional layers with different filter sizes and numbers of filters to extract features from the input data and make predictions. The specific details of the convolutional layers in MobileNet may vary depending on the specific implementation, but the basic equation described above remains the same.

There are several key equations that are used in the MobileNetV2 architecture for calculating the output of each layer.

3.1.1 Depthwise separable convolution

The depthwise separable convolution is a method applied in convolutional neural networks (CNNs) to minimize computational complexity and boost network efficiency. It involves dividing a standard convolution operation into two separate operations: depthwise convolution and pointwise convolution.

In depthwise convolution, each input channel is independently convolved by its own set of filters, resulting in a set of output channels. This allows the network to learn spatial relationships within each channel separately, rather than across all channels as in a standard convolution. This is the primary convolutional layer used in MobileNetV2 and is defined as follows:

$$depthwise\_convolution = input\_data * depthwise\_weights + depthwise\_biases$$

$$Output[i,j,k] = sum(Filter[i,j,l] * Input[stride[0] * i + dilation[0] * l, stride[1]$$
$$* j + dilation[1] * l, k] \ for \ l \ in[0, kernel\_size[0]]))$$

where Output is the output tensor, Filter is the convolutional kernel, Input is the input tensor, and stride and dilation are the stride and dilation parameters for the convolution.

3.1.2 Pointwise convolution

Pointwise convolution is then applied to combine the output channels produced by depthwise convolution into a single output feature map. This operation is similar to a 1×1 convolution, which is utilized to decrease the number of channels in a feature map. By dividing the standard convolution operation into these two separate steps, the overall computational complexity of the network is reduced, making it more efficient and easier to train. Depthwise Separable Convolution is often used in mobile and resource-constrained applications where efficiency is a key concern. This is a 1×1 convolutional layer used in MobileNetV2 to reduce the number of channels in the output tensor. It is defined as follows:

$$pointwise\_convolution = depthwise\_convolution * pointwise\_weights + pointwise\_biases$$

$$Output[i,j,k] = sum(Filter[i,j,l] * Input[i,j,l] \ for \ l \ in \ [0, num\_channels])$$

where Output is the output tensor, Filter is the convolutional kernel, and Input is the input tensor.

3.1.3 ReLU activation

The rectified linear unit (ReLU) activation function is frequently employed in deep learning networks, including MobileNet, and may be stated as follows:

$$f(x) = max(0, x)$$

In other words, the ReLU activation function outputs the input value x if it is positive and 0 if it is negative. The ReLU activation function has several advantages compared to other activation functions. It is simple and easy to compute, requiring only a single comparison and assignment operation. It is also nonsaturating, meaning it does not struggle with the problem of disappearing gradients that can occur with other activation functions such as sigmoid or tanh. This allows the network to continue learning and improving even when the inputs are large in magnitude. Overall, the ReLU activation function is an effective choice for the MobileNet architecture, as it allows the network to learn efficiently and effectively. It is a per-element activation function applied to the result of every layer in MobileNetV2 and is defined as follows:

$$\text{Output}[i] = \max(0, \text{Input}[i])$$

where Output is the output tensor and Input is the input tensor.

Algorithm 2: Breast cancer diagnosis using MobileNetV2

Pseudocode for MobileNet
Input: (batch_size, width, height, channels)
    Convolutional layer: output = (batch_size, width, height, filters)
    Pooling layer: output = (batch_size, width/2, height/2, filters)
    Depthwise Separable Convolutional layer: output = (batch_size, width/2, height/2, filters)
    Pointwise Convolutional layer: output = (batch_size, width/2, height/2, filters)
    average_pooling = (1/pool_size) * sum(input_data)
    global_average_pooling = (1/total_size) * sum(input_data)
    Repeat the above layers with decreasing width and height until the desired output size is reached.

3.2 Long short-term memory

LSTM is a popular RNN type for natural language analysis and time-series prediction. LSTMs possess the ability to recognize long-term relationships in sequential data, making them ideal for tasks that require the consideration of data order. In mammogram analysis, LSTM inputs a series of mammograms for the same patient over time, and the network's structure is crafted to learn relevant features for a specific objective, such as diagnosing breast cancer, through fine-tuning the preexisting weights on a mammogram data set.

3.2.1 LSTM equation for cell state update

The cell state in an LSTM cell is the "memory" of the cell, which stores information over multiple time steps. The cell state update equation is employed to refresh the cell state at each time step using both the previous cell state and a new input, which is managed by the forget gate. The following equation is used to update the cell's state:

$$\textbf{cell\_state} = \textbf{previous\_cell\_state} * \textbf{forget\_gate} + \textbf{input\_gat} * \textbf{input\_transformation}$$

The cell state update equation combines the previous cell state with the new input in a way that is controlled by the forget and input gates. The forget gate regulates the extent to which the previous cell state is discarded, while the input gate determines the portion of the current input to incorporate into the new cell state.

3.2.2 LSTM equation for hidden state update

The hidden state in an LSTM cell represents the output of the cell that is passed on to the next time step or to the output layer. The hidden state update equation calculates the final output of the LSTM by using the cell state and an output gate. The equation is expressed as

$$\textbf{hidden\_state} = \textbf{output\_gate} * \textbf{tanh}(\textbf{cell\_state})$$

where "hidden state" is the end result produced by the LSTM cell, and the "output gate" determines the extent to which the cell state contributes to this output through its activation. The "tanh(cell state)" is the cell state that has undergone transformation by the tanh activation function, resulting in values ranging from -1 to 1. The hidden state update equation creates the final output of the LSTM cell by applying a sigmoid activation (output gate) to the tanh transformed cell state. The output gate identifies the amount of data from the cell state that will be employed for the final output.

3.2.3 LSTM equation for input gate

The input gate in an LSTM cell determines the extent to which the new input contributes to the updated cell state. A weighted sigmoid activation function is applied to the past hidden state, the current input, and a bias factor to produce the activation of the input gate. The equation is expressed as

$$\textbf{input\_gate} = \textbf{sigmoid}(\textbf{previous\_hidden\_state} * \textbf{Wi} + \textbf{current\_input} * \textbf{Ui} + \textbf{bi})$$

where input_gate is the activation of the input gate, a scalar value between 0 and 1, previous_hidden_state is the hidden state of the LSTM cell at the previous time step, current_input is the input data at the current time step, Wi, Ui, and bi are weight matrices and bias terms that are learned during the training process.

3.2.4 LSTM equation for output gate

The equation for the output gate in an LSTM network is a mathematical formula that calculates the final output of the network. The formula uses a sigmoid function to squish the result into a range of $0-1$, which represents the gate's state of open or closed. The output gate is determined by the previous hidden state, the current input, and two matrices of weights (Wo and Uo) that learn from the data and a bias term (bo). The previous hidden state, current input, and biases are all multiplied by their respective weights and then summed together.

$$\textbf{output\_gate} = \textbf{sigmoid}(\textbf{previous\_hidden\_state} * \textbf{Wo} + \textbf{current\_input} * \textbf{Uo} + \textbf{bo})$$

The final result is then applied to the sigmoid activation function to produce the output gate. The output gate determines the information that will be allowed to pass through to the next layer in the network and ultimately affects the final prediction of the model.

3.2.5 LSTM equation for input transformation

LSTM networks calculate the updated cell state using the input transformation equation. The equation computes a candidate cell state, which is then filtered by the forget gate to generate the final cell state.

$$\textbf{input\_transformation} = \textbf{tanh}(\textbf{previous\_hidden\_state} * \textbf{Wc} + \textbf{current\_input} * \textbf{Uc} + \textbf{bc})$$

The weights (Wc, Uc, and bc) represent the relationship between the previous hidden state, current input, and bias term, and their impact on the input transformation. The tanh function ensures that the transformed values are between -1 and 1, which makes the cell state more stable and helps in preventing vanishing gradients during backpropagation.

Algorithm 3: Breast cancer diagnosis using LSTM

Pseudocode for LSTM
Input: (batch_size, sequence_length, input_size)
    LSTM layer: output = (batch_size, sequence_length, hidden_size)
    Dropout layer: output = (batch_size, sequence_length, hidden_size)
    Fully connected layer: output = (batch_size, output_size)

The given code outlines a neural network architecture utilizing LSTM, dropout, and fully connected layers for processing a 3D input tensor with shape (batch size, sequence length, input size). The LSTM layer is the first layer in the network and extracts features from the input data to produce an output tensor of shape (batch size, sequence length, hidden size). The dropout layer, which is designed to prevent overfitting by randomly dropping out neurons during training, processes the output of the LSTM layer to produce another output tensor with the same structure. The output of the dropout layer is processed by the final fully connected layer, which predicts the class labels with an output tensor of a certain shape (batch size, output size). The network's predictions can be transformed into class probabilities through the use of a softmax activation function.

---

4. Experimental results and discussion

The experiment was conducted using an online Colaboratory compiler, an Intel Core TM i5-8550U CPU with a clock speed of 1.99 GHz, and an 8 Gb RADEON (TM) 530 Graphics RAM. Due to the large amount of data involved in training the model for improved accuracy, the regular CPU may take a considerable amount of time to execute. To tackle this challenge, a GPU accelerator was utilized, resulting in a substantial reduction of time. The research was implemented using the Python Deep Learning framework. The performance of the MobileNetV2 model combined with the LSTM model was evaluated using various metrics, such as precision, recall, and F1-score outcomes, and statistical analysis, to assess the frequency of correctly classifying breast cancer. The authors standardized crucial characteristics across all

Table 5.2 Breast cancer diagnostic features table using MobileNetV2 and LSTM.

Attribute	Description
ID_Number	Identification number
Radius_Mean	Average distance between the center and points on the circumference
Texture_Mean	Standard deviation of gray-scale values based on texture
Perimeter_Mean	Average primary tumor size
Area_Mean	Average region size
Smoothness_Mean	Average of regional differences in radius lengths for smoothness
Compactness_Mean	Average of perimeter squared divided by area, used as a measure of compactness
Concavity_Mean	Average degree of concavity in the shape
Concave Points_Mean	Average number of concave areas on the contour
Infinite dimensionality	A measure of the "coastline approximation" with a value of 1

investigations to provide a fair comparison of different strategies in terms of implementation configurations. The parameters considered for classification using the proposed model are listed in Table 5.2.

4.1 Dataset collection

This study suggests the application of neural networks in automating the detection of breast cancer. The diagnosis of this disease can be difficult due to the variety of features found in abnormal regions and their similarity with dense breast tissue. Our goal is to provide a reliable solution for this issue by utilizing the latest advancements in deep learning techniques. To train these networks, we utilize the DDSM dataset, which is widely considered to be one of the most comprehensive datasets available for breast cancer research. The DDSM dataset includes 2700 cases, each featuring two mammogram views (Cranial Caudal and Medium Lateral Oblique) and accompanying patient information and photographs. This extensive dataset provides us with a wealth of information to train and evaluate our model. While our automatic segmentation systems are trained and tested using only the Cranial Caudal views, our mass lesion diagnosis model is trained and evaluated using both the Medium Lateral Oblique and Cranial Caudal views. This approach allows us to make the most of the available information and provides us with a more comprehensive understanding of the mammogram images. The task of determining whether a breast mass is benign or malignant is challenging due to the diverse features of abnormal regions and their overlap with dense breast tissue, as illustrated in the Deep Space

Sampler (DSSM) mass sample data. This makes it difficult for traditional diagnostic methods to achieve high levels of accuracy.

Our solution utilizes advanced neural network techniques, such as CNN, RNN, and LSTM networks. We also propose the use of the combination of MobileNetV2 and LSTM for improved performance. The evaluation of the algorithms is based on four key performance metrics, namely accuracy rate, precision, recall, and F1-score. The implementation of our solution is carried out using a Colaboratory online compiler, an Intel core TM i5-8550U CPU running at 1.99 GHz, and an 8 Gb RADEON (TM) 530 Graphics RAM. The regular CPU may take a long time to execute during the implementation stage while learning the model with massive quantities of data for improved accuracy. To address this issue, the model is built using a GPU accelerator, saving a significant amount of time. The Python Deep Learning framework is used to implement the deep learning method described in our research. In conclusion, this study proposes a solution for the automatic detection of breast cancer using advanced neural network techniques. Our approach makes use of the comprehensive DDSM dataset and utilizes both Cranial Caudal and Medium Lateral Oblique mammogram views for improved accuracy. The implementation of the solution is carried out using a GPU accelerator, reducing the time required to execute the model. The performance of the algorithms is evaluated based on four key performance metrics, providing a comprehensive understanding of the effectiveness of our solution.

4.2 Feature set extraction

This paper proposes using neural networks to automatically detect breast cancer. MobileNetV2 and LSTM models are combined for feature extraction and diagnosis. The MobileNetV2 is fine-tuned on mammogram data to optimize performance and extract deep features from the final convolutional layer. The LSTM model is then utilized for the analysis of the extracted features in terms of space and time. The LSTM model is fine-tuned on a specific dataset of sequences of mammograms to extract spatiotemporal features from the last hidden state of the LSTM. The extracted deep and spatiotemporal features are combined and fed into a fully connected layer for classification, which produces the final prediction through the application of a softmax activation function. The prediction is determined based on a specified threshold value. If the predicted value is greater than the threshold, the output is labeled as "abnormal" indicating that the patient has breast cancer. If the predicted value is less than the threshold, the output is labeled as "normal" indicating that the patient does not have breast cancer. In this study, a novel approach to detecting breast cancer is presented, which involves combining MobileNetV2 and LSTM neural networks. The MobileNetV2 model is used to extract deep features from mammograms, while the LSTM model performs spatiotemporal feature extraction. These two types of features are then combined and passed through a fully connected layer for classification using the softmax function. The accuracy of the model is continuously monitored using appropriate metrics, and adjustments are made as necessary. The data used for this study contains various attributes related to the features of abnormal regions in mammograms, and the combination of Mobile-NetV2 and LSTM offers a comprehensive approach to diagnosing breast cancer.

4.3 Performance evaluation

The evaluation of the performance of a classification model in diagnosing diseases like breast cancer typically involves the use of accuracy rate, precision, recall, and F1-score metrics. The accuracy rate reflects the model's overall accuracy in correctly identifying the observations as either "normal" or "abnormal." The model's correct identifications of "abnormal" cases are referred to as true positives, while its correct identifications of "normal" cases are referred to as true negatives. The total observations are the total number of cases that the model is being tested on.

Accuracy rate: (*True Positives* + *True Negatives*) / *Total Observations*

Precision assesses the model's accuracy in correctly identifying positive cases. In medical diagnosis, a high precision score is crucial as it shows the proportion of positive diagnoses that are accurate.

Precision: *True Positives* / (*True Positives* + *False Positives*)

The recall metric measures the model's capacity to recognize all positive cases. False negatives are instances in which the model wrongly identifies an observation as "normal" True positives are instances in which the model correctly classifies an observation as "abnormal" In medical applications, recall is crucial since it represents the proportion of genuine positive cases that the model is able to detect.

Recall: *True Positives* / (*True Positives* + *False Negatives*)

The F1-score is a weighted average of precision and recall that takes both the accuracy of the positive predictions and the model's ability to detect all positive cases into account. A higher F1-score suggests that the model has high precision and recall and makes accurate, well-balanced diagnoses.

*F*1 − *Score*: 2 ∗ (*Precision* ∗ *Recall*)/(*Precision* + *Recall*)

In classifying breast cancer using MobileNetV2 and LSTM, these metrics can be used to evaluate the performance of the combined model in making accurate diagnoses. By monitoring the accuracy rate, precision, recall, and F1-score, the model can be continuously refined and adjusted to ensure that it is making the most accurate diagnoses possible.

4.4 Exploring the efficacy of pretrained models in the training phase

Table 5.3 provides a novel perspective on the efficiency of machine learning algorithms by comparing their performance in classifying breast cancer using various

Table 5.3 Evaluation of mobile net V2 and LSTM in classifying breast cancer in the training model.

Algorithm	Accuracy rate	Precision	Recall	F1-score
CNN	0.92	0.88	0.91	0.89
RNN	0.93	0.89	0.91	0.90
LSTM	0.97	0.96	0.97	0.97
ANN	0.95	0.93	0.94	0.93
MobileNetV2 and LSTM	0.98	0.97	0.98	0.97

machine learning methods. In terms of accuracy rate, precision, recall, and F1-score, the combination of MobileNetV2 with LSTM exceeds all other methods. This highlights the importance of incorporating pretrained models with specialized tasks, like MobileNetV2, into the training process for improved performance in classifying complex medical datasets like breast cancer. The performance of various machine learning methods in identifying breast cancer is compared in Table 5.3. CNN, RNN, LSTM, ANN, and the combination of MobileNetV2 and LSTM are the techniques under consideration. The algorithms are evaluated using four main performance metrics: accuracy rate, precision, recall, and F1-score.

The definition of accuracy rate is the proportion of correct predictions to the total number of observations. In other words, it assesses the fraction of observations accurately categorized as normal or abnormal. The accuracy rate of CNN is 0.92, which means that 92% of the observations were classified correctly. Similarly, the accuracy rate of RNN, LSTM, ANN, and MobileNetV2 and LSTM is 0.93, 0.97, 0.95, and 0.98, respectively. The highest accuracy rate among the algorithms is achieved by MobileNetV2 and LSTM, which indicates that the combination of these two algorithms produces the most accurate results. Precision is the proportion of accurately predicted positive cases out of all positive cases correctly predicted by the model. In other terms, precision quantifies the fraction of accurate positive forecasts. The precision of CNN is 0.88, which means that 88% of the positive predictions made by the algorithm were correct. The precision of RNN, LSTM, ANN, and MobileNetV2 and LSTM is 0.89, 0.96, 0.93, and 0.97, respectively. The highest precision is achieved by MobileNetV2 and LSTM, which indicates that the combination of these two algorithms produces the most accurate positive predictions. The recall of 0.91 for the CNN algorithm means that 91% of the actual positive cases were correctly identified by the model. The recall of RNN, LSTM, ANN, and MobileNetV2 and LSTM is 0.91, 0.97, 0.94, and 0.98, respectively. The highest recall is achieved by MobileNetV2 and LSTM, which indicates that the combination of these two algorithms produces the most complete positive predictions.

The F1-score is a composite metric that takes both precision and recall into account. The F1-score provides a single, comprehensive score that indicates the overall performance of the model in terms of both accuracy and completeness of positive predictions. An F1-score of 0.89 for the CNN model indicates a good balance between precision and recall, with the model making accurate and comprehensive positive predictions. The F1-score of RNN, LSTM, ANN, and MobileNetV2 and LSTM is 0.90, 0.97, 0.93, and 0.97, respectively. The highest F1-score is achieved by LSTM and MobileNetV2 and LSTM, which indicates that these algorithms produce the most balanced and high-quality predictions.

In conclusion, the table shows that the combination of MobileNetV2 and LSTM produces the most accurate and complete predictions, with the highest accuracy rate, precision, recall, and F1-score. The high performance of MobileNetV2 and LSTM can be attributed to the strengths of both algorithms. MobileNetV2 is a deep learning algorithm that is efficient and effective in extracting.

4.5 Exploring the efficacy of models in the testing phase

This table compares the testing accuracy rate, precision, recall, and F1-score of different algorithms used for breast cancer classification. The algorithms compared in this table are CNN, RNN, LSTM, ANN, and MobileNetV2 with LSTM.

The accuracy rate metric measures the proportion of accurate predictions generated by the algorithm relative to the total number of predictions generated. The accuracy rate of the CNN algorithm, for instance, was 0.91, indicating that 91 out of 100 predictions were true. Precision measures the precision of the algorithm's positive predictions. The ratio of true positive predictions to the total number of positive predictions made by the algorithm indicates the proportion of positive predictions that are accurate. For instance, the RNN algorithm has a testing precision of 0.88, meaning that 88 out of 100 positive predictions provided by the algorithm are accurate. For instance, the ANN system has a testing recall of 0.93, which indicates that, out of 100 actual positive cases, it properly predicts 93 of them. The F1-score combines precision and recall into one score by taking the harmonic mean of both criteria. A high F1-score demonstrates a proper balance between precision and recall. For example, the MobileNetV2 with LSTM algorithm has a testing F1-score of 0.96, which indicates that it has a good balance of precision and recall in breast cancer classification. According to the results, the combination of MobileNetV2 and LSTM gives the best performance in terms of accuracy rate, precision, recall, and F1-score in comparison to other algorithms for breast cancer classification. It is essential to remember that these conclusions are dependent on a particular dataset and may not applicable to other datasets.

In conclusion, Table 5.4 provides a comparison of different algorithms used for breast cancer classification and their performance based on accuracy rate, precision, recall, and F1-score. The MobileNetV2 with LSTM algorithm provides the best performance among the algorithms compared, but it is important to evaluate the algorithms on different datasets to determine their generalizability.

Table 5.4 Evaluation of MobileNetV2 and LSTM in classifying breast cancer in the testing model.

Algorithm	Testing accuracy rate	Testing precision	Testing recall	Testing F1-score
CNN (convolutional neural network)	0.91	0.87	0.89	0.88
RNN (recurrent neural network)	0.92	0.88	0.90	0.89
LSTM	0.96	0.95	0.96	0.96
ANN (artificial neural network)	0.94	0.92	0.93	0.92
MobileNetV2 and LSTM	0.97	0.96	0.97	0.96

4.6 Discussion

The tables above compare five different machine learning algorithms for mammogram classification as either normal or abnormal. These algorithms include CNN, RNN, LSTM, ANN, and a combination of MobileNetV2 and LSTM. The first table shows the accuracy rate, precision, recall, and F1-score metrics for each algorithm during the training process, while the second table displays these metrics for testing.

Accuracy measures the proportion of correct predictions made by the algorithm. Precision assesses the accuracy of positive predictions, with a higher score indicating fewer false positive predictions. Recall measures the ability of the algorithm to detect positive cases, with a higher score indicating more correct positive detections. The F1-score combines precision and recall into a single score to provide a comprehensive evaluation of the algorithm's performance.

In both tables, the MobileNetV2 and LSTM algorithm has the highest accuracy rate, precision, recall, and F1-score. The ANN and RNN algorithms have a similar performance, with a slightly higher performance by the ANN algorithm. The CNN algorithm has the lowest performance of all the algorithms. This indicates that the MobileNetV2 and LSTM algorithms are the most effective for classifying mammograms as normal or abnormal. It is important to note that the performance of the algorithms during training and testing may not be the same. This can be seen in the comparison of the two tables, where the testing accuracy rate, precision, recall, and F1-score of the algorithms are slightly lower than the training metrics. This is a common phenomenon in machine learning and is referred to as overfitting. Overfitting occurs when an algorithm fits the training data too closely, leading to poor generalization to unseen data. In conclusion, the two tables show that the MobileNetV2 and LSTM algorithms are the most effective for classifying mammograms as normal or abnormal. While the performance of the algorithms during testing is slightly lower than during training, they still perform well overall as shown in

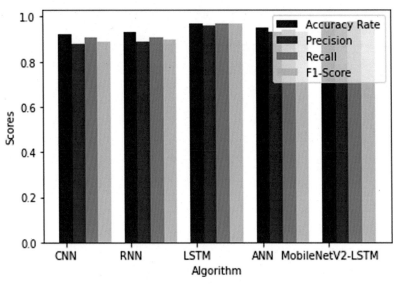

FIGURE 5.1

Comparison of accuracy, precision, recall, and F1-score for different algorithms in breast cancer diagnosis.

Fig. 5.1. The results of this comparison can be used to guide the selection of algorithms for future mammogram classification projects.

In conclusion, deep learning algorithms have shown promising results in the field of breast cancer diagnosis and have the potential to assist radiologists in making more accurate and consistent diagnoses, ultimately improving patient outcomes. However, further research is needed to optimize their performance and ensure their generalizability.

5. Conclusion

Breast cancer is a significant health concern among women worldwide, and its early detection can lead to better treatment outcomes. However, the traditional methods of diagnosing breast cancer such as biopsies are expensive, time consuming, and prone to conflicts among professionals. This is where the use of deep learning techniques can be highly beneficial. Deep learning algorithms have the ability to analyze and understand patterns in large amounts of data, making it a promising method for the diagnosis of breast cancer.

A deep learning-based framework is presented in this study for diagnosing breast cancer using images of the affected area. The framework combines MobileNetV2 and LSTM to detect the presence of tumors. MobileNetV2 is an efficient image classification model, while LSTM is a type of RNN that is commonly used to process sequential data. The performance of the model is evaluated using the DDSM dataset of mammograms. The proposed method was compared to various deep learning models such as InceptionV3, DenseNet121, Res-Net50, VGG16, and CNN. The comparison showed that the proposed method outperforms the others with a higher classification accuracy of 97%, precision of 96%, recall of 97%, and F1-score of 96%.

The combination of MobileNetV2 and LSTM in this study offers several advantages over traditional diagnostic methods. The deep learning model is able to learn complex patterns quickly, reducing the computational effort required to diagnose breast cancer. Additionally, the accuracy of the model in detecting affected regions is faster and more precise than traditional methods, leading to more timely diagnoses for patients. In conclusion, the proposed deep learning-based framework for the diagnosis of breast cancer offers a promising alternative to traditional diagnostic methods. The application of deep learning methods has the potential to produce more precise and effective diagnoses, ultimately leading to better treatment outcomes for patients. However, it is important to note that further research and evaluation is necessary to establish the reliability of this method in a clinical setting.

References

[1] Rock CL, Thomson CA, Sullivan KR, Howe CL, Kushi LH, Caan BJ, et al. American Cancer Society nutrition and physical activity guideline for cancer survivors. CA: A Cancer Journal for Clinicians 2022;72(3):230–62.

[2] Massafra R, Comes MC, Bove S, Didonna V, Diotaiuti S, Giotta F, et al. A machine learning ensemble approach for 5-and 10-year breast cancer invasive disease event classification. PLoS One 2022;17(9). e0274691.

[3] Zhang Y, Chan S, Park VY, Chang K-T, Mehta S, Kim MJ, et al. Automatic detection and segmentation of breast cancer on MRI using mask R-CNN trained on non–fat-sat images and tested on fat-sat images. Academic Radiology 2022;29:S135–44.

[4] Gao F, Wu T, Li J, Zheng B, Ruan L, Shang D, et al. SD-CNN: a shallow-deep CNN for improved breast cancer diagnosis. Computerized Medical Imaging and Graphics 2018; 70:53–62.

[5] Dafni Rose J, VijayaKumar K, Singh L, Sharma SK. Computer-aided diagnosis for breast cancer detection and classification using optimal region growing segmentation with MobileNet model. Concurrent Engineering 2022;30(2):181–9.

[6] Barnett AJ, Schwartz FR, Tao C, Chen C, Ren Y, Lo JY, et al. A case-based interpretable deep learning model for classification of mass lesions in digital mammography. Nature Machine Intelligence 2021;3(12):1061–70.

[7] Begum A, Kumar VD, Asghar J, Hemalatha D, Arulkumaran G. A combined deep CNN: LSTM with a random forest approach for breast cancer diagnosis. Complexity 2022;2022.

[8] Alshayeji MH, Ellethy H, Gupta R. Computer-aided detection of breast cancer on the Wisconsin dataset: an artificial neural networks approach. Biomedical Signal Processing and Control 2022;71:103141.

[9] Boudouh SS, Bouakkaz M. Breast cancer: using deep transfer learning techniques alexnet convolutional neural network for breast tumor detection in mammography images. In: 2022 7th international conference on image and signal processing and their applications (ISPA). IEEE; May 2022. p. 1–7.

[10] Tsochatzidis L, Costaridou L, Pratikakis I. Deep learning for breast cancer diagnosis from mammograms—a comparative study. Journal of Imaging 2019;5(3):37.

[11] Jasti VDP, Zamani AS, Arumugam K, Naved M, Pallathadka H, Sammy F, et al. Computational technique based on machine learning and image processing for medical image analysis of breast cancer diagnosis. Security and Communication Networks 2022;2022:1—7.

[12] Rahman ASA, Belhaouari SB, Bouzerdoum A, Baali H, Alam T, Eldaraa AM. Breast mass tumor classification using deep learning. In: 2020 IEEE international conference on informatics, IoT, and enabling technologies (ICIoT). IEEE; February 2020. p. 271—6.

[13] Al-Haija QA, Adebanjo A. Breast cancer diagnosis in histopathological images using ResNet-50 convolutional neural network. In: 2020 IEEE international IOT, electronics and mechatronics conference (IEMTRONICS). IEEE; September 2020. p. 1—7.

[14] Albashish D, Al-Sayyed R, Abdullah A, Ryalat MH, Almansour NA. Deep CNN model based on VGG16 for breast cancer classification. In: 2021 International conference on information technology (ICIT). IEEE; 2021. p. 805—10.

[15] Saleh H, Alyami H, Alosaimi W. Predicting breast cancer based on optimized deep learning approach. Computational Intelligence and Neuroscience 2022;2022.

[16] Khanna P, Sahu M, Singh BK, Bhateja V. Early prediction of pathological complete response to neoadjuvant chemotherapy in breast cancer MRI images using combined Pre-trained convolutional neural network and machine learning. Measurement 2023; 207:112269.

[17] Mobark N, Hamad S, Rida SZ. Coronet: deep neural network-based end-to-end training for breast cancer diagnosis. Applied Sciences 2022;12(14):7080.

[18] Dar RA, Rasool M, Assad A. Breast cancer detection using deep learning: datasets, methods, and challenges ahead. Computers in Biology and Medicine 2022:106073.

[19] Akingbade O, Nguyen KT, Chow KM. Effect of mHealth interventions on psychological issues experienced by women undergoing chemotherapy for breast cancer: a systematic review and meta-analysis. Journal of Clinical Nursing 2022;32:3058—73.

[20] Suchodolska G, Senkus E. Mobile applications for early breast cancer chemotherapy-related symptoms reporting and management: a scoping review. Cancer Treatment Reviews 2022:102364.

[21] Ahmadi M, Shahrokhi SN, Khavaninzadeh M, Alipour J. Development of a mobile-based self-care application for patients with breast cancer-related lymphedema in Iran. Applied Clinical Informatics 2022;13(05):935—48.

[22] Moon Z, Zuchowski M, Moss-Morris R, Hunter MS, Norton S, Hughes LD. Disparities in access to mobile devices and e-health literacy among breast cancer survivors. Supportive Care in Cancer 2022;30:117—26.

[23] Scholz S, Teetz L. Smart health via mHealth? Potentials of mobile health apps for improving prevention and adherence of breast cancer patients. Digital Health 2022;8. 20552076221074127.

[24] Akingbade O, Nguyen KT, Chow KM. Effect of mHealth interventions on psychological issues experienced by women diagnosed with breast cancer receiving chemotherapy: a systematic review and meta-analysis. medRxiv 2022. https://doi.org/10.1111/jocn.16533. 2022-02.

[25] Roh S, Lee Y-S. Developing culturally tailored mobile web app education to promote breast cancer screening: knowledge, barriers, and needs among American Indian women. Journal of Cancer Education 2023:1—10.

Autoencoder-based dimensionality reduction in 3D breast images for efficient classification with processing by deep learning architectures

J. Vijayaraj[1], B. Satheesh Kumar[2], R. Manikandan[3]

[1]*Department of Artificial Intelligence and Data Science, Easwari Engineering College, Chennai, Tamil Nadu, India;* [2]*Department of Computer Science and Engineering, School of Computing Science and Engineering, Galgotias University, Greater Noida, Uttar Pradesh, India;* [3]*School of Computing, SASTRA Deemed University, Thanjavur, Tamil Nadu, India*

1. Introduction

Most prevalent cancer in women as well as a significant contributor to cancer-related deaths globally is breast cancer [1]. Over the past few decades, mammography has significantly decreased the mortality from breast cancer. Due to its inexpensive cost and quick collection time, mammography has been used in many nations to create population-based screening programs. However, a 4D imaging approach called dynamic contrast-enhanced magnetic resonance imaging (DCE-MRI) offers higher sensitivity than mammography. Breast MRI has been introduced to the majority of screening programs for women in whom mammography alone proved to be insufficient. The vast majority of these women's malignancies are currently found in practice by screening breast MRI. In women who have a higher-than-average risk of developing breast cancer, a more aggressive screening plan is used. The American Cancer Society's recommendations [2] state that women with a lifetime risk of 20% or above should undergo yearly breast MRI screenings, which should be supplemented by mammography starting at the age of 30. The sensitivity of MRI is roughly double that of mammography in high-risk screening. In actuality, less than 10% of all cases of cancer are found by mammography alone, and these cases are mainly DCIS and low-grade lesions. Additionally, it has been demonstrated that thick breasts reduce mammography's sensitivity [3]. The DENSE trial examined the value of MRI screening in women with the highest density (ACR category 4) and demonstrated its impact in lowering the incidence of cancer between scans. However, interpretation is time consuming and difficult due to the HD of breast DCE-MRI data. With the aim of reducing interpretation time and supervision errors, computer-aided detection (CADe) tools have been created to assist and support radiologists in the analysis of breast MRIs. As can be seen, the first step makes use of available methods to acquire internal tissue dynamics of the breast to express them in an image; the second step uses algorithms to carry out fundamental operations on images to segment, which is the detection of regions of interest (ROI); and finally, the third step quantifies differences between images that have abnormalities from the perspective of images. Rapid development of novel methods is more precisely record dynamics of breast tissues has led to numerous developments in all the aforementioned disciplines. In this aspect, all of the approaches still have the extremely desirable attribute of achieving the goal of detecting all abnormalities without raising false alarms [4]. There is still a need for an article that outlines key methods utilized to obtain breast images as well as stages of processing essential to establish a diagnosis [5]. Recent research has studied a number of methodologies for feature categorization and interpretation.

2. Related works

Deep learning (DL) has recently been utilized in DBT pictures to increase the accuracy of lesion or cancer identification. To detect as well as categorize breast cancer, numerous researchers have used a variety of deep learning architectures [6]. For mass identification as well as segmentation in DBT pictures, work [7] proposed a deep learning system based on a 3-D version of Mask-RCNN method. A U-Net architecture-based DBT mass automated segmentation technique was reported in Ref. [8]. Major screening technique for breast cancer is mammography. Only imaging procedure that lowers breast cancer mortality is mammography [9]. Cancer cannot be prevented by mammography. However, mammography can be used to detect cancer in its early stages. Mammography is thought to have a sensitivity between 77% and 95%. Mammography-based estimates of specificity range from 92% to 97% [10]. Mammography, however, performs poorly on breasts with dense tissue. As a result, about 38% of cancers go unnoticed or receive an incorrect diagnosis. Radiation exposure and discomfort for the subject are other drawbacks of mammography. Mammogram interpretation takes time and is prone to mistakes [11]. Researchers recently switched from using conventional CAD to CNN-based techniques [12]. SVM, RF, and other statistical methods were well-liked before deep learning and CNNs, but their reliance on labor-intensive feature engineering has driven researchers toward CNN-based techniques, or hybrid techniques that combine statistical models with CNN models [13]. The increased capabilities of a new generation of graphical processor units, on other hand, have made it possible to execute high-load computational methods faster than in a multicore processor [14], which led to the development of new NN methods that perform feature extraction as well as quantification. For instance, according to a study [15], ultrasound images containing suspicious areas can be extracted and categorized into four different groups: skin, glandular tissue, masses, and fat. To expedite the procedure, they alter convolutional filters. Additionally, the author [16] uses a combination of CNN methods to directly extract questionable zones. To expedite the training process, they simply change the last layers. Ref. [17] create the characteristics from mammograms using a deep neural network. They adopt a modified architecture in which the network's outputs and inputs are used to update the model's parameters while it is being trained. In Refs. [18,19], similar strategies are outlined.

3. System model

This section discusses a novel method in 3D breast image-based classification and processing using autoencoder in dimensionality reduction with DL. Here, input has been collected as 3D breast images and has been processed for noise removal, smoothening, and dimensionality reduction. Dimension of processed input image has been reduced utilizing autoencoder-based Kernel Principal analysis. Then the classification has been carried out using regressive convolutional AlexNet architecture and the classified output has been given as detected 3D breast cancer. The proposed architecture is given in Fig. 6.1.

3.1 Autoencoder-based Kernel Principal analysis in dimensionality reduction

To make things easier to distinguish, we add a new letter WX to Eq. (6.1), making $\mathbf{h}_i = g(\mathbf{x}_i \mathbf{W_X})$. We take away symbol b from Eq. (6.1) and add an additional dimension to sample x i as follows: $\mathbf{x}_i = [x_{i1}, x_{i2}, \dots, x_{in-1}, 1]$. Outcome of this procedure will be identical to that of utilizing the bias term in Eq. (6.1).

DIAE decoding procedure has two parts as opposed to the traditional AE model's single part. One is input data reconstruction utilizing hidden representation $h\ i$ Eq. (6.1):

$$\widehat{\mathbf{x}}_i = \mathbf{g}\left(\mathbf{h}_i \mathbf{W}_x^T\right) \tag{6.1}$$

The second component is prediction of the input data utilizing h "i" Eq. (6.2):

$$\widehat{\mathbf{y}}_i = \mathbf{g}(\mathbf{h}_i \mathbf{W}_T) \tag{6.2}$$

Moreover, we incorporate a symmetric matrix $\mathbf{L} \in \mathbb{R}^{r \times r}$ into the DIAE model to reflect the structural information among various fault types. Then Eq. (6.3) can be used to rewrite prediction of the input data:

$$\widetilde{\mathbf{y}}_i = \widehat{\mathbf{y}}_i \mathbf{L} = \mathbf{g}(\mathbf{h}_i \mathbf{W}_T)\mathbf{L} \tag{6.3}$$

We deduce both the structural information and the discriminant data among various fault types from Eq. (6.4). We create the loss function of DIAE using Eq. (6.5), as shown below:

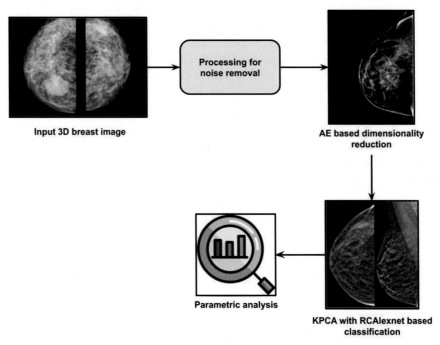

Input 3D breast image

Processing for noise removal

AE based dimensionality reduction

Parametric analysis

KPCA with RCAlexnet based classification

FIGURE 6.1

Proposed architecture.

$$J = \frac{\alpha}{2n}\sum_{i=1}^{n}\left(\mathbf{x}_i - \widetilde{\mathbf{x}}_i\right)^2 + \frac{1}{n}\sum_{i=1}^{n}(-\kappa_\sigma(\mathbf{y}_i, \widehat{\mathbf{y}}_i\mathbf{L})) + \frac{\beta}{2}\|\mathbf{L} - \mathbf{L}^T\|_F^2 + \frac{\lambda}{2}\left(\|\mathbf{W}_x\|_F^2 + \|\mathbf{W}_T\|_F^2\right)$$

(6.4)

where the regularization parameters are α, β, λ. We can determine the ideal $\mathbf{W}_x^*, \mathbf{W}_T^{\cdot}$, and $\mathbf{L}^-$ by reducing J because matrices $\mathbf{W}_X, \mathbf{W}_r$, and L are all randomly started. We provide the following explanations of each term in Eq. (6.6) for clarity:

- The reconstruction error in the first term $\frac{1}{2n}\sum_{i=1}^{n}\left(\mathbf{x}_i - \dot{\mathbf{x}}_i\right)^2$ is the same as the traditional AE. Features' capacity to represent input data can be enhanced.
- The second term is intended to reflect the discriminant information: $\frac{1}{n}\sum_{i=1}^{n}(-\kappa_\sigma(\mathbf{y}_i, \widehat{\mathbf{y}}_i\mathbf{L}))$. This term's reduction would enhance the retrieved characteristics' capacity for discrimination. $\kappa_\sigma(\mathbf{y}_i, \mathbf{y}_i\mathbf{L})$ is the Gaussian kernel in this Eq. (6.5):

$$\kappa_\sigma(\mathbf{y}_i, \widehat{\mathbf{y}}\mathbf{L}) = 1 / 1(\sqrt{2\pi}\,\sigma)(\sqrt{2\pi}\,\sigma)\exp\left(-(\mathbf{y}_i - \widehat{\mathbf{y}}_i\mathbf{L})^2 / 2\sigma^2\right) \qquad (6.5)$$

As indicated by $\frac{1}{n}\sum_{i=1}^{n}\kappa_\sigma(a, b)$, one can interpret the following as a rough computation of the correntropy $V_\sigma(A, B)$ Eq. (6.6):

$$V_\sigma(A, B) = E(\kappa_\theta(A, B)) = \kappa_\theta(A, B)dF_{AB}(a, b) \tag{6.6}$$

where $A = [a_1, a_2, ..., a_N]^T$ and $B = [b_1, b_2, ..., b_N]^T$ are two stochastic variables, $E(\kappa_\sigma(A, B))$ is expectation of $\kappa_\sigma(A, B)$, $\kappa_\sigma(\cdot, +)$ is Mercer kernel, and $F_{AB}(a, b)$ is joint PDF. As PDF of finite samples is unknown, as in many real applications, the above approximation of the entropy is typically used.

As a convenience, we reduce the complexity of Eq. (6.7) by expressing it in matrix form as follows:

$$\min_{w_x, w_1 \mathbf{L}} J_1 + J_2 + J_3 + J_4 \tag{6.7}$$

where:

$$J_1 = -\text{tr}\left(K_\sigma\left(\widehat{\mathbf{Y}}\mathbf{L}, \mathbf{Y}\right)\right)$$

$$J_2 = \frac{\alpha}{2} \| \dot{\mathbf{X}} - \mathbf{X} \|_F^2$$

$$J_3 = \frac{\beta}{2} \| \mathbf{L} - \mathbf{L}^T \|_F^2 \tag{6.8}$$

$$J_4 = \frac{\lambda}{2} \left(\| \mathbf{W}_X \|_F^2 + \| \mathbf{W}_1 \|_F^2 \right)$$

To resolve the equation, we use an alternative optimization technique (Eq. 6.9). Starting with initializing $\mathbf{W}_X, \mathbf{W}_r, \mathbf{L}$, these three steps run sequentially until convergence. First, fix $\mathbf{W}_\gamma$ and L, then use the gradient descent method to optimize $\mathbf{W}_X$; second, fix $\mathbf{W}_X$ and L, use the gradient descent method to optimize $\mathbf{W}_I$; to optimize L. Following equations are utilized to evaluate the values of $\mathbf{W}_X, \mathbf{W}_r, \mathbf{L}$.

$$\mathbf{W}_\tau = \mathbf{W}_T + \eta \frac{\partial J}{\partial \mathbf{W}_T} \mathbf{L} = \mathbf{L} + \eta \frac{\partial J}{\partial \mathbf{L}} \tag{6.9}$$

Partial derivatives of J is $\mathbf{W}_x, \mathbf{W}_T, \mathbf{L}$, i.e., $\frac{d}{W_x}, \frac{d}{W_\tau}, \frac{A}{\pi}$ by Eq. (6.10)

$$\frac{\partial J}{\partial W_X} = \frac{\partial J_1}{\partial W_X} + \frac{\partial J_2}{\partial W_X}$$

$$\frac{\partial y}{\omega_T} = \frac{\partial_1}{W_T} + \frac{\partial W_T}{W_T}$$

$$= \frac{c}{\sigma^2} \mathbf{H}^T \left(\left(\mathbf{D}\left(\widetilde{\mathbf{Y}}\mathbf{L} - \mathbf{Y} \right)\mathbf{L}^T \right) \odot dg(\mathbf{H}\mathbf{W}_T) \right) + \lambda \mathbf{W}_T \tag{6.10}$$

$$\frac{\partial J}{\partial \mathbf{L}} = \frac{\partial J_1}{\partial \mathbf{L}} + \frac{\partial J_4}{\partial \mathbf{L}} = \frac{C}{\sigma^2} \widehat{\mathbf{Y}}^T \mathbf{D}\left(\widehat{\mathbf{Y}}\mathbf{L} - \mathbf{Y} \right) + \beta\left(\mathbf{L} - \mathbf{L}^T \right)$$

If there are K anchor vectors in the N-dimensional unit sphere, they are represented by the notation $\mathbf{a}_k \in R^N, k = 1, \cdots, K$.

$$\mathbf{y} = (y_1, \cdots, y_k, \cdots, y_K)^T$$

$$y_k(\mathbf{x}, \mathbf{a}_k) = \max(0, \mathbf{a}_k^T \mathbf{x}) \equiv \mathrm{Rec}(\mathbf{a}_k^T \mathbf{x}) \tag{6.11}$$

Propagate error back to hidden layer when there is an error $[n]E$ at output layer "L." Meaning that derivative of $[n]E$ with regard to the output layer's weights $W:, k$ is the quantity of interest. To be more precise, we use an Eq. (6.12) to return from every output unit k to the hidden unit j of the hidden layer L 1

$$\frac{\partial^{[n]} E}{\partial W_{jk}^{\{L\}}} = \frac{\partial^{[n]} E}{\partial^{[n]} \widehat{y}_k} \cdot \frac{\partial^{[n]} \widehat{y}_k}{\partial^{[n]} a_k^{\{L\}}} \cdot \frac{\partial^{[n]} a_k^{\{L\}}}{\partial W_{jk}^{\{L\}}}, j = 1, \dots, N_{L-1}, k = 1, \dots, N_L \tag{6.12}$$

Nontrivial derivative of $\frac{\partial^{[n]} E}{\partial W_{jk}^{[L]}}$ decomposes into multiple much simpler parts only by using the chain rule. You can check Eq. (6.13)'s right side term by phrase, from right to left:

$$\frac{\partial^{[n]} E}{\partial^{[n]} \widehat{y}_k} = -([n]y_k - [n]\widehat{y}_k) \tag{6.13}$$

The bias at each stratum is a second matter of importance. The bias at the output layer follows a similar pattern according to Eq. (6.14)

$$\frac{\partial^{[n]} E}{\partial b_k^{\{L\}}} = \frac{\partial^{[n]} E}{\partial^{[n]} \widehat{y}_k} \cdot \frac{\partial^{[n]} \widehat{y}_k}{\partial^{[n]} a_k^{\{L\}}} \cdot \frac{\partial^{[n]} a_k^{\{L\}}}{\partial b_k^{\{L\}}} = -(y_k - \widehat{y}_k) \cdot \sigma'(a_k^{\{L\}}) \cdot 1 \tag{6.14}$$

because Eq. (6.14) are unaffected by the bias. This implies that the two interest derivatives is given in terms of one another. $\frac{\partial [[n]_E}{\partial W_{jk}^{(L)}} = \frac{\partial [n]_E E}{\partial b_k^{[L]}} \cdot [n]z_j^{\{L-1\}}$

A function $K : S \times S \to C$ is a positive definite (p.d.) kernel on S if, from Eq. (6.15)

$$\sum_{i,j=1}^{N} \bar{c}_i c_j K(v_i, v_j) \geq 0 \tag{6.15}$$

for all $|\{x_i\}_{i=1}^{N} \subset S, \{c_i\}_{i=1}^{N} \subset \mathbb{C}$, and $N \in \mathbb{N}$

Given a p.d. kernel, a mapping: $S H(K)$ such that from Eq. (6.1), a reproducing kernel Hilbert space $\Phi: S \to H(K)$ and a p.d (6.16)

$$K(x, y) = \langle \Phi(x), \Phi(y) \rangle_{\mathscr{H}(K)} \tag{6.16}$$

A feature map Φ is the term used for the function in the problem. Additionally, the reproducing property listed in Eq. (6.17) is true:

$$f(x) = \langle K_x, f \rangle_{\mathscr{H}(K)} \tag{6.17}$$

$$\mathrm{span} \{K_x := K(., x) \tag{6.18}$$

$H(K)$-inner product is given by Eq. (6.19)

$$\langle \sum c_i K_{x_i}, \sum d_j K_{x_j} \rangle_{\mathscr{H}(K)} := \sum \bar{c}_i d_j K(x_i, x_j) \tag{6.19}$$

3.2 Regressive convolutional AlexNet architecture-based classification

Instead of the more traditional method of classifying every pixel in the image, we use our linear-regression CNN model to infer the radius specification of the vessel wall at 100 evenly spaced radial positions in each polar image. Radial distances from the catheter's center in polar space were used to parameterize the lumen segmentation. Fig. 6.2 depicts the general flow of the suggested CNN method.

Our network has a straightforward design with three fully linked layers, the final output layer, and four convolutional layers. Before being windowed for input, all polar images were padded in a circle to the left and right. Each unique radial point was the center of a window with a size of 488 × 128 pixels, producing 100 inputs as well as 100 evaluated radial distances per image. Table 6.1 lists the specifics of network design. A layer that is completely connected has 512 nodes. To increase the network's robustness, dropout with a keep probability of 0.75 was added to fully connected layers FC1 and FC2. The last layer produces a single number that represents the radial separation between the catheter's center and the lumen border for the radial position being assessed.

We utilized the widely known AlexNet architecture, which was pretrained with more than one million photos and has an incredible 60 million specifications with almost 650,000 neurons. AlexNet is organized into eight learned layers, the first three of which are fully connected layers, followed by five convolutional, pooling blocks. In essence, convolutional as well as pooling blocks aim to extract significant

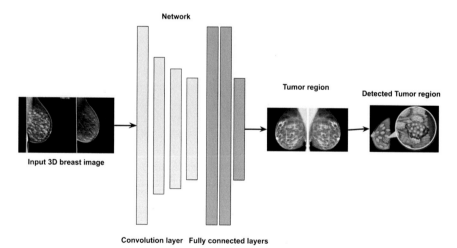

FIGURE 6.2

Overview of linear-regression CNN classification.

Table 6.1 Lumen segmentation at every windowed image utilizing a linear-regression CNN architecture. The output is the radial separation between the catheter's center and the lumen's edge. FC, fully connected layer; CN, convolutional layer.

Layer	Weights	In	Out	Pooling
CN1	$1 \times 5 \times 5 \times 24$	$488 \times 128 \times 1$	$244 \times 32 \times 24$	2×2
CN2	$24 \times 5 \times 5 \times 24$	$244 \times 32 \times 24$	$122 \times 16 \times 24$	2×2
CN3	$24 \times 5 \times 5 \times 24$	$122 \times 16 \times 24$	$61 \times 8 \times 24$	2×2
CN4	$24 \times 5 \times 5 \times 24$	$61 \times 8 \times 24$	$61 \times 8 \times 24$	–
FC1	11712×512	$11,712$	512	–
FC2	512×512	512	512	–
Out	512×1	512	1	

characteristics. The intermediate layers are capable of capturing acceptable representations for AlexNet that are neither either particular or overly general. Due to the size of feature maps produced by the convolutional layers, pooling is crucial to reducing the size of features. We carried out an additional overlapping max-pooling after C3 and C4 in particular because max-pooling layers were only carried after C1, C2, and C5. Computed feature maps were from three different convolutions as well as pooling layer block combinations.

Standard mean-squared error is the goal function used for network training. Network was stochastically trained at a base learning rate of 0.005 with a mini-batch size of 100. For every 50,000 runs, the learning rate was reduced by half. At 400,000 runs, where convergence was seen, the training was halted. The network's learned weights and biases—a total of about 6.3 million parameters—are then applied to the test sets to forecast the lumen contour. In numerous deep learning architectures, the dropout layer has demonstrated its potential as a regularization strategy. It offers the regularization process that occurs throughout learning. This method, which uses stochastic dropping out of some neurons during training, is used to avoid overfitting. Before the final regression layer, in the fully connected (FC) layer, there is just one neuron that represents the estimated orientation angle value. Consequently, FC is in charge of remapping feature map to focus on a specific neuron value. In supervised learning, the regression layer uses a loss function during the training phase to determine optimal parameters.

Deep network performance depends heavily on hyper-parameters. Each convolutional layer (CL) is made up of K kernels, each with a different size (KS). Utilizing the learning rate value, the smallest loss value is attained with a comparatively quick convergence rate. We start by choosing hyperspecifications are been applied in previous studies. We choose learning rate based on an estimation deep network that was followed in a similar direction. In a related remote sensing application, less batch size value was examined. Minimum batch size 64 has outperformed other minimum batch sizes in terms of average accuracy. The network training iterations are determined by the epochs number. About 30 epochs are used in our estimation issue to arrive at the least loss value. The KS was set to a lower value of 3, dropout percentage was set to 20% of neurons, as well as initial learning rate was set to 0.001. As a result, these are initial hyperspecifications we established for our proposed DCRN, which are displayed in Table 6.2. Second, we use an exhaustive grid search strategy during the trials to look into the optimum hyperparameters for our suggested architecture. The parameters that gave our architecture optimum performance were the KS set to 7, the dropout rate set to 30% of neurons, and the initial learning rate set to 0.004.

4. **Performance analysis**

Python 3.6, Keras 2.0.6, and a Tensor flow backend were utilized to execute each CNN. We employed a Windows 10 computer with an Intel(R) CPU and Tesla P100-PCIE-12GB graphic processor (3.50 GHz). To get a suitable scale, we applied the well-known Xavier initialization to convolutional layer weights. With Adam optimization, we trained with a batch size of 10 across 200 iterations. All tests have used a fourfold cross-validation technique to assess the proposed DNN. Genuine and fictitious components for five different frequencies were used as inputs to train the U-net. In our dataset, which had 600 phantoms, 150 samples were included in each fold of fourfold cross-validation. The testing set was made up of 150 held-out samples, while the training set used 450 cases for each fold. Thus, when they were not in the training set, all 600 cases were used as test examples.

Table 6.2 Hyper-parameters setting for our proposed network.

Layer name	Optimized hyperparameters	Initial hyperparameters
CLs in convolutional unit 1	KS[7, 7], K = 256	KS[3, 3], K = 256
CLs in convolutional unit 2	KS[7, 7], K = 128	KS[3, 3], K = 128
	KS[7, 7], K = 64	KS[3, 3], K = 64
	KS[7, 7], K = 512	KS[3, 3], K = 512
CLs in convolutional unit 3	KS[7, 7], K = 1024	KS[3, 3], K = 1024
	KS[7, 7], K = 1024	KS[3, 3], K = 1024
Activation layers	Drop percentage = 30%	Drop percentage = 20%
	KS[2, 2], stride = 2	KS[2, 2], stride = 2
Dropout layers	Relu unit	Relu unit

4.1 Dataset description

A review of our institution's medical records from March 2013 to December 2016 yielded the cohort for the study. The following criteria were used for inclusion: (1) The same MR procedure was used for scanning the images. (2) The pathology reports for the lesion were complete (biopsy or surgery), with the exception that three lesions were deemed benign following a 3-year follow-up. (3) Two breast radiologists with a combined experience of 12 years diagnosed the imaging reports with definite BI-RADS categories. In the event of a disagreement, a senior radiologist with a combined experience of 15 years was contacted. (4) With the same BI-RADS and pathological findings, the lesions were either (a) solitary in one breast or (b) present in both breasts. BPE in the bilateral breasts that is normal or typical was disregarded. A 1.5T system with an eight-channel breast coil was used for the breast MRI. Patients were inspected while lying on their backs with their two breasts in the coil cavity. The following parameters were used in conventional plain scans: nonfat axial T1WI 3D suppressed. These studies were performed on patients between the ages of 13,954 and between the dates of August 26, 2014, and January 29, 2018.

Table 6.3 shows input 3D breast image of various dataset processing using proposed classification and dimensionality reduction with tumor detection. Here, the processing and classification of 3D breast image of tumor have been shown.

Table 6.4 shows the analysis for various 3D breast cancer datasets. Datasets compared are LCC, RCC, LMLO, and RMLO. Parameters analyzed are training accuracy, validation accuracy, precision, recall, F-1 score, RMSE, and dice coefficient.

Fig. 6.3A−G shows the analysis for the LCC tumor dataset. Proposed technique attained a training accuracy of 75%, validation accuracy 85%, precision 68%, recall 62%, F-1 score 81%, RMSE 73%, and dice coefficient 63%; CNN attained a training accuracy of 71%, validation accuracy 81%, precision 64%, recall 58%, F-1 score 77%, RMSE 65%, and dice coefficient 55%; SVM attained a training accuracy of 73%, validation accuracy 83%, precision 66%, recall 59%, F-1 score 79%, RMSE 71%, and dice coefficient 59%.

Fig. 6.4A−G shows the analysis for RCC tumor dataset. Proposed technique attained a training accuracy of 82%, validation accuracy 86%, precision 75%, recall 65%, F-1 score 85%, RMSE 75%, and dice coefficient 65%; CNN attained a training accuracy of 77%, validation accuracy 82%, precision 71%, recall 61%, F-1 score 79%, RMSE 68%, and dice coefficient 59%, SVM attained a training accuracy of 79%, validation accuracy 84%, precision 73%, recall 63%, F-1 score 83%, RMSE 73%, and dice coefficient 63%.

Fig. 6.5 shows the analysis for LMLO tumor dataset. Proposed technique attained a training accuracy of 88%, validation accuracy 89%, precision 77%, recall 68%, F-1 score 88%, RMSE 77%, and dice coefficient 68%; CNN attained a training accuracy of 79%, validation accuracy 85%, precision 72%, recall 63%, F-1 score 81%, RMSE 73%, and dice coefficient 61%; SVM attained a training accuracy of 85%, validation accuracy 88%, precision 75%, recall 66%, F-1 score 85%, RMSE 75%, and dice coefficient 63%.

Table 6.3 Various 3D breast image analysis using proposed classification dimensionality reduction.

Dataset	Input 3D breast image	Preprocessed image	Classified image	Detected tumor
LCC				
RCC				
LMLO				
RMLO				

Fig. 6.6 shows the analysis for RMILO tumor dataset. Proposed technique attained a training accuracy of 93%, validation accuracy 94%, precision 81%, recall 75%, F-1 score 86%, RMSE 81%, and dice coefficient 73%; CNN attained a training accuracy 89%, validation accuracy 89%, precision 75%, recall 71%, F-1 score 82%, RMSE 76%, and dice coefficient 65%; SVM attained a training accuracy of 91%, validation accuracy 92%, precision 79%, recall 73%, F-1 score 84%, RMSE 79%, and dice coefficient 69%.

Table 6.4 Analysis based on various 3D breast cancer dataset.

Dataset	Techniques	Training accuracy	Validation accuracy	Precision	Recall	F1_Score	RMSE	Dice coefficient
LCC	CNN	71	81	64	58	77	65	55
	SVM	73	83	66	59	79	71	59
	ADR_3DBI_DLA	75	85	68	62	81	73	61
RCC	CNN	77	82	71	61	79	68	59
	SVM	79	84	73	63	83	73	63
	ADR_3DBI_DLA	82	86	75	65	85	75	65
LMLO	CNN	79	85	72	63	81	73	61
	SVM	85	88	75	66	85	75	63
	ADR_3DBI_DLA	88	89	77	68	88	77	68
RMLO	CNN	89	89	75	71	82	76	65
	SVM	91	92	79	73	84	79	69
	ADR_3DBI_DLA	93	94	81	75	86	81	73

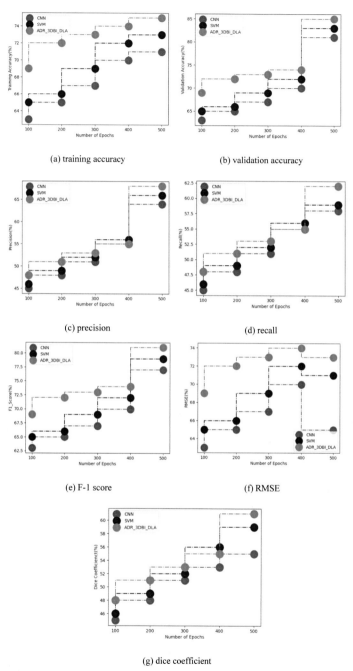

(a) training accuracy

(b) validation accuracy

(c) precision

(d) recall

(e) F-1 score

(f) RMSE

(g) dice coefficient

FIGURE 6.3

Analysis for LCC tumor dataset in terms of training accuracy, validation accuracy, precision, recall, F-1 score, RMSE, and dice coefficient.

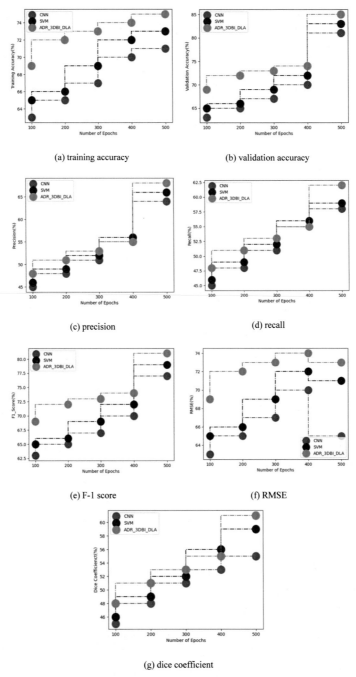

(a) training accuracy

(b) validation accuracy

(c) precision

(d) recall

(e) F-1 score

(f) RMSE

(g) dice coefficient

FIGURE 6.4

Analysis for RCC tumor dataset in terms of training accuracy, validation accuracy, precision, recall, F-1 score, RMSE, and dice coefficient.

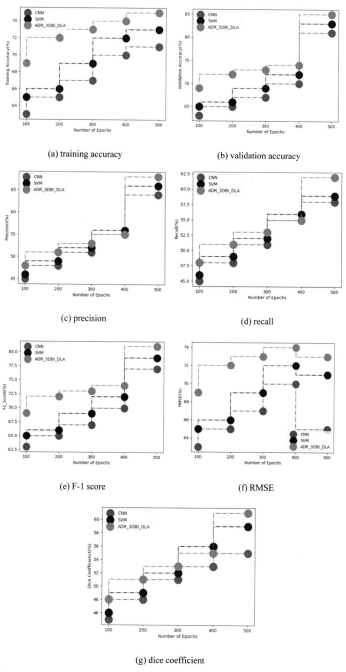

(a) training accuracy

(b) validation accuracy

(c) precision

(d) recall

(e) F-1 score

(f) RMSE

(g) dice coefficient

FIGURE 6.5

Analysis for LMLO tumor dataset in terms of training accuracy, validation accuracy, precision, recall, F-1 score, RMSE, and dice coefficient.

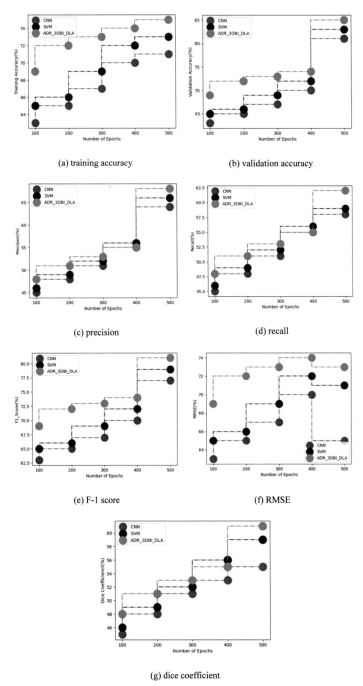

FIGURE 6.6

Analysis for RMILO tumor dataset in terms of training accuracy, validation accuracy, precision, recall, F-1 score, RMSE, and dice coefficient.

5. Conclusion

This research proposes a novel technique in 3D breast image-based classification and processing using autoencoder in dimensionality reduction with deep learning techniques. The dimensionality reduction is done using autoencoder-based Kernel Principal analysis, classification has been done using regressive convolutional Alex-Net architecture, and the classified output has been given as detected 3D breast cancer. To lessen the impact of class imbalance, we employ the data level strategy, which involves resampling the data. Tenfold cross-validation is used for evaluation. Proposed technique attained a training accuracy of 93%, validation accuracy of 94%, precision of 81%, recall of 75%, F-1 score of 86%, RMSE of 81%, and dice coefficient of 73%.

References

[1] Hamed G, Marey MAER, Amin SES, Tolba MF. Deep learning in breast cancer detection and classification. In: The international conference on artificial intelligence and computer vision. Cham: Springer; April 2020. p. 322–33.

[2] Ak MF. A comparative analysis of breast cancer detection and diagnosis using data visualization and machine learning applications. Healthcare April 2020;8(2):111 [MDPI].

[3] Mohammed SA, Darrab S, Noaman SA, Saake G. Analysis of breast cancer detection using different machine learning techniques. In: International conference on data mining and big data. Singapore: Springer; July 2020. p. 108–17.

[4] Dar RA, Rasool M, Assad A. Breast cancer detection using deep learning: datasets, methods, and challenges ahead. Computers in Biology and Medicine 2022:106073.

[5] Priyanka KS. A review paper on breast cancer detection using deep learning. In: IOP conference series: materials science and engineering, Vol 1022. IOP Publishing; 2021. p. 012071.

[6] Surendhar SPA, Vasuki RJMTP. Breast cancers detection using deep learning algorithm. Materials Today: Proceedings 2021.

[7] Allugunti VR. Breast cancer detection based on thermographic images using machine learning and deep learning algorithms. International Journal of Engineering in Computer Science 2022;4(1):49–56.

[8] Alanazi SA, Kamruzzaman MM, Islam Sarker MN, Alruwaili M, Alhwaiti Y, Alshammari N, et al. Boosting breast cancer detection using convolutional neural network. Journal of Healthcare Engineering 2021;2021.

[9] Zheng J, Lin D, Gao Z, Wang S, He M, Fan J. Deep learning assisted efficient AdaBoost algorithm for breast cancer detection and early diagnosis. IEEE Access 2020;8: 96946–54.

[10] Chugh G, Kumar S, Singh N. Survey on machine learning and deep learning applications in breast cancer diagnosis. Cognitive Computation 2021;13(6):1451–70.

[11] Ragab DA, Sharkas M, Marshall S, Ren J. Breast cancer detection using deep convolutional neural networks and support vector machines. PeerJ 2019;7:e6201.

[12] Sadhukhan S, Upadhyay N, Chakraborty P. Breast cancer diagnosis using image processing and machine learning. In: Emerging technology in modelling and graphics. Singapore: Springer; 2020. p. 113–27.

[13] Vaka AR, Soni B, Reddy S. Breast cancer detection by leveraging machine learning. ICT Express 2020;6(4):320–4.

[14] Das A, Mohanty MN, Mallick PK, Tiwari P, Muhammad K, Zhu H. Breast cancer detection using an ensemble deep learning method. Biomedical Signal Processing and Control 2021;70:103009.

[15] Wang X, Ahmad I, Javeed D, Zaidi SA, Alotaibi FM, Ghoneim ME, Eldin ET. Intelligent hybrid deep learning model for breast cancer detection. Electronics 2022;11(17): 2767.

[16] Mahmood T, Li J, Pei Y, Akhtar F, Imran A, Rehman KU. A brief survey on breast cancer diagnostic with deep learning schemes using multi-image modalities. IEEE Access 2020;8:165779–809.

[17] Suh YJ, Jung J, Cho BJ. Automated breast cancer detection in digital mammograms of various densities via deep learning. Journal of Personalized Medicine 2020;10(4):211.

[18] Witowski J, Heacock L, Reig B, Kang SK, Lewin A, Pysarenko K, Geras KJ. Improving breast cancer diagnostics with deep learning for MRI. Science Translational Medicine 2022;14(664):eabo4802.

[19] Rautela K, Kumar D, Kumar V. A systematic review on breast cancer detection using deep learning techniques. Archives of Computational Methods in Engineering 2022: 1–31.

Further reading

[1] Bhise S, Gadekar S, Gaur AS, Bepari S, Deepmala Kale DSA. Breast cancer detection using machine learning techniques. International Journal of Engineering Research and Technology 2021;10(7).

Prognosis of breast cancer using machine learning classifiers

7

Anurag Dutta[1], John Harshith[2], A. Ramamoorthy[3], A. Stephan Antony Raj[4], M. Balamurugan[5], D. Vinodhini[6]

[1]*Department of Computer Science and Engineering, Government College of Engineering and Textile Technology, Serampore, Kolkata, India;* [2]*Department of Computer Science and Engineering, Vellore Institute of Technology, Vellore, Tamil Nadu, India;* [3]*Department of Mathematics, Velammal Engineering College, Anna University, Chennai, Tamil Nadu, India;* [4]*Department of Mathematics, THE TIPSGLOBAL INSTITUTE, Coimbatore, Tamil Nadu, India;* [5]*Department of Mathematics, Vel Tech Rangarajan, Dr. Sagunthala R&D Institute of Science and Technology, Chennai, Tamil Nadu, India;* [6]*Department of Mathematics, Amrita School of Agricultural Sciences, Amrita Vishwa Vidyapeetham, Coimbatore, Tamil Nadu, India*

1. Introduction

A distinct aberrant state known as illness, which is unrelated to trauma [1], is one that negatively impacts the body's structure or function overall or in part. An illness is a term used to describe a medical condition that has specific indications and symptoms. Both internal and external causes, such as microorganisms, can contribute to illness. Internal immune system failure [2], for instance, has been linked to several illnesses, such as allergies, autoimmune diseases, hypersensitivity, and other types of immunodeficiency. In *Homo sapiens*, the terminology "illness" is quite regularly utilized to describe any scenario that leads to suffering, social issues, or death for the person who has it. In a larger sense, it may occasionally include trauma [3], disability [4], disease [5], syndrome [6], illness [7], specific symptoms, deviant conduct, and structural and functional atypical variants, albeit in other contexts and for different reasons, these may be viewed as separate categories. People can be affected by illnesses not just physically but also mentally, as acquiring and dealing with the condition can alter a person's perspective on life.

An anomaly or functional failure is a disorder. Mental, physical, genetic, emotional, Behavioral, and functional diseases are different types of medical conditions. As it is often thought to be more impartial and less biased than illness or disease, the term "disability" may be preferred. The phrase "mental disorder" in mental health refers to the intricate interplay between biological, social, and psychological elements that contribute to mental illness. However, the word "disability" is also used in a variety of different medical contexts, most frequently to describe physical impairments that are not brought on by an infectious agent.

Computational Intelligence and Modelling Techniques for Disease Detection in Mammogram Images
https://doi.org/10.1016/B978-0-443-13999-4.00003-1

129

A set of illnesses known as cancer [8] involve cells that develop abnormally and invade or spread to other bodily areas. As opposed to these, benign [9], nonmetastatic tumors [10]. Lumps [11], irregular bleeding [12], a persistent cough [13], unexplained weight loss [14], and stoma alterations [15] are all potential warning signs and symptoms. These signs and symptoms could be cancer-related, but there are other possible explanations as well. Humans are affected by more than a 100 different cancers. Quitting smoking, maintaining a healthy weight, abstaining from alcohol, eating plenty of fruits, vegetables, and whole grains, consuming resistant starch, and being vulnerable to specific infections all reduce the risk of acquiring certain types of cancer. Vaccination and the eating of processed and red meat can both help to lower it. Cervical [16] and colorectal [17] cancer can be effectively caught early with screening. Breast cancer screening has burdensome advantages. Radiation therapy, surgery, chemotherapy, and targeted therapy are frequently combined to treat cancer. A key aspect of care is controlling symptoms and pain. The sort of cancer you have and how advanced it is when therapy is started will determine your chances of survival. In affluent nations, the average 5-year survival rate for patients who were younger than 15 at the time of diagnosis is 80%. In the United States, the average 5-year survival rate for cancer is 66%. Lung, prostate, colon, and stomach cancers are the most prevalent types of cancer in men. The most prevalent cancers in women are cervical, breast, colon [18], and lung cancers [19]. When cancer first starts to spread, there are no symptoms. As the tumor swells or ulcerates, symptoms appear. Results vary depending on cancer's nature and location. There are not many distinct symptoms. In folks with other medical issues, many are typical. Cancer is challenging to identify and is occasionally considered a "huge copycat" disease.

Genetic alterations brought on by environmental and way-of-life variables account for 90%—95% of cancer instances. The final 5%—10% come via inheritance. The way we live, economic and Behavioral variables, and pollution are all environmental factors. These causes are not genetic. Common environmental variables that raise the death risk from most cancers include smoking [20], being overweight or obese, catching infectious diseases, exposure to radiation, insufficient exercise, and pollution. Even while it no longer appears to be a risk factor for cancer development, psychological stress can still hurt people who already have the disease. It is frequently challenging to identify the particular origin of cancer since different causes may not always have visible symptoms. The reason a persistent smoker develops lung cancer, for instance, may be related to tobacco use, but lung cancer can develop in anyone who is exposed to radiation or air pollution [21]. Cancer is often not contagious, unless in extremely rare circumstances involving organ donation and infection; nonetheless, several conditions can cause cancer can.

The majority of malignancies are first identified through screening or the emergence of symptoms. None of these result in a conclusive diagnosis that calls for a pathologist to examine tissue samples. A person who may have cancer is examined by a physician. These frequently consist of blood tests, X-rays, contrast-enhanced CT scans, and endoscopies. The term "cancer prevention" refers to taking preventive steps to lower the chance of developing cancer. Environmental risk factors are a major contributor to many cancer incidences. Many of these environmental influences are regulated by lifestyle decisions. As a result, cancer can usually be avoided. Most common cancers are preventable since environmental factors are to blame for 70% −90% of them. Avoiding the risk factors listed below will help prevent nearly 30% of cancer-related deaths: smoking, being overweight or obese, eating poorly, not exercising enough, drinking alcohol, having STDs [22], and being exposed to air pollution. Furthermore, it is possible to see poverty as a veiled risk factor for cancer in people. Not all environmental causes, such as naturally occurring background radiation and malignancies brought on by inherited genetic disorders, are outside our control and can thus not be avoided. Depending on the type of cancer and its stage at diagnosis, the survival rate might range from high to zero 5 years following diagnosis. The prognosis is typically poorer if the cancer has spread. Except for carcinoma in situ and nonmelanoma skin cancer, around half of the patients treated for invasive cancer pass away as a result of the disease or the medication. Metastasis [23] from the initial tumor is a major cause of cancer-related fatalities. The reason why survival rates are lower in underdeveloped nations is partial since the most prevalent cancers there are more challenging to treat than in wealthy nations.

In our work, we would work on the Breast Cancer Mammography Dataset and devise the most efficient machine learning algorithm that would be efficient enough to predict carcinogenic tissues in the breast. For the dataset, we have considered 250 healthy females, and 350 carcinogenic. The dataset is made publicly available on GitHub. Data, code, and supporting materials are publicly available via https://github.com/Anurag-Dutta/Stratifying-Mastographic-dossier-for-Cancer-Prognosis.

2. Breast cancer

Breast tissue can grow into cancer in cases of breast cancer. Breast lumps, abnormal nipple shape, dimpling, milk refusal, drainage flowing from the tip, an arching breast, or a red or crusted region of the epidermis are all signs of melanoma. Patients who are afflicted may have nerve damage, swollen lymph nodes, difficulty breathing, or yellow skin. Fatness, a low level of physical activity, alcohol addiction, hormonal replacement therapy throughout postmenopausal, ionizing radiation, having kids later in life or none at all, being older, having an earlier risk of the disease, and possessing cancer in the family are all significantly more likely to develop breast cancer. Inherited genetic biases, including BRCA1 and BRCA2, account for 5% −10% of cases. Breast cancer most commonly manifests itself in the cells that surround lactiferous ducts and the temporalis that feed milk to these ducts. Ductal melanomas are malignancies that begin in the channels, while lobular melanomas begin in the lobules. There are approximately 18 different forms of breast cancer. Some develop from preinvasive abnormalities, such as invasive ductal carcinoma. Breast cancer is confirmed by conducting a test on questionable tissue. For diagnosis, more tests are undertaken to determine whether cancer has spread beyond the mammary and to find the most viable therapies. It is controversial if breast cancer has better than adverse effects. Considering that many women who screened positive for malignancy do not have it, a 2013 Cochrane analysis concluded that it was uncertain whether screening mammography testing is more harmful than beneficial. In 2009, the United States Preventive Services Task Force did a review and discovered evidence of benefit in people aged 40–70. The organization recommends evaluating women aged 50 and 74 every 2 years. Individuals who are at high risk of contracting breast cancer can avoid it by taking the medicines tamoxifen or raloxifene. Surgical removal of both breasts is another prophylactic treatment for some high-risk women. Numerous therapies, such as surgery, radiation therapy, chemotherapy, hormone therapy, and targeted therapy, may be utilized on cancer patients. The various surgical procedures range from mastectomy to breast-conserving surgery. Breast reconstruction can happen right away during surgery or later. Treatments are mostly focused on enhancing comfort and quality of life for patients whose cancer has spread to other body parts. The type of breast cancer, the severity of the condition, and the patient's age all affect the outcome. Between 80% and 90% of people survive 5 years in the United States and England. The 5-year mortality rates are greater in poorer nations. With 25% of all occurrences, breast carcinoma is the most common type of cancer among women worldwide. There were 2 million additional cases and 627,000 fatalities as a result of 2018. It is more prevalent in industrialized nations and affects women over 100 times more frequently than it does men. A lump that sounds weird than the tissue that surrounds it is the most common indication of breast cancer. More than 80% of cases are identified when an individual feels a bump of this kind with their fingertips. Mammography, on the other hand, can detect early breast tumors. Lumpiness in the underarm lymphatic system may also be an indication of breast cancer. Other than a bump, indicators of breast cancer include a nipple shifting positions, shape, or becoming inverted, skin puckering or discoloration, dermatitis on or near a nip, drainage from a lip, back fatigue in an area of the breast or underarm, and enlargement underneath the upper arm or near the clavicle. Mastodynia may be a sign of other breast health problems but is not a reliable indicator of the presence or absence of breast cancer. Paget's disease of the breast is another symptom combination associated with breast cancer. Skin changes resembling eczemas, such as redness, discoloration, or slight flaking of the nipple skin, are how this syndrome manifests. Breast Paget's disease symptoms might include tingling, itching, increasing sensitivity, burning, and pain as it progresses. There might also be nipple discharge. Most women who are diagnosed with Paget's disease of the breast also have a breast lump. An uncommon but aggressive type of breast cancer known as "inflammatory breast cancer" is characterized by the formation of swollen, red patches on the top of the breast. The visual features of inflammatory breast cancer are brought on by the cancer cells blocking the lymphatic arteries. Obese, African American and younger women are more likely to be diagnosed with this kind of breast cancer. There may occasionally be a delay in diagnosis since inflammatory breast cancer does not typically appear as a lump. A rare variety of secretory carcinomas that only affects the breast is known as mammary secretory carcinoma (MSC). Although it typically affects adults, children are also affected by it frequently: 80% of all cases of breast cancer in children are caused by MSC. MSC lesions are often small, painless, slow-growing ductal breast tumors that have infiltrated the tissue near their ducts of origin and frequently migrated to axillary and/or sentinel lymph nodes. Rarely, however, have they metastasized to distant regions. These tumors frequently exhibit unusual microscopic characteristics, and the tumor cells contain a balanced genetic translocation in which a portion of the NTRK3 gene is fused to a portion of the ETV6 gene to create the fusion gene ETV6-NTRK3. ETV6-NTRK3 is a chimeric protein that is encoded by this fusion gene. The NTRK3 portion of the ETV6-NTRK3 protein exhibits upregulated tyrosine kinase activity that activates the PI3K/AKT/mTOR and MAPK/ERK signaling pathways, which support cell proliferation and survival and may help MSC grow. The most popular treatments for treating adults with MSC have been conservative surgery, modified radical mastectomy, and radical mastectomy, whereas the best treatments for treating children with MSC have been simple mastectomy, local excision with sentinel lymph node biopsy, and complete axillary dissection. Long-term, that is, >20 years, follow-up exams are advised in all circumstances. Chemotherapy and radiation therapy have had little to no effect on the extremely infrequent cases of MSC that have spread to other tissues. Entrectinib, an antityrosine kinase medication that blocks the tyrosine kinase activity of the ETV6-NTRK3 fusion protein, produced positive partial responses in three patients with metastatic illness. MSC has a 20-year survival rate of 93.16% due to its sluggish growth and low rate of metastasis to distant tissues. In a few rare instances, a tumor called a phyllode may first look like a fibroadenoma. Phyllodes tumors develop within the breast stroma and are made up of both glandular and stromal tissue. Phyllodes tumors are not staged in the traditional sense; instead, they are categorized as benign, borderline, or malignant based on how they appear under a microscope. Metastatic tumors, or secondary tumors that spread outside of the original tumor's place of origin, can develop from malignant tumors. The site of the metastasis will determine the symptoms brought on by metastatic breast cancer. Bone, liver, lung, and the brain are common locations for metastasis. A cancer is in stage 4 when it has become so invasive. Cancers in this stage frequently result in death. Unexpected weight loss, joint and bone pain, jaundice, and neurological symptoms are all common signs of stage 4 cancer. Due to the possibility that they could represent signs of numerous different conditions, these symptoms are referred to as nonspecific symptoms. Rarely, breast cancer can spread to extremely rare locations, such as the peripancreatic lymph nodes, which can clog the bile duct and make diagnosis challenging. The majority of breast problem symptoms, including the majority of lumps, turn out not to be underlying breast cancer. For instance, less than 20% of lumps are malignant, and benign breast diseases including mastitis and fibroadenoma of the breast are more frequently to blame for the symptoms of breast disorders.

The majority of breast cancer types can be quickly identified through microscopic examination of a biopsy sample taken from the breast's afflicted area.

Additionally, some forms of breast cancer necessitate specialist lab tests. Mammography and physical exams of the chest by a medical professional can both provide an approximate possibility that a tumor is cancerous and may also pick up on other abnormalities, such as a straightforward cyst. When these tests are unhelpful, a medical professional may take a sample of the lump's fluid for microscopic inspection to help make a diagnosis. Aspiration with a needle can be done in a doctor's office or clinic. If the lump is not under the skin, a local anesthetic may not be required to freeze the breast tissue to prevent discomfort during the treatment. A bulge that contains clear fluid is extremely unlikely to be carcinogenic, however, bloody fluids may be submitted for a microscope examination to check for cancerous cells. Breast cancer can be accurately diagnosed with fine needle aspiration cytology (FNAC), mammography, and medical examination of the chest altogether. Other biopsy alternatives include excisional biopsy, which involves removing the entire lump, core biopsy, which involves removing a part of the breast mass, and vacuum-assisted breast biopsy. The results of a physical examination by a healthcare professional, a mammogram, and any additional tests that might be carried out under unique conditions are frequently sufficient to support excisional biopsy as the primary diagnostic and therapeutic approach. Machine learning algorithms have been additionally proposed as a diagnostic measure. Machine learning approach allows a trained model to predict the possible presence of a tumor in the breast, which is having a much higher accuracy than that of the primarily evident mastographic tumor evaluation strategies.

3. Machine learning

Machine learning is indeed a field of study that is concerned with understanding and creating "learning" techniques, or methodologies which use data to improve performance on a specific set of tasks. It is an aspect of artificial intelligence. Machine learning algorithms develop a model utilizing data samples, also known as training examples, to form hypotheses or assessments without even being explicitly instructed to do so. Algorithms for machine learning are used in a spectrum of uses, like machine vision, natural language processing, phishing emails, pharmacology, and farming, where existing algorithms are difficult or prohibitive to build. Computational statistics is strongly connected to a subgroup of cognitive computing, however, not all algorithms are empirical learning. The methods, methodology, and implementation techniques derived from the investigation of mathematical programming enrich the science of machine learning. Unsupervised machine learning for information extraction is the subject of data mining, a comparable field of study. Some applications for machine learning use information based on biology in ways that are similar to how biological brains work. When it comes to solving real-world problems, machine learning is also referred to as predictive analytics. UC Berkeley divides a machine learning algorithm's learning system into three major components.

1. Algorithms for machine learning are typically used to make recommendations or categorizations. Based on the data, the algorithms will provide an approximation of a trend in the inputs, which can be tagged or unlabeled.
2. An evaluation metric is used to evaluate the model's projection. If documented cases exist, a functional form can contrast them to determine the model's correctness.
3. If the models can perfectly accommodate the pieces of data in the training data set, weights are modified to reduce the distance between both the known case and the model prediction. The algorithms will continue this "analyze and optimize" method, with parameters changed dynamically until a preset level of precision is reached.

Machine learning models fall into three primary categories.

1. *Supervised machine learning*: The process of teaching computers to properly perform classification or predict consequences using tagged samples are referred to as "supervised learning," [24] which is also utilized to refer to machine learning supervised learning. As input data is entered into the model, its weights are adjusted until it is well suited. This occurs throughout the cross-validation method to ensure that the prototype does not fit too well or too inadequately. One popular way supervised learning helps businesses is by identifying spam in a separate folder from your email.
2. *Unsupervised machine learning*: Unsupervised learning, also known as unsupervised machine learning, employs machine-learning algorithms to analyze and group unlabeled datasets. Without the intervention of a human, these algorithms detect underlying knowledge or data clusters. Because it may uncover parallels and differences in information, this method is excellent for hypothesis testing, cross-selling tactics, buyer behavior, and imagery and pattern detection.
3. *Reinforcement machine learning*: Even though the technique is not taught on a data sample, reinforced algorithm is a controlled learning-like predictive model. This model learns by generating errors along the way. By encouraging a run of favorable outcomes, the best suggestion or technique for a particular complaint will be established.

4. Machine intelligence-aided mammography

This section puts forth the prediction of breast cancer by studying the X-ray scan images. The creation of an analysis that "corresponds too closely or exactly to a given set of data, and may, therefore, fail to fit further data or forecast future observations accurately" is known as overfitting in mathematical modeling. To get rid of such an issue in future data, in the modeling part, we undertook augmentation of data (Figs. 7.1 and 7.2).

It is a classification-based problem in machine learning. Here, in this article, we would make use of standard classification-based machine learning algorithms for

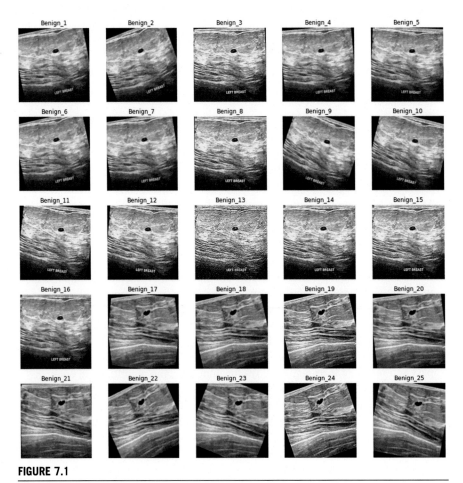

FIGURE 7.1

An instance of the augmented dataset, used for modeling using machine intelligence. It contains a set of 25 benign breast cancer images.

modeling and prediction. Now, since we are making use of multiple machine learning algorithms, we would compare them based on some metric. Here, we are making use of the confusion matrix to compare them.

A confusion matrix, alternatively referred to as just an error matrix, is indeed a type of tabular structure that permits the depiction of the effectiveness of an approach, most typically a learn-and-gain knowledge one, in the machine learning field and, more specifically, the problems of prediction. Both forms of the matrix are reported in the literature, where each row contains cases in a class label and each column uniquely identifies occurrences in a forecasting class. The name alludes to how easy it is to tell if the machine is merging two types. The dimensions of this

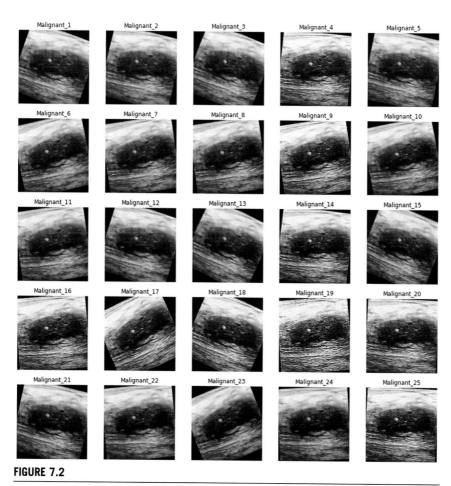

FIGURE 7.2

An instance of the augmented dataset, used for modeling using machine intelligence. It contains a set of 25 malignant breast cancer images.

contingency table are accurate and anticipated, with equivalent sets of categories in each degree (Fig. 7.3).

Further, we have also considered the accuracy, precision, recall, and their F1 score.

The accuracy of a classification test is a preparation to find how well it incorporates or eliminates a circumstance. In another sense, the accuracy is indeed the proportion of correct predictions (including true positives and false negatives) over all examples examined. As a consequence, it compares projected probability before and after the assessment.

Mathematically,

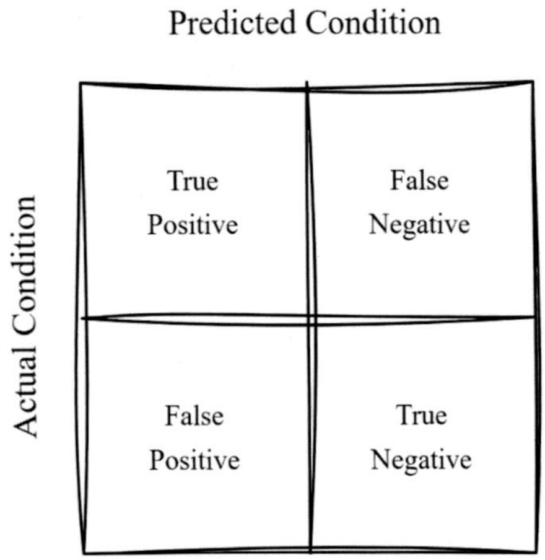

Predicted Condition

FIGURE 7.3

Confusion matrix. In true positive (TP), the actual scenario and the model's prediction were YES. In false negative (FN), although the model indicated NO, the actual result was YES. In false positive (FP), although the model expected YES, the actual result was NO. In true negative (TN), the model predicted NO, and the actual or true value likewise indicated NO.

$$\text{Accuracy} = \frac{TP + TN}{TP + FN + FP + TN} \tag{7.1}$$

In a classification task, the precision for a class is calculated as the number of true positives, or the proportion of elements that are correctly classified as belonging to the positive class, divided by the total number of elements that are classified in this way, or the sum of true positives and false positives, or the proportion of elements that are classified in this way but are not members of the class. Mathematically,

$$\text{Precision} = \frac{TP}{TP + FP} \tag{7.2}$$

Recall is defined in this context as the sum of true positives and false negatives, or items that should have been classified as belonging to the positive class but were not and is calculated as the number of true positives divided by the total number of elements that belong to the positive class. Mathematically,

$$\text{Recall} = \frac{TP}{TP + FN} \tag{7.3}$$

The F-measure commonly referred to as the F score, is indeed a test accuracy parameter used in data analysis of dichotomous categorization. It is calculated using the precision and recall of the test. The harmonic mean of the precision and recall is the F1 score. In the more general F_β score, extra weights are applied, favoring either precision or recall over the other. An F-score can have a maximum value of 1.0, which denotes perfect precision and recall, and a minimum value of 0, which occurs when either precision or recall is zero. Mathematically,

$$\text{F1 score} = \left(\frac{2 \times \text{Precision} \times \text{Recall}}{\text{Precision} + \text{Recall}}\right) = 2 \times \left(\frac{\left(\frac{\text{TP}}{\text{TP} + \text{FP}}\right) \times \left(\frac{\text{TP}}{\text{TP} + \text{FN}}\right)}{\left(\frac{\text{TP}}{\text{TP} + \text{FP}}\right) + \left(\frac{\text{TP}}{\text{TP} + \text{FN}}\right)}\right) \quad (7.4)$$

Also, we have considered the training and testing scores for each of the classifiers.

The classifiers that have been put into action are as follows:

1. Logistic regression
2. Support vector machine
3. *k*-Nearest neighbors
4. Random forest classifier
5. Gaussian Naïve Bayes
6. Decision tree

The results corresponding to these classifiers are as follows:

```
========================
[ Logistic Regression ]
========================

--------
Training
--------
Training Score: 0.9755083179297597
-------
Testing
-------
Testing Score: 0.8731527093596059
Precision: 0.8608374384236454
Recall: 0.8825757575757576
F1 Score: 0.8715710723192021
```

Figs. 7.4 and 7.5 show the precision–recall curve and confusion matrix for the model modeled using logistic regression.

```
===========================
[ Support Vector Machine ]
===========================

--------
Training
--------
Training Score: 0.9741219963031423
-------
Testing
-------
```

```
Testing Score: 0.9347290640394089
Precision: 0.9319899244332494
Recall: 0.9343434343434344
F1 Score: 0.9331651954602774
```

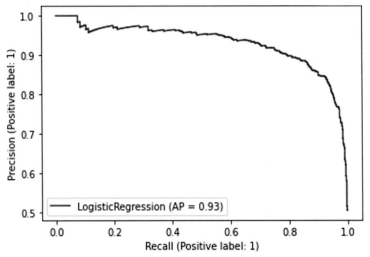

FIGURE 7.4

Precision–recall curve for the logistic regression model with precision along the Y-axes and recall along the X-axes.

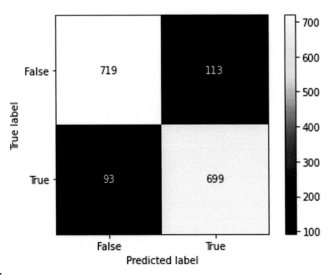

FIGURE 7.5

Confusion matrix for the logistic regression model. The respective quadrants have their usual meaning, which has been described earlier.

Figs. 7.6 and 7.7 show the precision recall curve and confusion matrix for the model modeled using support vector machine.

```
===============================
[ k — Nearest Neighbours ]
===============================

--------
Training
--------

Training Score: 0.9799753542821935
-------
Testing
-------

Testing Score: 0.9538177339901478
Precision: 0.9324487334137516
Recall: 0.976010101010101
F1 Score: 0.9537322640345465
```

Figs. 7.8 and 7.9 show the precision—recall curve and confusion matrix for the model modeled using k-nearest neighbors.

```
===============================
[ Random Decision Forest ]
===============================

--------
Training
--------

Training Score: 1.0
-------
Testing
-------

Testing Score: 0.9538177339901478
Precision: 0.9674054758800521
Recall: 0.9368686868686869
F1 Score: 0.9518922386144965
```

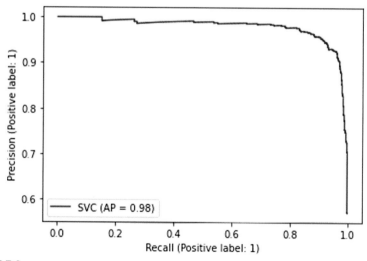

FIGURE 7.6

Precision—recall curve for the support vector machine with precision along the Y-axes and recall along the X-axes.

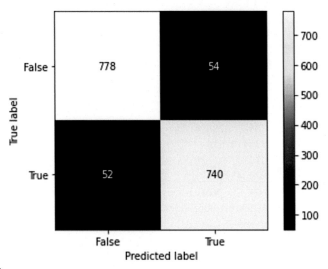

FIGURE 7.7

Confusion matrix for the support vector machine model. The respective quadrants have their usual meaning, which has been described earlier.

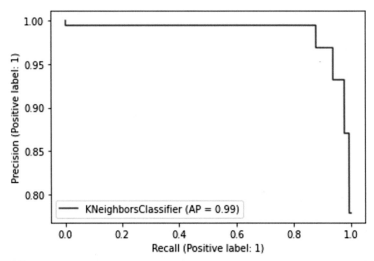

FIGURE 7.8

Precision—recall curve for the k nearest neighbors classifier with precision along the Y-axes and recall along the X-axes.

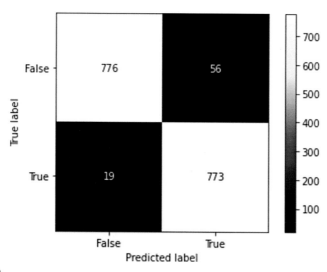

FIGURE 7.9

Confusion matrix for the *k* nearest neighbors model. The respective quadrants have their usual meaning, which has been described earlier.

Figs. 7.10 and 7.11 show the precision–recall urve and confusion matrix for the model modeled using random forest.

```
===========================
[ Gaussian Naîve Bayes ]
===========================

--------
Training
--------
Training Score: 0.6592729513247073
-------
Testing
-------
Testing Score: 0.6416256157635468
Precision: 0.6058467741935484
Recall: 0.7588383838383839
F1 Score: 0.6737668161434979
```

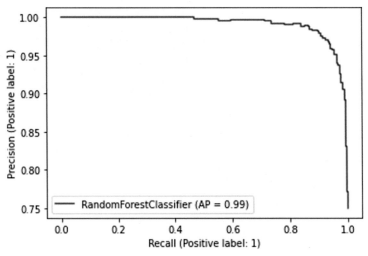

FIGURE 7.10

Precision–recall curve for the random forest classifier with precision along the Y-axes and recall along the X-axes.

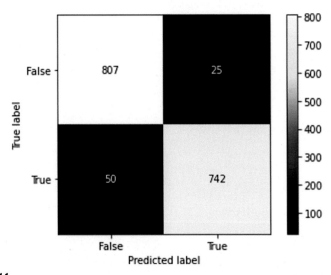

FIGURE 7.11

Confusion matrix for the random forest model. The respective quadrants have their usual meaning, which has been described earlier.

Figs. 7.12 and 7.13 show the precision—recall curve and confusion matrix for the model modeled using Gaussian Naïve Bayes.

```
====================
[ Decision Tree ]
====================
--------
Training
--------
Training Score: 1.0
-------
Testing
-------
Testing Score: 0.8454433497536946
Precision: 0.8351920693928129
Recall: 0.851010101010101
F1 Score: 0.8715710723192021
```

Figs. 7.14 and 7.15 show the precision—recall curve and confusion matrix for the model modeled using decision tree.

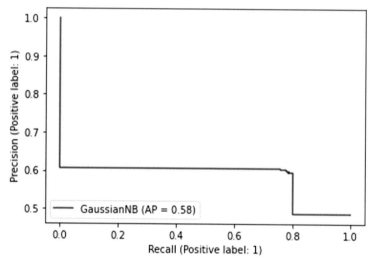

FIGURE 7.12

Precision—recall curve for the Gaussian Naïve Bayes with precision along the Y-axes and recall along the X-axes.

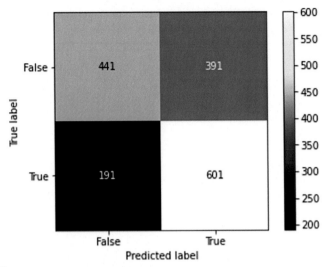

FIGURE 7.13

Confusion matrix for the Gaussian Naïve Bayes model. The respective quadrants have their usual meaning, which has been described earlier.

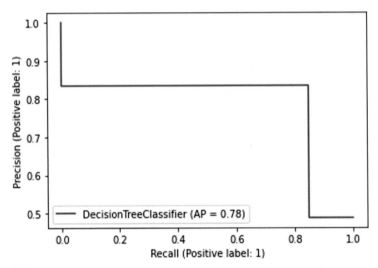

FIGURE 7.14

Precision–recall curve for the decision tree with precision along the Y-axes and recall along the X-axes.

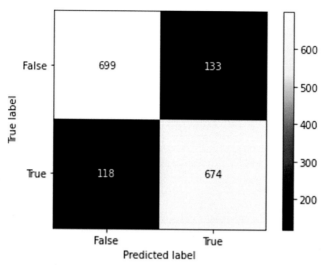

FIGURE 7.15

Confusion matrix for the decision tree model. The respective quadrants have their usual meaning, which has been described earlier.

Comparing All, we could conclude that the k-nearest neighbors is best in modeling mastographic classifier. It gives an F1 score of 0.9537, which is the best we could get of these standard classifiers. Probably, we would get a more boost in efficiency if we incorporate the neural networks for modeling.

5. Conclusion

Medical science is evolving daily, and so is the wrath of infectious diseases. This chapter focused on combining medical science with the latest technologies in the world of informatics, like machine intelligence. Mammography, as a diagnosis paradigm, is quite an important domain these days. This chapter considers a popular machine learning algorithm for prognosis of breast cancer. For the classification, of whether the carcinogenetic nature of the same is benign, or malignant, we utilized popular ML classifiers like logistic regression, support vector machine, k-nearest neighbors, random forest classifier, Gaussian Naïve Bayes, and decision tree. These paradigms were compared based on their F1 score. The k-nearest neighbors turned out to be overruling for all other algorithms, with an F1 score of 0.9537, precision of 0.9324, and a recall of 0.9760. Table 7.1 summarizes the result from all machine learning classifiers.

Table 7.1 Result summarization of the classifier algorithms for mammography.

Algorithms	F1 score	Precision	Recall
Logistic regression	0.8716	0.8608	0.8826
Support vector machine	0.9332	0.9320	0.9343
k-nearest neighbors	0.9537	0.9324	0.9760
Random forest classifier	0.9519	0.9674	0.9369
Gaussian Naïve Bayes	0.6738	0.6058	0.7588
Decision tree	0.8716	0.8352	0.8510

References

[1] Brunson KL. Mechanisms of late-onset cognitive decline after early-life stress. Journal of Neuroscience Oct. 2005;25(41):9328–38. https://doi.org/10.1523/jneurosci.2281-05.2005.

[2] Matzinger P. The danger model: a renewed sense of self. Science Apr. 2002;296(5566): 301–5. https://doi.org/10.1126/science.1071059.

[3] Jeronimus BF, Ormel J, Aleman A, Penninx BWJH, Riese H. Negative and positive life events are associated with small but lasting change in neuroticism. Psychological Medicine Feb. 2013;43(11):2403–15. https://doi.org/10.1017/s0033291713000159.

[4] Shakespeare T, Watson N. The social model of disability: an outdated ideology? Research in Social Science and Disability 2001;2:9–28. https://doi.org/10.1016/s1479-3547(01)80018-x.

[5] Marcantonio M, Pascoe EL, Baldacchino F. Sometimes scientists get the Flu. Wrong…! Trends in Parasitology Jan. 2017;33(1):7–9. https://doi.org/10.1016/j.pt.2016.10.005.

[6] Ghaemi SN. Nosologomania: DSM and Karl Jaspers' critique of Kraepelin. Philosophy, Ethics, and Humanities in Medicine 2009;4(1):10. https://doi.org/10.1186/1747-5341-4-10.

[7] White MP, et al. Spending at least 120 minutes a week in nature is associated with good health and wellbeing. Scientific Reports Jun. 2019;9(1). https://doi.org/10.1038/s41598-019-44097-3.

[8] Wu S, Powers S, Zhu W, Hannun YA. Substantial contribution of extrinsic risk factors to cancer development. Nature Dec. 2015;529(7584):43–7. https://doi.org/10.1038/nature16166.

[9] Inoki K, Corradetti MN, Guan K-L. Dysregulation of the TSC-mTOR pathway in human disease. Nature Genetics Jan. 2005;37(1):19–24. https://doi.org/10.1038/ng1494.

[10] Bakhoum SF, et al. Chromosomal instability drives metastasis through a cytosolic DNA response. Nature Jan. 2018;553(7689):467–72. https://doi.org/10.1038/nature25432.

[11] Bancroft LW, Kransdorf MJ, Peterson JJ, O'Connor MI. Benign fatty tumors: classification, clinical course, imaging appearance, and treatment. Skeletal Radiology Oct. 2006;35(10):719–33. https://doi.org/10.1007/s00256-006-0189-y.

[12] Edelman A, et al. Association between menstrual cycle length and coronavirus disease 2019 (COVID-19) vaccination. Obstetrics and Gynecology Jan. 2022;139:481. https://doi.org/10.1097/aog.0000000000004695.

[13] Goldsobel AB, Chipps BE. Cough in the pediatric population. The Journal of Pediatrics Mar. 2010;156(3):352−8. https://doi.org/10.1016/j.jpeds.2009.12.004.

[14] Müller TD, Blüher M, Tschöp MH, DiMarchi RD. Anti-obesity drug discovery: advances and challenges. Nature Reviews Drug Discovery Nov. 2021:1−23. https://doi.org/10.1038/s41573-021-00337-8.

[15] Thorpe G, Arthur A, McArthur M. Adjusting to bodily change following stoma formation: a phenomenological study. Disability & Rehabilitation Mar. 2016;38(18): 1791−802. https://doi.org/10.3109/09638288.2015.1107768.

[16] zur Hausen H. Papillomaviruses and cancer: from basic studies to clinical application. Nature Reviews Cancer May 2002;2(5):342−50. https://doi.org/10.1038/nrc798.

[17] Ionov Y, Peinado MA, Malkhosyan S, Shibata D, Perucho M. Ubiquitous somatic mutations in simple repeated sequences reveal a new mechanism for colonic carcinogenesis. Nature Jun. 1993;363(6429):558−61. https://doi.org/10.1038/363558a0.

[18] Abdulamir AS, Hafidh RR, Bakar FA. The association of Streptococcus bovis/gallolyticus with colorectal tumors: the nature and the underlying mechanisms of its etiological role. Journal of Experimental and Clinical Cancer Research Jan. 2011;30(1). https://doi.org/10.1186/1756-9966-30-11.

[19] Ramada Rodilla JM, Calvo Cerrada B, Serra Pujadas C, Delclos GL, Benavides FG. Fiber burden and asbestos-related diseases: an umbrella review. Gaceta Sanitaria Jun. 2021. https://doi.org/10.1016/j.gaceta.2021.04.001.

[20] Axéll T, Hedin CA. Epidemiologic study of excessive oral melanin pigmentation with special reference to the influence of tobacco habits. Scandinavian Journal of Dental Research Dec. 1982;90(6):434−42. https://doi.org/10.1111/j.1600-0722.1982.tb00760.x.

[21] Lelieveld J, Evans JS, Fnais M, Giannadaki D, Pozzer A. The contribution of outdoor air pollution sources to premature mortality on a global scale. Nature Sep. 2015;525(7569): 367−71. https://doi.org/10.1038/nature15371.

[22] Holmes KK, Johnson DW, Trostle HJ. An estimate of the risk of men acquiring gonorrhea by sexual contact with infected Females1. American Journal of Epidemiology Feb. 1970;91(2):170−4. https://doi.org/10.1093/oxfordjournals.aje.a121125.

[23] Ettinger DS, et al. NCCN clinical practice guidelines occult primary. Journal of the National Comprehensive Cancer Network Dec. 2011;9(12):1358−95. https://doi.org/10.6004/jnccn.2011.0117.

[24] Smith MR, Martinez T. Improving classification accuracy by identifying and removing instances that should be misclassified. The 2011 International Joint Conference on Neural Networks Jul. 2011:2690−7. https://doi.org/10.1109/ijcnn.2011.6033571.

Breast cancer diagnosis through microcalcification

Sindhu P. Menon[1], Pramodkumar Naik[2], Ramya Srikanteshwara[3]

[1]*School of Computing and Information Technology, Reva University, Bengaluru, Karnataka, India;*
[2]*Department of Computer Science & Engineering, Dayananda Sagar University, Bengaluru,*
Karnataka, India; [3]*Department of Computer Science & Engineering, Nitte Meenakshi Institute of*
Technology, Bengaluru, Karnataka, India

1. Introduction

The most prevalent reason today causing cancer mortality for women is breast cancer, which succeeds lung cancer. Statistics shared by WHO states that more than a 3,000,000 people will be diagnosed with breast cancer in 2023 [1]. Numbers also show that this may reach 4.5 million by 2030. According to the ACS, 287,000 women in the United States are anticipated to receive an invasive breast cancer diagnosis in 2005 (from preliminary to final stages). Moreover, 61,475 women are anticipated to have had in situ breast (meaning that they have not spread) cancer diagnoses, which are a very early stage of the illness. In Europe, breast cancer affects more women than any other type of cancer.

According to estimates, 14 million cases of breast cancer will be diagnosed in Europe by 2040, while 859,000 people died from the disease in 2022. In Europe, breast cancer contributes to 27.3% of all new instances of cancer in women and 19.2% of cancer fatalities. ACS reported that there were 1,821,450 new cases of cancer in the country in 2021 and 806,880 cancer-related deaths. In the United States, there will be 342,170 cases of breast cancer in women that year and 52,360 fatalities are anticipated. In addition, it is anticipated that China will see 312,000 fresh cases each year and roughly 70,000 deaths. Breast cancer has been identified as the malignancy with the quickest rate of growth in China [2].

Till date, there has been no discovery to prevent breast cancer, and its specific etiology is unknown.

Early detection could lead to better survival rates of breast cancer, according to a vast number of research [3]. Most common procedures to diagnose are through imaging diagnosis, histological biopsy, and clinical breast examination. In addition, thermography, magnetic resonance imaging, tissue sampling, mammography, ultrasound, and computed tomography are frequently utilized breast evaluation techniques.

Despite more than 2 decades of research on computer-aided mammography, the fuzzy character of microcalcification, poor contrast, and limited capacity to identify

Computational Intelligence and Modelling Techniques for Disease Detection in Mammogram Images
https://doi.org/10.1016/B978-0-443-13999-4.00008-0
151

them from their surroundings makes the automated characterization of microcalcifications as exceedingly challenging. The size of these average out to be 0.3 mm, and sometimes can be less than 0.1. Mammography can reveal early tumors that are difficult to see visually. It reacts strongly to intramammary calcification, especially if it is minor [4]. Mass, microcalcification, modifications to the areola and nipple, structural distortion, and enhanced vascular deformation are typical findings from X-rays in the case of breast cancer. From the perspective of cancer detection and diagnosis, bulk and microcalcification are the most significant and evident pathological indicators of this form of cancer.

Breast lesions, both benign and malignant, may have microcalcification. Microcalcification is frequently the only symptom of breast cancer, particularly the early "occult" form [5]. Mammography, however, is more responsive to microcalcification.

A more accurate clinical diagnosis can now be made by doctors thanks to the introduction of computer-aided diagnosis into the medical profession with the development of computer-based technology. Computer diagnostics can speed up the diagnostic process for doctors and improve the precision of the results.

A significant and frequent indication of early breast cancer is characteristic calcification. In mammography, the microcalcification point typically appears as a little brilliant spot.

Typically, it demonstrates that a number of microcalcifications are grouped together, much like the distribution of clusters. It is challenging to locate microcalcification spots with the naked eye since each microcalcification has high brightness, excellent grayscale consistency, a strong contrast to the breast tissue in the background, and few edge blurs. Doctors frequently overlook it as a result, which could result in a cancer misdiagnosis. A magnifying glass may occasionally be used by professionals to inspect the mammography. Although a significant amount of time and commitment is utilized, this method is not precise. Computers have been utilized by researchers in recent years to start diagnosing breast calcification.

Three classification techniques were used for the purpose of detecting calcifications in mammography: (1) In Ref. , the authors have presented a quantum noise model based on the image enhancement technique by taking the average of the grayscale, which could help us in detecting microcalcifiers. This method would increase the local contrast. To compare the results of the microcalcification spots that were discovered, enhancement techniques [12] are often used, such as histogram modeling, convolution filters, region-based enhancement techniques, morphological methods, and contrast stretching. (2) Another method was proposed [13] to filter high-frequency sounds and signals utilizing various images, which was based on the multiscale decomposition method. The distribution of microcalcification was then extracted using neural network classification after using wavelet reduction to suppress sounds. By using a conjugate gradient approach to change the weight, a cost function [14] was introduced. After the wavelet decomposition, they reduced the cost function and found microcalcifications in the initial layers of high-frequency subbands. (3) The SVM algorithm was employed by Lee et al. [15] based on the machine learning technique.

Because of the variation in how they form, it is exceedingly hard to discern between the infected and uninfected microcalcification clusters. A picture region could refer to as either malignant or benign, and a choice for a detection result may either be true or false. Therefore, one of four possible categories will be chosen for a detection result: true positives, true negatives, false positives, and false negatives. FN and FP stand for different types of errors. A false positive error happens when a normal region was mistakenly recognized as abnormal, while an FN error indicates that an actual abnormality was not found. An accurate assessment of an existing anomaly is a TP decision, while a TN decision indicates that it was incorrectly labeled. The following indices are used to gauge and document a diagnostic system's performance: true positive rate and true negative rate. True positive and true negative rates are interchangeable with words sensitivity and specificity, that is, true positive rate = TP/(TP + FN) and true negative rate = TN/(TN + FP), respectively. The values of these performance parameters in a clinical diagnosis will rely on the specific threshold used to divide the data into various choices for a fixed discrimination capability. If the confidence level is modified, both sensitivity and specificity will vary. Various extraction techniques are discussed next.

1.1 Feature extraction

This is a significant phase in the classification of microcalcifications. The following qualities [4] should be present in the extracted features: (1) feature should be reliable if its values are consistent across all patterns in the same class; independent if its values are as uncorrelated as possible; and discriminatory if its values are considerably different across patterns in various classes.

Additionally, the number of features should be kept to a reasonable minimum because a CAD system's complexity rises quickly as the number of features is used. The techniques used for retrieving the image and the attributes of the image determine how the features are distinguished. Therefore, the features can be categorized as follows: features such as foreground, compactness, elongation, eccentricity, area, perimeter, thickness, line, orientation, contrast, direction, backdrop, and distance are examples of features that can be directly retrieved from mammograms. They are simple to extract and come from radiologists' experience (2) features extracted from the cooccurrence vector for spatial gray values; (3) features extracted from the GLRL and GLD matrices; (4) features extracted from the wavelet-based subimages; (5) features extracted from the fractal structure of an image; and (6) microcalcification distribution, cluster size, and quantity are characteristics used to describe an area. We will provide a summary of these various feature extraction techniques in the paragraphs that follow.

1.2 Shape feature extraction

Numerous researchers directly define individual microcalcification using form data taken from mammograms; 15 features are chosen in Ref. [27]. These include the standard deviation and max size, area of the cluster, count of the calcifications in each cluster, the average and the maximum compactness value, size of microcalcifications, the largest calcification in a cluster determined by its size, the SD of the size of calcifications in each cluster, the count of calcified values that are present in each pixel defined by its size and its area, the grayscale, scattering, the standard deviation of the mean of the calcifications in each cluster, and the average and largest value of SD of gray scales in each cluster.

Authors in Ref. [6] classified the microcalcification clusters using a total of 16 features, which can be divided into two categories: distribution features, which are based on how each feature is distributed in a cluster, and secondly, the features corresponding to the position and shape of the cluster, which describe where the clusters are located on a mammogram.

Cluster can be defined using the standard deviation, mean, orientation, compactness, eccentricity, and area. Cluster shape is defined by the eccentricity, area, and orientation of the cluster. The distance from the breast edge to the cluster and the pectoral muscle to the cluster are two cluster location features that are finally examined. Scattering density of blobs within the region of interest, circularity, elongation, density, irregularity, and defective degree of microcalcifications were the features extracted in Ref. [7] using shape curvature and orientation details.

Microcalcifications are classified in Ref. [8] using shape parameters such as the ratio of circularity and suspicious areas between two regions of interest in various ipsilateral imaging views.

Based on a study of the shape, Foggia et al. [9] derive four features: local contrast, gradient strength of the border, roughness, and compactness; 22 features were extracted by Lo et al. The standard deviation and mean of the various features were calcification area and distance, eccentricity, orientation, Fourier descriptor, moment, the log of the number of calcifications, and its count.

1.3 Statistical feature extraction

A comparison of how well the local area dependence method and other traditional statistical texture analysis techniques, such as the spatial graylevel dependence method (SGLDM), graylevel difference method (GLDM), and graylevel run length method (GLRLM), perform at spotting clustered microcalcifications in digital mammograms was shown in Ref. [4]. These methods extract textural elements that are then utilized to categorize ROIs into positive ROIs that contain clustered microcalcifications and negative ROIs that contain normal tissues. These features include second angular moment, entropy, contrast, inverse difference moment, and mean.

Features used in Ref. [11] using the SRDM were grid and diagonal, horizontal, and vertical weighted sum.

The usage of texture analysis in Ref. [12] enables the assessment of textural changes brought on by a developing cancer in breast tissue. They take the following characteristics out of the SGLD matrix: second angular moment denoted by energy, correlation, inertia, inverse difference moment, entropy, average, sum and variance difference, and measure of correlation.

In [5] SGLDM is used where 10 features—namely variance, contrast, correlation, inverse difference moment, entropy, second angular momentum, entropy and variance difference, and two information correlation measures—are computed to quantify texture.

In [13], the authors have used a statistical procedure that combines covariance, contrast, and heterogeneity to assess the texture parameters for different distances between pixels, from the medical images rather than using the cooccurrence matrix method.

The primary benefit of this approach is that it has a calculation cost that is only dependent on the image size being processed and has no dependence on the count of gray levels.

Texture is defined by using the moments of the graylevel histogram in Ref. [14]. Each object's statistical parameters like mean, skewness, variance, and kurtosis are computed from the central moments. Additionally, they estimate the number of things discovered within a 2 cm radius.

In locations that have been discovered, a number of texture and spatial and morphological properties were extracted in Ref. [8]. The texture attributes comprise the region's mean gradient and the ratios of intensity and intensity dividing line between two sections in various image perspectives.

Textural characteristics were used for classification as listed in Ref. [9]. The region of interests average square intensity was the energy in the region of interest, its background average square intensity is given by the energy in the background. The ROIs intensity is the intensity obtained through standard deviation. The regularity of the spread of the gray values in the region of interest is measured by the entropy of the first-order histogram.

1.4 Multiscale texture features extraction—wavelet-based method

For multiresolution analysis and texture analysis, wavelet theory offers a potent paradigm. The wavelet transform might distinguish between little objects like microcalcifications and large objects like substantial background structures by making use of its multiresolution capacity.

By using a multiresolution analysis and the wavelet transform, the authors in Ref. [15] were able to identify signals with a size of less than 1 mm. The wavelet transform was used in Ref. [16] to decompose a picture using the Haar and Daubechies 4 wavelets, only keeping the largest coefficients at the first order of decomposition. To find calcifications, wavelet analysis was used in Ref. [17]. In wavelet processing, this is accomplished by simply eliminating the approximation coefficients with the lowest resolution and doing a reverse wavelet transform [10].

To enable binary detections of calcifications, the reverse wavelet transform output is then thresholded suitably.

The Daubechies 6, 10, and 12 wavelets are three distinct wavelets used in the study [27]. Their approach to feature extraction is focused on breaking down the photos and figuring out the energy and entropy of each subband [28]. Additionally, they use the multiwavelet transformation, which employs a number of scaling functions and mother wavelets, using GHM, CL, and SA4 multiwavelets. They gather the first- and second-order statistical data as well as the measurements of the run lengths of the gray levels to calculate their wavelet transformation coefficients.

To improve the microcalcifications, the wavelet transformation using the lowest asymmetric Daubechies' with dimensions 20, 8, and 12 was performed in Ref. [18].

1.5 Cluster features extraction

All wavelet architectures according to Ref. [18] employ the Daubechies 4-TAP wavelet filter from every signal and to determine their wavelet transformation coefficients, the first- and second-order statistical data along with the measurements of the run lengths of the gray values are being collected. Wavelet transformation utilizing the least asymmetric Daubechies' wavelets was performed in Ref. [19] with dimensions 20, 8, and 12 to increase the microcalcifications. Finally, they look into two cluster position characteristics: the distance a cluster is from the pectoral muscle and the distance it is from the breast border.

Three different types of features are extracted in Ref. [9]: the shape of the clusters, the dispersion of the microcalcifications inside the cluster, and the microcalcification presence within asymmetrical and irregular shapes. The more asymmetrical and irregular the shapes of the microcalcifications are, the higher the likelihood that a cluster is malignant.

To differentiate between the microcalcification clusters, a set of 12 features was used in Ref. [20], including the standard deviation, maximum number of areas, the count of calcifications, the average numbers, and the range of microcalcification areas within the cluster.

1.6 Classifiers

In the deployment of computer-aided mammography diagnostics, classifiers are crucial. Classifiers use these criteria, or a subset of them, to categorize microcalcifications as benign or malignant. Below is a summary of different classifier types.

1.6.1 Neural networks

The nonparametric pattern recognition systems known as artificial neural networks (ANNs) are capable of learning from instances and extracting general principles. They are especially helpful in situations when there are ambiguous choice rules and no formal understanding of the probability density functions guiding sample distributions. The localized functions, distributed representation, and asymmetric processing are the main traits of ANN. The ANN may be quickly changed to incorporate new features as needed.

Kim and Park used textural properties as the input to the artificial neural network [4]. A comparison of the accuracy of classification for the GLRLM, GLDM, SRDM, and SGLDM was conducted. The study's conclusion is that the SRDM outperforms the other traditional approaches in terms of classification accuracy and computing complexity. The A_z values of the SGLDM (10 hidden neurons), GLRLM (7 hidden neurons), GLDM (which contained seven neurons in the hidden layer), and SRDM (which contained five neurons in the hidden layer) are, respectively, 88%, 74%, 68%, and 93%. When training two distinct three-layer ANNs which contained the patient clusters, an A_z value of 64% was achieved [20]. To reduce the amount of clusters that were false, a second BPANN classifier was employed after the first to identify whether there was calcification in each region of interest. The best A_z is 69%, and the best partial A_z, denoted by the range of sensitiveness denoted through average specificity was from 90% to 100%, is only 21%. As a result, they produce a very poor outcome. Similar to this, they have achieved a sensitivity of greater than 90% with a lower false positive rate of 67% per region of interest [11] using four hidden neurons and a three-layer BPANN classifier.

In [12], the authors trained a BPANN classifier to distinguish between the good and bad calcified clusters and evaluate it using the leave-one-case-out technique. Analysis of the ANN's performance reveals that when 6 out of a total of 22 texture features were considered, the classifier can obtain receiver operating characteristic (ROC) curve of 88%

A variety of feature extraction strategies were used in Ref. [21] to extract a mixture of three features. Standard deviation, entropy, and pixel count are combined and found to be the optimum method for differentiating between benign patterns from a cancerous one.

Data is then classified using a feedforward BPANN. There is only one hidden layer in the BPANN architecture. The Nijmegen mammographic database has the classification in place, and the TP is 88.9%. Form Cognitron is an ANN developed in Ref. [7] that uses universal feature planes for shape identification (SCognitron). The Nijmegen mammography database is used to evaluate the system, and the experimental findings indicate that the system's specificity and sensitivity can both approach 74.1% and 86.1%, respectively. The classification approach implemented in Ref. [22] utilized a Kohonen layer followed by a multilayer feedforward network. The neural network produced 91% TP when one hidden layer was used, compared to 82% TP when no hidden layer was used.

In [14], the authors have used two distinct topologies with 5 and 10 hidden nodes, to test the performance of a multilayer perceptron. In comparison to the second system, which succeeds with an average of 27% false positives and a sensitivity of 94.7% in each image, a sensitivity of 84.2% was obtained on a system with five hidden neurons with an average of 26% false positives in each image.

Moreover, 191 cases of breast cancer that are challenging to diagnose were classified [20] using a convex-set-based neurofuzzy method. The backpropagation and neurofuzzy pattern classifier techniques are assessed using a 20-dimensional collection of challenging to diagnose mammographic microcalcifications. The comparison results demonstrate that the convex-set-based method outperformed backpropagation in all tests where knowledge inference was concerned, making it appropriate for implementation in practical systems. When 90% of the features are used, the maximum mean ROC area is 88.37%.

A three-layer supervised ANN classifier was created in which the neural network is trained using the backpropagation approach in conjunction with Kalman filtering. The BP algorithm's adaptive convergence rate is achieved, thanks to the Kalman filter gain added to the algorithm. The Kalman filter-based backpropagation method demonstrates quick convergence using the delta learning rule. With three FP discoveries, the method obtains 90% sensitivity.

In [23], the authors achieved an accuracy of >97% true positives with the false positive rate of three clusters per image using a feed-forward, three-layered, back propagation ANN to categorize the identified clusters of microcalcifications using the standard deviation and mean as inputs. They make use of the Nijmegen database for mammography. Utilizing two separate ANNs, microcalcification analysis was carried out in Ref. [24]. The first one identifies regions that contain segmented images using a feed-forward neural network. The second ANN, which categorizes ROIs using the principle component technique, is fed with this ANN's result as its input. The method achieves a 92% sensitivity and 92% specificity in the classification of clusters.

The hierarchical pyramid neural network (HPNN) is implemented in Ref. [25] and uses multilayer perceptrons that contain four to eight hidden units present in one hidden layer. Each network in the HPNN hierarchy also receives input from the integrated feature hierarchy, and the hidden unit receives input from networks that are below it. Depending on the design, networks are trained either fine to coarse or coarse to fine. Before teaching the following network in the hierarchy, the networks below in the hierarchy are trained initially until convergence, to a fine HPNN, at which point all of its variables are kept constant. Because the positions of the little items are clearly defined as the resolution is reduced, coarse-to-fine training is possible. Extended objects for the fine-to-coarse HPNN lack a clear placement at high resolution. While the FP findings are just 21%, the ROC curve's area is 94%. An expert system is used to categorize a cluster of microcalcifications [9]. The cluster expert classifies the cluster as a whole, while the mC-Expert classifies each microcalcification. The final output is obtained by combining their results and applying an appropriate judgment procedure based on an evaluation of the correctness of each classification. BPANN is applied to both classification systems. The ANN has a three-layer architecture with a number of input neurons, two output neurons that denote the malignant and benign classes, and a hidden layer that contains 25 neurons. The model yields 89% sensitivity and 70% FP results as measured by the ROC curve.

1.6.2 k-nearest neighbor classifiers

A classifier called k nearest neighbor (KNN) separates new instances depending on how closely they resemble existing instances. This method calculates the separations between an unknown instance and each known instance, choosing the k nearest samples as the classification foundation. Among the k-nearest samples, the class with the most samples receives the unidentified pattern. The decision criterion used in Ref. [5] uses a KNN classifier as a measure of malignancy.

The multiwavelet approach is used to extract the input features, and for CL multiwavelet, the area under the ROC curve (AUC) reaches 89%. With this method, TP findings are 85% and FP findings are 10%.

A classification method with two classification steps is used in Ref. [6]. The clusters are classified using the KNN approach using the first-step classifier. This categorization stage results in the assignment of a likelihood to each cluster. The classifier averages the classification results of the respective clusters to provide the final patient-based classification result. In A_z, 83% is generated by the system.

1.6.3 Nearest neighbor classifiers based on Euclidean distance

To verify the classification, Ferreira and Borges [16] create a closest neighbor classifier employing the Euclidean distance between the corresponding wavelet coefficients as a metric. The calcifications were classified using the Euclidean distance calculation, and the TP was close to 80% [17].

1.6.4 Support vector machines

SVMs are learning tools used in regression prediction and pattern-matching tasks. They originate from difficulties in SLT (statistical learning theory), which provide helpful limitations on the ability of machines to generalize for learning tasks. In the input space, the SVM algorithm creates a separating hypersurface. It functions as follows: (1) it uses some nonlinear mapping selected beforehand (Kernel) to transform the input space into a high dimensional feature space; and (2) it creates the maximal margin hyperplane in this feature space. Over the past 20 years, numerous studies have looked at the automated classification of breast microcalcifications. Although significant progress has been made in improving these techniques, there are still problems to be solved and areas for further work. The first is the area of developing better feature selection and detection methods to more effectively distinguish between cancerous and noncancerous clusters, and the second is the incorporation of classifiers to lower the values of false Negatives and False Positives. Various characteristics, perhaps combined with clinical information, should be employed to decrease false positives. It goes without saying that an algorithm takes longer to run the more sophisticated it is. Therefore, it is crucial that researchers choose algorithms that strike a compromise between accuracy and computing complexity.

Utilizing a common database for the algorithms' evaluation is another crucial issue. Table 8.1 lists the various classifiers which could detect clustered microcalcifications resulting in better detection and lesser FPs. Additionally, the classification

Table 8.1 Summary of classifiers for microcalcification classification.

Classification method used.	Dataset	Database	Feature extraction	Performance
Decision tree [13]	260 images	Human development index 3000 database	Statistical features	Specificity: 85.4% Sensitivity: 94.6%
SVM (support vector machine) [15]	45 images	Nijmegen database	Scale and location based features	False positive: 60% Sensitivity: 95%
k-nearest neighbor classifier [6]	168 mammograms	Nijmegen database	Cluster shape	A_z: 83%
[5]	42 mammograms		Multi wavelet and wavelet features	True positive: 85% False positive: 10% A_z: 89%
Artificial neural networks [4]	120 mammograms, 172 ROIs were selected from the above database	NA	SGLDM GLRLM SRDM GLDM	A_z: 68% A_z: 88% A_z: 74% A_z: 93%
[13]	292 cases	Digital database for screening mammography	Morphological features	
[20]	272 images	Source N/A	Features of the cluster	A_z: 64%
[12]	86 images of mammograms		SGLD	A_z: 88%
[11]	120 images of mammograms, 140 regions of interest		SRDM	False positive: 67% for each ROI Sensitivity: 90%
[7]	40 mammograms, 102 ROIs	Mammograms obtained from University of Netherlands (Nijmegen)	Features extracted directly	Specificity: 74.1% Sensitivity: 86.1%

Reference	Data	Database	Features	Results
[21]	18 microcalcification regions	Images of digital mammograms retrieved from University of Nijmegen (The Netherlands)	Number of pixels Standard deviation and entropy.	True positive: 88.9%
[21]	191 images which were difficult to diagnose	NA	Wavelet features	A_z: 89.67%
[22]	30 mammograms, 50 ROIs	NA	Features related to morphology	True positive: 91%
[25]	32 mammograms, 138 ROIs	NA	Energy of the orientation, maximum energy, pixel location	False positives: 21% A_z: 94%
[14]	25 mammograms, database by the	Images obtained from MIAS (Mammographic image analysis society)	Textural features	False positives: 26% with five hidden neurons Sensitivity: 84.2% False positives: 27% with 10 hidden neurons Sensitivity: 94.7%
[8]	40 mammograms	NA	Shape features	Sensitivity: 90%
Nearest neighbor [26]	84 images	Database of ISSSTEP hospital	Features extracted directly	True positive: 93.2%
[17]	25 mammograms	NA	Wavelet features extracted based on Euclidean distance	True positive: 80%

rates are significantly influenced by the feature vector that is employed. Higher A_z values in the classification process are shown by using statistical feature extraction methods.

2. Proposed method

2.1 Dataset

Datasets used for this work were from the Curated Breast Imaging Subset of DDSM and Breast Cancer Digital Repository, which contains data of 1734 patients. We applied the same preprocessing techniques outlined in Ref. [20] for clipping, normalizing, and enhancing the pictures. Using the dataset's given ground truth, we first cut fixed-sized zones of interest and scaled them to 227×227 as we were using AlexNet model. Then, we employed global contrast normalization, where each image was normalized by dividing by the standard deviation of its parts and subtracting the mean. To create batches of tensor data, augmentation techniques were applied to it like changing the height and width by a fraction of 0.28, by rotating images randomly with rotations varying from 0 to 50°, by randomly transforming the images, or by flipping them.

CBIS-DDSM: The digital database for screening mammography (DDSM) [33] has been upgraded and modernized as curated breast imaging subset of DDSM (CBIS-DDSM) [21]. After being decompressed, the images were changed to DICOM format. Mammograms are provided alongside updated ROI segmentation and pathologic diagnoses, which was not the case for DDSM [21], where annotations for the anomalies were not very accurate. The database includes annotations at the pixel level for the ROIs, such as pathology classification (benign or malignant), finding calcified masses for classification, and ACR class density classification. By merging the training and test sets for both classes suggested in Ref. [32], we employed 3684 ROI lesions derived from the mammograms of 1566 patients.

BCDR: 344 patients with their biopsy proved data are included in the BCDR-F03. We used all 736 ROIs in the dataset, each of which had accurate labels describing its density, disease, and conclusions.

2.2 Feature selection

Selection of features follows a specific process. Breast cancer's primary warning indicators, and often even the only signs, are not very sensitive or effective. In this study, we seek to identify the ideal feature combination for the quick and precise categorization of breast tumors.

Using a procedure akin to the coordinate ascent, our technique chooses a feature to be added to the feature collection. The most discriminating feature is the only one initially included in the feature set. Then, features are picked one by one since the performance of classification can be greatly enhanced by including a feature. This cycle is repeated until adding a feature either has no effect on performance or even has the opposite effect.

All of the features were chosen to represent different dimensions of microcalcifications, including one-dimensional features like the diameter, 2D morphological features like area, fractal features like volume ratio, circularity proportion, density, solidity, spiculation, and sandy microcalcification followed by statistics features like linear microscopic features, linear microscopic features, and microcalcification number. We chose 38 features using the method of texturing features in MATLAB to broaden the variety of features and improve experimental conditions. GLRLM and GLCM are two widely used techniques for estimating texture features. By counting the instances in which the orientation of adjacent pixels is the same, the GLCM was determined. These characteristics can accurately describe the scattering of calcification.

2.3 Transfer learning

A machine learning technique called transfer learning allows a network that has already been trained to learn additional features for classifying fresh images. Transfer learning allows for quick and simple fine-tuning compared to network initialization with randomly initialized weights.

2.3.1 Using AlexNet for transfer learning

The AlexNet architecture is a 25-layer deep CNN made up of 8-layer deep CNN with three fully connected layers and five convolutional layers. This divides items into 1000 different categories after being trained on the imageNet dataset. The input is an image of size 227×227. By replacing the completely connected CNN layer with differing values of learning rate factors for weights and bias, we were able to fine-tune AlexNet with the help of the new images.

To convert the ImageNet CNN trained for a 1000 class assignment to a mammography task which is of two classes, two fully connected layers with 100 and 2 nodes each were added to the network's end.

A similar strategy was proposed in Ref. [29] by reducing the number of nodes in the last completely linked layer by adding extra layers. Stages 1 and 2 of transfer learning utilized the same CNN structure. On a Tesla K40 GPU, the multinomial logistic regression objective was maximized by employing minibatch stochastic gradient descent optimization with a batch size of 128.

The ROC was employed as a performance metric, and the output value of each sample from the CNN's softmax layer was used as a decision variable for ROC analysis [30]. To maximize efficiency, the trapezoidal rule was utilized to estimate the AUC during iterations. In Ref. [31] the authors have told that a momentum of 0.9 and an initial learning rate of 0.001 would result in a stable plateau for the task under study after 200 epochs. Therefore, for the current investigation, we used these parameters (Table 8.2).

Table 8.2 Tabular representation of various layers of CNN.

Layer	Count of the neurons	Size of filter	Number of nodes
L1	61,504	11 by 11	64
L2	43,200	5 by 5	192
L3	18,816	5 by 5	384
L4	12,544	3 by 3	256
L5	12,544	3 by 3	256
Fl1			2048
Fl2			2048
Fl3			1000
Fl4			100
Fl5			2

Here L1 indicates that the L1 layer was frozen during pretraining. L2-Fl3 indicates that layers from L2 to Fl3 were frozen during the transfer learning process.

Transfer learning—Initial phase:

We first carried out experiments with the dataset mentioned earlier to determine the appropriate transfer learning technique for improving the ImageNet-trained CNN's performance at identifying breast masses. Six transfer networks in total were assessed. From the six transfer layers, the best one was chosen and the average results of 10 experiments to aid in the work of mammography transfer learning. To choose the next transfer layer for a mammography-trained CNN, which is obtained from a CNN trained on ImageNet, the mammography images were divided into training and test. Following the selection, the whole mammography dataset was utilized to examine the effects of training sample size on stage 1 transfer learning, and the DBT training set was used as a validation set to assess the inference capacity for the mammography-trained CNNs.

3. Results and discussion

Calcified mammographic images of the breast were taken and their numeric values were computed. Information on microcalcifications and worrisome breast masses was obtained. To categorize features and acquire a thorough characterization of the microcalcifications and masses, statistical and textural features were both applied. Eleven quantitative measures altogether were kept on file for every patient.

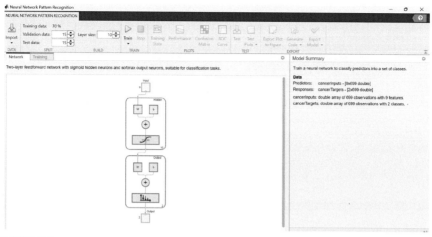

FIGURE 8.1

Two layered feedforward network showing the predictors and responses.

Initially, the network was trained using neural network with 10 layers. The predictors and responses are shown in Fig. 8.1.

In total, 70% of the data were used for training and 15% each for testing and validation. The cross entropy error for training, validation, and testing was observed to be 0.050, 0.057, and 0.0387, respectively. The gradient was observed to be 0.019 during the 11th epoch as shown in Fig. 8.2.

Fig. 8.3 is used to represent the training state plot. Figs. 8.4 and 8.5 show the confusion matrix and the ROC curve for training, testing, validation, and the combined one on a neural network.

In the next step, we applied various machine learning algorithms as listed in Table 8.3. From the results shown in Table 8.3, it can be observed that SVM produced the best result with an accuracy of 98.1% but CNN gave an accuracy of 98.9%.

The confusion matrix is shown in Figs. 8.6 and 8.7, respectively.

Tables 8.3 and 8.4 show the classification results of each classifier. The former uses only masses whereas the latter uses microcalcification and mass features for classifying. The proposed CNN showed the best results in performance. The best measurements were highlighted in bold. Scatter plot for SVM showing the benign and malignant tumors based on microcalcification is shown below. The red dots denote the malignant tumors and the blue dot represents the benign tumors.

Image segmentation in the first case resulted in 15 features. For the SVM, KNN, LDA, decision tree, Naïve Bayes, and CNN, the overall accuracies were 98.1%, 96.3%, 95.6%, 93.2%, 94.2%, and 98.9%, respectively. In the second case, image

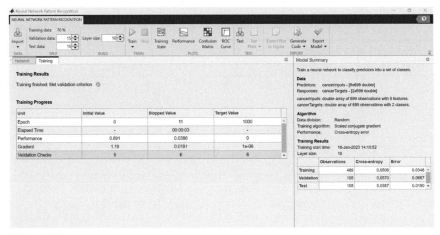

FIGURE 8.2

Statistics of training data indicating the epoch, gradient, and the validation checks.

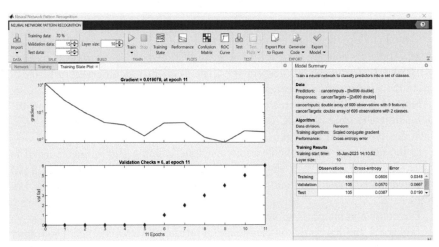

FIGURE 8.3

Training state plot.

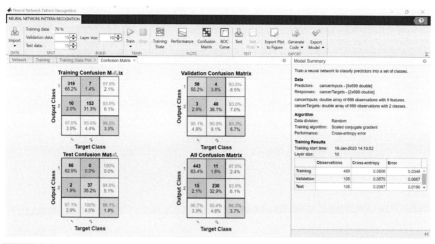

FIGURE 8.4

Confusion matrix of neural network.

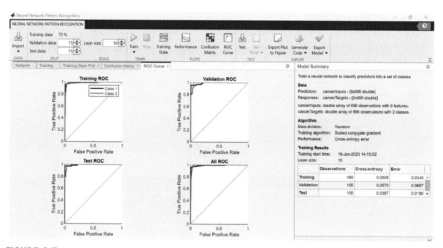

FIGURE 8.5

ROC for neural network.

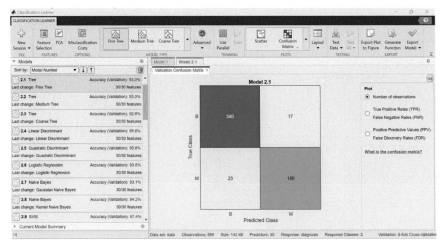

FIGURE 8.6

Confusion matrix for fine tree.

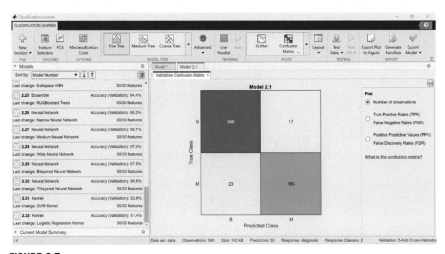

FIGURE 8.7

Confusion matrix for validation data of medium tree.

Table 8.3 Classification results of each classifier through mass features.

Classifier	Classification type	Test data			
		Accuracy	Sensitivity	Specificity	AUC
Decision tree	Fine tree	93.0	0.89	0.32	0.9288
	Medium tree	93.0	0.89	0.19	0.9288
	Coarse tree	92.6	0.88	0.21	0.9136
Support vector machines (SVM)	Linear SVM	97.4	0.98	0.12	0.9636
	Cubic SVM	**98.1**	**1.00**	**0.18**	**0.9799**
	SVM (medium Gaussian)	97.9	0.98	0.27	0.9712
	SVM (fine Gaussian)	80.0	0.83	0.38	0.7986
	SVM (coarse Gaussian)	95.6	0.96	0.29	0.9433
k-Nearest neighbor (KNN)	Fine KNN	95.1	0.96	0.78	0.9401
	Medium KNN	95.8	0.96	0.76	0.9489
	Coarse KNN	93.1	0.93	0.65	0.9213
	Cosine KNN	96.3	0.97	**0.90**	0.9589
	Cubic KNN	95.8	0.97	0.33	0.9476
Ensemble classifiers	Boosted trees	94.4	0.94	0.45	0.9412
Neural network	Narrow NN	96.3	0.97	0.33	0.9513
	Neural network (medium)	96.7	0.97	0.36	0.9516
	Neural network(wide)	97.0	0.98	0.28	0.9617
	Neural network (trilayered)	96.5	0.97	0.46	0.9701
	Bilayered neural network	97.5	0.98	0.45	0.9654
Linear discriminant	Linear discriminant	95.6	0.96	0.52	0.9486
Logistic regression	Quadratic discriminant	93.5	0.94	0.57	0.9219
Naïve Bayes	Gaussian Naïve Bayes	93.1	0.94	0.18	0.9209
	Kernel Naïve Bayes	94.2	0.95	0.29	0.9301

segmentation produced 26 features based solely on breasts. Masses. Figs. 8.8 and 8.9 show the scatter plot and ROC for SVM.

For SVM, KNN, LDA, decision tree, and Naïve Bayes, respectively, the AUC were 0.97, 0.94, 0.94, 0.92, and 0.93, respectively. It was observed that the model

Table 8.4 Performance of different models through microcalcifications and mass features (10 features were considered).

Classifier	Classification type	Test data			
		Accuracy	Specificity	Sensitivity	AUC
Decision tree	Fine tree	91.0	0.90	0.23	0.9188
	Medium tree	94.0	0.89	0.17	0.9288
	Coarse tree	92.3	0.88	0.21	0.9136
Support vector machines (SVM)	Linear SVM	96.4	0.98	0.15	0.9636
	Cubic SVM	**98.4**	**1.00**	**0.12**	**0.9799**
	SVM (fine Gaussian)	80.0	0.82	0.38	0.7986
	SVM (medium Gaussian)	97.2	0.98	0.27	0.9712
	SVM (coarse Gaussian)	95.1	0.96	0.29	0.9423
k-Nearest neighbor (KNN)	Fine KNN	95.1	0.96	0.78	0.9421
	Medium KNN	95.8	0.96	0.76	0.9439
	Coarse KNN	93.4	0.93	0.65	0.9273
	Cosine KNN	96.7	0.98	**0.90**	0.9569
	Cubic KNN	94.8	0.98	0.33	0.9476
Ensemble classifiers	Boosted trees	93.4	0.93	0.45	0.9412
Neural network	Narrow NN	95.3	0.97	0.33	0.9513
	Medium NN	95.3	0.98	0.36	0.9516
	Wide NN	96.8	0.97	0.28	0.9617
	Bilayered NN	97.5	0.98	0.45	0.9654
	Trilayered NN	96.5	0.99	0.46	0.9701
Linear discriminant	Linear discriminant	95.6	0.96	0.52	0.9486
Logistic regression	Quadratic discriminant	93.5	0.94	0.57	0.9219
Naïve Bayes	Gaussian Naïve Bayes	93.1	0.91	0.18	0.9209
	Kernel Naïve Bayes	93.2	0.92	0.29	0.9301
CNN	Using transfer learning	**98.9**	0.99	0.15	0.9812

performed better under CNN when compared to other algorithms. Despite this discovery, the model's sensitivity was close to 100%, suggesting that all of the patients who tested positive had breast masses. As a result, this approach might make it easier

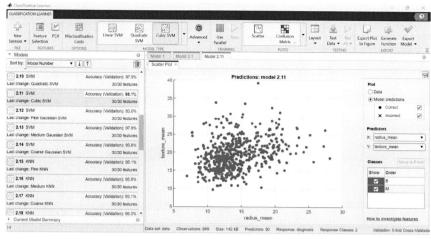

FIGURE 8.8

Scatter plot showing the benign and malignant tumors.

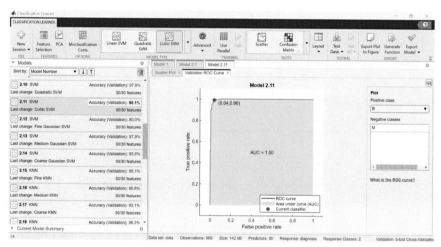

FIGURE 8.9

ROC for SVM.

to make a diagnosis in benign situations, but it might not be a reliable diagnostic tool in actual clinical settings.

Fig. 8.10 shows a parallel coordinate plot. A parallel coordinate plot maps each tuple in the input table as a line. A point in the line denotes each attribute of a row. The parallel coordinate plots resemble line charts as a result, but the data are converted into a plot in a very different method. This determines how decisions are

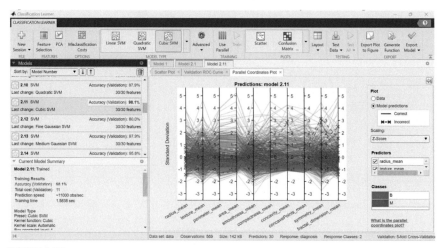

FIGURE 8.10

Parallel coordinate plot for SVM showing all features.

being taken inside the network. The plot displays the standard deviation with respect to the various features. Each feature is represented using a vertical line. The points are then represented as a series of lines connected along the respective axis.

It can be observed from the figure that benign tumors have smaller radius mean and perimeter mean but wider fractal dimensions. The objective of using a parallel coordinate plot is to study the various features with respect to certain statistical parameters.

4. Conclusion

The current study has covered medical image processing and computer-aided calcification diagnosis. More knowledge has been gained about the field of medical image processing, and the brightness image characteristics of breast images have been discovered. SVM initially performed better than other machine learning algorithms in its category, with an accuracy of 98.2%, but when CNN was used utilizing transfer learning, its performance was far higher. It is envisaged that in the future, demographic characteristics such as weight, age, and height will also be taken into account when identifying the characteristics of breast calcification. AI systems have made it easier for doctors to make correct diagnoses. In conclusion, deep learning is the most efficient and precise approach to discover if we have a lot of datasets.

References

[1] American Cancer Society: Breast Cancer Facts and Figures. http://www.cancer.org/.

[2] Breast cancers. In: Health Statistics-Atlas on Mortality in the European Union. Eurostat, http://epp.eurostat.cec.eu.int/cache/ITY_OFFPUB/ATLAS/EN/ATLAS-EN.HTM.

[3] Woods K, Sallam MY, Bowyer KW. Evaluating detection algorithms. In: Image-processing techniques for tumor detection. CRC Press; 2002. p. 28–54.

[4] Kim JK, Park HW. Statistical textural features for detection of microcalcifications in digitized mammograms. IEEE Transactions on Medical Imaging 1999;18(3):231–8.

[5] Rebecca L, Kimberly D, Jemal A. Cancer statistics. CA: A Cancer Journal for Clinicians 2019;69(1):7–34.

[6] Veldkamp WJH, Karssemeijer N, Hendriks JHCL. Experiments with radiologists and a fully automated method for characterization of microcalcification clusters. International Congress Series 2001;1230:586–92.

[7] Lee SK, Chung PC, Chang CI, Lo CS, Lee T, Hsu GC, et al. Classification of clustered microcalcifications using a shape cognitron neural network. Neural Networks 2003; 16(1):121–32.

[8] Sun X, Qian W, Song D. Ipsilateral-mammogram computeraided detection of breast cancer. Computerized Medical Imaging and Graphics 2004;28:151–8.

[9] Foggia P, Sansone C, Tortorella F, Vento M. Automatic classification of clustered micro-calcifications by a multiple classifier systems. International Workshop of Multiple Classifier Systems 2001:208–17.

[10] Cai S, Liu PZ, Luo YM, Du YZ, Tang JN. Breast microcalcification detection algorithm based on contourlet and ASVM. Algorithms 2019;12(7):135.

[11] Kim JK, Park JM, Song KS, Park HW. Detection of clustered microcalcifications on mammograms using surrounding region dependence method and artificial neural network. VLSIVideo 1998;18:251–62.

[12] Chan HP, Sahiner B, Petrick N, Helvie MA, Lam KL, Adler DD, et al. Computerized classification of malignant and benign microcalcifications on mammograms: texture analysis using an artificial neural network. Physics in Medical and Biology 1997;42: 549–67.

[13] Kuo WJ, Chang RF, Moon WK, Lee CC, Chen DR. Computer-aided diagnosis of breast tumors with different US system, vol. 9. Academic Radiology; 2002. p. 793–9.

[14] Nishikawa RM, Giger ML, Vyborny CJ, Schmidt RA. Computer-aided detection of clustered microcalcifications on digital mammograms. Medical, & Biological Engineering & Computing 1995;33(2):174–8.

[15] Bazzani A, Bevilacqua A, Bollini D, Campanini R, Lanconelli N, Riccardi A, et al. Automatic detection of clustered microcalcifications using a combined method and an SVM classifier. In: ESANN'2000 proceedings - European symposium on artificial neural networks, Belgium; 2000. p. 195–200.

[16] Ferreira CBR, Borges DL. Analysis of mammogram classification using a wavelet transform decomposition. Pattern Recognition Letters 2003;24(7):973–82.

[17] Sentelle S, Sentelle C, Sutton MA. Multiresolution-based segmentation of calcifications for the early detection of breast cancer. Real-Time Imaging 2002;8:237–52.

[18] Zhang L, Hao C, Wu Y, Zhu Y, Ren Y, Tong Z. Microcalcification and BMP-2 in breast cancer: correlation with clinicopathological features and outcomes. OncoTargets and Therapy 2023;12.

[19] Yoshida H, Doi K, Nishikawa RM, Muto K, Tsuda M. Application of the wavelet transform to automated detection of clustered microcalcifications in digital mammograms, vol. 17. Academic Reports of Tokyo Institute of Polytechnics; 1994. p. 24–37.

[20] Sorantin E, Schmidt F, Mayer H, Becker M, Szepesvari C, Graif E, et al. Computer aided diagnosis of clustered microcalcifications using artificial neural nets. Journal of Computing and Information Technology 2000;2:151–60.

[21] Duggento A, Aiello M, Cavaliere C, Cascella GL, Cascella D, Conte G, et al. An Ad Hoc random initialization deep neural network architecture for discriminating malignant breast cancer lesions in mammographic images. Contrast Media and Molecular Imaging 2019;9:5982834.

[22] Rubio Y, Montiel O, Sepúlveda R. Quantum inspired algorithm for microcalcification detection in mammograms. Information Sciences 2019;480:305–23.

[23] WHO. Fact Sheet World Health Organization.https://www.who.int/news-room/fact-sheets/detail/cancer.

[24] American College of Radiology, Mammography and Breast Imaging Resoruces. https://www.acr.org/ClinicalResources/Breast-Imaging-Resources (accessed on 15 July 2021).

[25] Ribli D, Horváth A, Unger Z, Pollner P, Csabai I. Detecting and classifying lesions in mammograms with deep learning. Scientific Reports 2018;8:1–7.

[26] Hadjidj I, Feroui A, Belgherbi A, Bessaid A. Microcalcifications segmentation from mammograms for breast cancer detection. International Journal of Biomedical Engineering and Technology 2019;29:1–16.

[27] Soltanian-Zadeh H, Rafiee-Rad F. Comparison of multiwavelet, wavelet, Haralick, and shape features for microcalcification classification in mammograms. Pattern Recognition October 1, 2004;37(10):1973–86.

[28] Sannasi Chakravarthy S, Rajaguru H. Detection and classification of microcalcification from digital mammograms with firefly algorithm, extreme learning machine and nonlinear regression models: a comparison. International Journal of Imaging Systems and Technology 2020;30:126–46.

[29] Mabrouk MS, Afify HM, Marzouk SY. Fully automated computer-aided diagnosis system for micro calcifications cancer based on improved mammographic image techniques. Ain Shams Engineering Journal 2019;10:517–27.

[30] Panachakel JT. Contourlet transform and iterative noise free filtering based bilayer filter for enhancing echocardiogram. In: Proceedings of the 2012 international conference on green technologies (ICGT), Trivandrum, India; 2012. p. 18–20.

[31] Menon SP. A survey on algorithms in deep learning. In: Industrial internet of things and cyber-physical systems: transforming the conventional to digital. IGI Global; 2020. p. 339–50.

[32] Khanmohammadi S, Adibeig N, Shanehbandy S. An improved overlapping k-means clustering method for medical applications. Expert Systems with Applications 2017; 67:12–8.

[33] Yoshida H, Zhang W, Cai W, Doi K, Nishikawa RM, Giger ML. Optimizing wavelet transform based on supervised learning for detection of microcalcifications in digital

mammograms. In: Proceedings of the international conference on image processing. Washington, DC, USA; 1995. p. 152–5.

Further reading

[1] Menon SP, Vaishaali K, Sathvik NG, Gollapalli SPA, Sadhwani SN, Punagin VA. Brain tumor diagnosis and classification based on AutoML and traditional analysis. In: IEEE global conference on computing, power and communication technologies (GlobConPT) . 20221-7. IEEE; September 2022.

[2] Menon SP. Deep learning for prediction of amyotrophic lateral sclerosis using stacked auto encoders. International Journal of Biomedical Data Mining 2020;1(2):119–34.

Scrutinization of mammogram images using deep learning

S.R. Reeja[1], Tulasi Thotakura[1], Ishfaq Yaseen[2]

[1]*VIT-AP University, Amaravati, Andhra Pradesh, India;* [2]*Prince Sattam Bin Abdul Aziz University, Al-Kharj, Saudi Arabia*

1. Introduction

Breast cancer is a frequent disease with a high fatality rate, particularly among women. As a result, the greatest strategy for combating this condition is early examination. For visualizing the breasts, a set of modalities are available, notably mammography, tomosynthesis, molecular breast imaging, and magnetic resonance imaging. Mammography is the most prevalent analytical procedure among them because of its ubiquity and inexpensive cost. Mammography pathognomonic is the most often used accurate technology for initial identification of carcinoma. Deep learning, pertinently deep convolutional neural networks (CNNs), which may be easily implemented with cutting-edge performance, plays an essential role in improving outcomes in the domain of biomedical engineering. Many people still utilize mammography as their primary approach. It examines the human breast with low-dose X-rays. On mammography, malignant tumors and calcium accumulation appear prominent. Mammography now serves as the global standard strategy to mitigate breast cancer mostly in the beginning phases, prior tumors become clinically conspicuous.

Breast malignancies are broadly categorized into two types: benign and malignant. Benign lesions are noninvasive (noncancerous), whereas malignant lesions are pathogenic (cancerous). The infectious stage of a condition determines whether it is benign or malignant. Tumors serve as the most predominant clinical criterion of carcinomas, which are present as grayish to white in pixel luminosity readings on mammograms. Throughout the breast region, breast lumps differ in intensity, dispersion, structure (pedunculated, anomalous, spherical, elliptical), and boundaries (spiculated, inadequately, constricted), enhancing the hazard of misinterpretation are shown in Fig. 9.1.

Various levels, comprising pooling layers and convolutional and fully linked layers, are characteristic of the CNN design for mammography [1]. The convolutional layers utilize filtration to acquire distinctive features from mammography images, whereas the pooling layers down-sample the feature mappings to substantially minimize computation and preserve merely the most significant traits. The

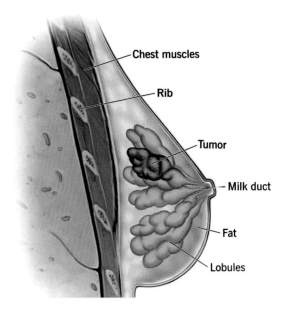

FIGURE 9.1

This diagram depicts the anatomy of a woman's breast. The regions within the lobes are where epithelial cancers or lesions develop.

completely linked layers incorporate the preceding layers' characteristics and create the ultimate categorization or detection decision. CNN is supervised on a tremendous database of mammography pictures designated as carcinogenic or noncarcinogenic in breast cancer diagnosis. The model begins to identify cancer-related structures and characteristics in visuals. CNN is determined to assess abnormal sections in mammography pictures that require additional inspection in lesion segmentation.

The goal of this research is to look at approaches to give a deep learning (DL) method that can adequately distinguish malignant neoplasm so that premature medications might indeed be prevented because of incorrect negatives and inefficient treatments can be eliminated owing to wrongful convictions. A database of 1312 images comprises both sorts of photographs, that is, the labels benign and malignant have been employed. The models VGG16, DenseNet201, AlexNet, and ResNet50 are used. Without any transfer learning, we attain the best accuracy for our basic model [2]. Augmentation is a technique for rescaling data and reducing overfitting. When evaluated on a widely used accessible dataset, the presented approach yields good results. Pretrained techniques are also helpful in healthcare image processing, as they execute exactly with a low loss rate. These methodologies are developed on vast datasets of mammography pictures and can evolve to distinguish cancer-related structures and attributes.

Numerous DL approaches, notably CNN, have subsequently achieved spectacular outcomes in a variety of fields, involving malignancy prediction, skin carcinoma analysis, and breast cancer screening. CNN is the most extensively utilized deep learning approach, allowing for automatic mass categorization, pattern recognition, and segmentation with minimal training datasets and no human interaction. CNNs are built in a layer structure. Furthermore, Deep Learning algorithms have demonstrated remarkable advancements in the medical arena, providing researchers confidence that patients may now securely interface with an activated medical equipment and participate in a significant role in enhancing patient healthcare. Owing to the inherent constraints of traditional approaches, we demonstrate a method that utilizes deep learning in this study, exhibiting the capabilities of CNNs to identify. Breast microcalcifications are segmented to aid radiologists in mammography analysis. Step by step procedure is shown in Fig. 9.2.

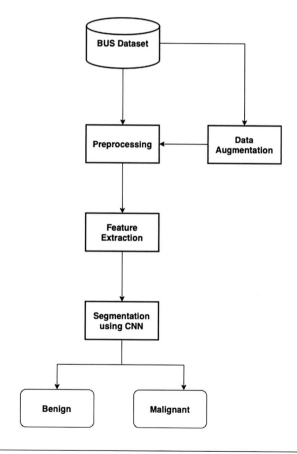

FIGURE 9.2

Step-by-step methodology we employed in our method to deploy the algorithms to the dataset.

2. Literature review

A deep learning mammography literature review would entail looking for and assessing research papers, articles, and studies that have utilized deep learning algorithms to interpret mammogram images. Deep learning applications in mammography include breast cancer categorization, lesion identification, and picture segmentation. The review might examine the performance of various deep learning approaches and compare them to standard methods.

2.1 Risk factors

Breast cancer is a complicated illness with unknown origins. However, there are various variables that might raise a person's chance of having the condition. Among the most profound hazard concerns is age, with the likelihood of having breast cancer increasing with age. Women are also at a higher risk of having breast cancer than males. A person's risk increases if their mother, sister, or daughter has had the condition in the past [3]. Certain genetic abnormalities, such as BRCA1 and BRCA2, can also raise a person's chance of getting breast cancer. Women who had their first child beyond the age of 30, have never given birth, or had a late menopause, as well as women who use hormonal therapy for menopause or take oral contraceptive pills for a prolonged period of time, are at elevated risk. Alcohol consumption, as well as being overweight or obese, can all raise a person's likelihood of acquiring breast cancer. It is important to emphasize that possessing one or more of these risk factors does not guarantee that a person will get breast carcinoma, and many women who get the illness have no known risk factors.

2.2 Related work

A comprehensive analysis was done to analyze the results achieved in computer-aided breast cancer prognosis since the implementation of the project. The prospective study was prepared to be employed in a broad range of publications in the field by using a diverse set of technological data sources as an allusion. Nonetheless, this article's focus was confined to research and academic writings, with no regard for profit entities. The findings of this study include a description of the prevailing state of prognosis techniques in terms of the image modalities utilized and the classifiers relying on deep learning that are implemented [4].

Deep learning-based categorization of breast tissues from histological pictures, in compliance to Yang et al., has low efficiency due to a shortage of training data and information about architectural and textual data that really can span numerous levels. In this publication, the ensemble of the multiresolution CNN (EMCN) substructure is employed to classify hematoxylin-eosin-stained breast microscope pictures into four classifications: a benign tumor, a healthy cell, an intrusive lesion, and a malignancy. Each picture is transcribed to different scales before training the pre-trained models, such as DenseNet-161, ResNet-152, and ResNet-101 [5]. During each scale, the acquired training bits are utilized and enhanced. The EMS-net technique outperforms the other three algorithms examined in terms of accuracy [6].

The literature the notion of a multi view CNN, a 6-input convolutional neural networks that includes the mass area, microcalcifications zone, and perspectives including both Mediolateral oblique and CC viewpoints of each breast. The 6-input convolution neural networks amalgamated to create a combination line that is employed to build a classification algorithm. The IN-breast data set was utilized for experimental analysis, yielding a precision of 80.10% and an AUC score of 0.78. The author [7] classified photos from the Mammographic Image Analysis Society (MIAS) dataset using a CNN. The photos are preprocessed using the watershed segmentation technique. When compared to classic machine learning approaches such as SVM, the approach generated an accuracy of 98%.

The new algorithm was tested using a digital mammography dataset. Superresolution (SR) and nonsubsampled contourlet transform (NSCT) were coupled with Can AdaBoost classifier to construct the suggested technique [8]. After computerizing the identification of intersect region, NSCT and SR enhanced the image quality (ROI). The prediction accuracy was 91.43%.

Another research, by Lo et al., assisted us acknowledge that the amount of network levels utilized should be governed by the sophistication of the sequence to be assessed and that two or three network layers are generally sufficient for the simplicity of mammography abnormality patterns. This article also helped us grasp how one-hot encoding with multilayer neural network output nodes allow for stochastic output with a CNN [9]. Lo et al. opted to have a handful of those output neurons symbolizing the location where the probabilistic conclusion is ambiguous. While Lo et al. extracted features using 2×2 average pooling, we chose to keep the 2×2 architecture but employ utmost pooling in the aim of recovering the most essential characteristics at each pooling step. Both Lo et al.'s article and Sahiner et al.'s journal recommended using a ROC (receiver operating characteristic) [10,11] curve to analyze the reliability of the results, particularly given that the allocation of class labels in our dataset (as in numerous medical datasets) is skewed further toward baseline characteristics than malignant samples.

All of these factors had a role in the growth of our classifiers and the examination of the findings. Although not every proposal was included in our final models, several of the ideas presented in these publications prompted us to make judgments that resulted in improved computing performance and good findings.

3. Methodologies

3.1 Deep learning in mammography

Deep learning has been utilized to mammogram screening in recent years with the objective of enhancing the precision and effectiveness of breast cancer diagnosis. Deep learning is a sort of cognitive computing that uses artificial neural networks to model complicated connections in huge datasets [12]. Deep learning models employed in mammography may evaluate medical pictures and discover symptoms of breast cancer, such as lumps or calcifications, that are not evident to the naked eye.

The following are some of the advantages of employing deep learning in mammography.

- Enhanced accuracy: Research has demonstrated that deep learning models are more precise at categorizing breast cancer than conventional mammography analysis strategies.
- Automated exploration: Deep learning techniques can quickly evaluate mammography pictures, improving efficiency and labor-intensive manual interpretation.
- Enhanced patient metrics: Deep learning in mammography does have the potential to increase patient outcomes and mortality rates by incentivizing more precise and effective breast cancer identification.

Overall, deep learning has demonstrated remarkable promise in mammogram screening and has the capability to have a large influence on patient prognoses and breast cancer diagnosis. To address the difficulties of integrating deep learning into clinical practice, further study and development are required.

3.2 Deep learning approaches for mammographic image processing

Mammogram picture analysis may be accomplished using a number of efficient deep learning techniques, including the following [12].

- CNNs: CNNs are being audition to mammography images for breast malignancy categorization and lesion identification. CNNs are often employed for image classification activities.
- Transfer learning: Transfer learning entails adopting a deep learning model that has already been built and optimizing it for the particular job of mammography interpretation. When there is a lack of readily available annotated data, this method has demonstrated good outcomes in practice and is a practical, affordable option.
- Generic mammography pictures may be produced using generative adversarial networks (GANs) that will be applied to increase the training dataset and enhance the effectiveness of deep learning algorithms.
- Recurrent neural networks (RNNs): RNNs are capable of analyzing the chronological information included in sequential data, such as mammography pictures collected over time to monitor breast tissue changes.

3.3 **CNN methodologies in mammography**

Mammography uses CNNs [13,14], a prominent deep learning paradigm. CNNs may be applied to mammography to perform tasks including picture segmentation, lesion identification, and breast cancer classification. Convolutional, pooling and fully linked layers are among the layers that make up a typical CNN architecture for mammography. While the pooling layers down-sample the feature maps to speed up computation and save just the most crucial characteristics, the convolutional layers apply filters on the mammography images to learn local features. The ultimate classification or detection choice is made by the fully linked layers, which incorporate the characteristics from the preceding layers. CNN is trained on a sizable dataset of mammography pictures classified as either malignant or benign to classify breast cancer. The model gains the ability to spot patterns and characteristics in the photos that point to malignancy [15]. The CNN is taught to identify problematic areas in mammography pictures that need more investigation during lesion detection. In general, CNNs [16] in mammography have demonstrated encouraging results and have the potential to increase the precision and effectiveness of breast cancer diagnosis and treatment. Deep learning models for medical imaging are still in the early phases of development, therefore additional study is required to determine their clinical relevance and obtain regulatory approval. The five-layered architecture of convolutional neural network (CNN) is shown in Fig. 9.3.

3.4 **Comprehensive architecture of CNN in mammography**

The design of a CNN for mammography might vary in accordance with the objective and the data employed [17]. However, a typical CNN design for mammography analysis can comprise the following components.

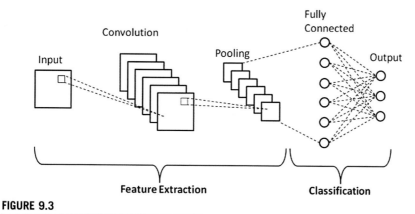

FIGURE 9.3

Five-layered architecture of convolutional neural network (CNN).

- Input layer: The input layer is where the mammography pictures are sent after they have typically been preprocessed to uniformize their size and brightness.
- Convolutional layers: By applying filters to the input pictures, convolutional layers train themselves to extract pertinent information from the images. These filters are developed throughout the training phase and modified based on the model's accuracy. To deepen the model and capture more complicated characteristics, several convolutional layers can be layered.
- Pooling layers: Pooling layers allow the most crucial data to be preserved while reducing the spatial resolution of the characteristic maps created by the convolutional layers. This aids in lowering the model's computational complexity and minimizes overfitting, a major issue with deep learning models.
- Fully linked layers: The characteristics are compressed into a 1D vector after the pooling layers and put into a string of fully linked layers. These layers train a nonlinear function that connects the input forecasts to the features.
- Output layer: This layer generates the model's ultimate forecasts, which may include multiclass assumptions like different forms of breast cancer or binary classifications like normal versus abnormal.
- This is only a generalized design for CNN for mammogram interpretation is shown in (Fig. 9.4); actual implementations may differ. To avoid overfitting, certain models may include extra layers, such as washout layers, or recurrent layers, to capture temporal relationships in the information. To boost the models' performance, it is also possible to change the size and quantity of filters in each layer in addition to the activation functions that are employed (Fig. 9.4).

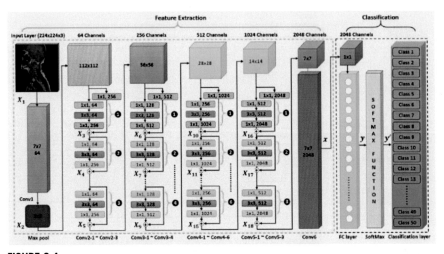

FIGURE 9.4

Extensive CNN mammography infrastructure.

4. **Resources and procedures**

The research for this study involved the gathering, processing, and explication of clinical findings. The subsequent sections give in-depth examinations of various components. The execution of any technique is dependent on the nature of the data; before presenting a way of analysis for medical imaging, it is required to study the databases.

4.1 **Data exploration**

Dr. Aly Fahmy and his group obtained the mammographic images dataset, which was implemented in this thesis [20]. Breast ultrasound pictures were gathered at the start of the study from women aged 25 to 75. This information was collected during 2018. There are total 600 female patients. The dataset comprises 780 images, each measuring 500×500 pixels. The photos are stored in PNG files. Original photos are shown with ground truth photographs. The photos are divided into three categories: normal, benign, and malignant. To improve accuracy, the photos are downsized to 128×128 pixels. All photos are parameterized for better performance by partitioning to the RGB range 255 and splitting labels. Similarly, mask pictures are important aspects in the model, and my calculations produced proper results in accordance with the model. Because typical photos are simple to categorize, it has not been examined further. After executing data augmentation with the appropriate flow, there are a total of 3057 photos. To extract significant characteristics from photos, I used multiple convolution layers with 16, 32, and 64 filters, correspondingly, with an input feature structure of $128 \times 128 \times 1$, and MaxPooling2D for dimensionality compression. To transform the retrieved features into a 1D array, I inserted a flattened layer. Dense layers of 128, 64, and 32 units are afterward added. A dropout layer with a rate of 0.3 is used to shrink overfitting. Adaptive moment approximation is used to optimize the activation function. Similarly, feature development and hyperparameter tweaking are used to produce superior parameters. Transfer learning models are also used to validate the findings.

4.2 **Techniques and methodologies**

Deep learning implementations in the medical sector have grown substantially because they may be used to rapidly uncover patterns relevant to anticipating distinct patient risk categories. Deep learning's potential advantages in healthcare are anticipated to rise as the discipline evolves and additional data becomes accessible. Mammography may be converted into a more precise and effective tool for breast cancer identification and administration by harnessing the power of deep learning.

4.2.1 *ResNet50*

ResNet50 is a CNN framework built by Microsoft Research that earned the Large-Scale ImageNet Visual Accolade Challenge in 2015. It is known as "ResNet50" because it contains 50 layers and belongs to the ResNet (Residual Network) community of CNN technologies. ResNet50 is meant to tackle the obstacle of disappearing gradients in very deep networks, when the error gradients become exceedingly tiny and the network is unable to train effectively. The ResNet architecture overcomes this issue by leveraging residual connections, which enables the network to bypass multiple stages of computation and immediately transport data from the lower layers to the upper layers. ResNet50 has been prevalently applied in computer vision applications such as picture classification, object identification, and semantic segmentation. Furthermore, it has been developed and fine-tuned for a wide range of medical imaging applications, including mammography, where it has been demonstrated to perform effectively in identifying breast cancer. Overall, ResNet50 is a robust and extensively used CNN architecture that has shown high performance in a range of computer vision tasks, including mammography.

However, the fundamental building parts of a ResNet50 design are as follows.

- Convolutional layers: These layers perform a convolutional implementation on the input picture to extract regional information from the image.
- Rectified linear unit (ReLU) activation: To add nonlinearity to the network, the ReLU activation function is employed element-by-element to the outcome of the convolutional layer.
- Batch normalization layers: By bringing each layer's action potentials within the normal range, these layers serve to stabilize the training procedure and minimize overfitting.
- Pooling layers: By downsampling the spatial configurations of the feature mappings, these layers serve to lower the network's computational expense.
- Residual connections: These connections circumvent a few processing steps and let the network send data straight from lower layers to higher layers.

Depending on the particular job and dataset being used, the ResNet50 architecture's precise specifications, as well as the number of layers, the quantity of filters in all the layer, the dimension of the filters, and the precise types of pooling and activation functions employed, can change. The interpretation of all layers is shown in Fig. 9.5, where the framework of ResNet50 is shown with all layers.

4.2.2 DenseNet201

A CNN architecture called DenseNet201 was first described in the Huang et al. study "Densely Connected Convolutional Networks." Because it is a member of the DenseNet family of CNN designs and contains 201 layers, it is known as "DenseNet201." In very convolutional models, where the amount of parameters rises sharply as the complexity of the layers rises, DenseNet201 is made to handle the problem of parameter efficiency. This problem is solved by the thick connections used in the DenseNet design, which combines all of the outputs from earlier levels into a single input for each new layer. This decreases the amount of parameters needed by the network and enables the network to reuse features learned by preceding layers in an effective manner. DenseNet201 has been extensively implemented in

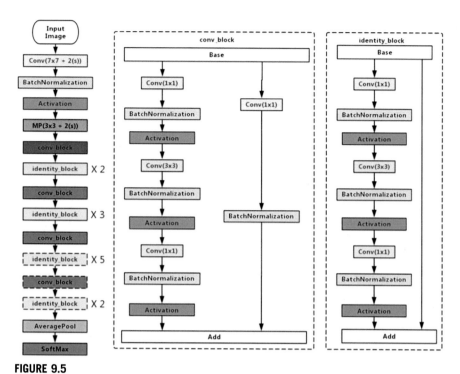

FIGURE 9.5

Integrated architecture of ResNet50.

assorted computer vision tasks, including object identification, picture categorization, and semantic segmentation. Additionally, it has been adapted and fine tuned for various medical imaging tasks, including mammography, where it has been demonstrated to perform well in identifying breast cancer. All things considered, DenseNet201 is a potent and popular CNN design that has proven to be accomplished in a number of image category function, including mammography.

The infrastructure of DenseNet201 is made up of many compact blocks, each of which has a number of convolutional layers. Transitional layers that link the dense blocks together increase the number of channels while lowering the spatial resolution of the feature maps. The results of every layer before it are concatenated and used as inputs by every layer after it in a dense block. As a result, the network's need for parameters is decreased and it is able to reuse features learned by prior layers more effectively. A ReLU activation function, a batch normalization layer, and a 3×3 convolutional filter are all included in each convolutional layer of a dense block. As the network gets deeper, there are more filters in each layer, which helps to capture more complicated characteristics. In DenseNet201, the transitional layers are 1×1 convolutional filters that are employed to scale back the number of channels and boost the spatial resolution of the feature maps. This aids in lowering the network's computational expense and enhancing its capacity to learn fine-grained information.

Finally, a worldwide conclusion of the ultimate dense block is applied with an average pooling layer. combining the feature maps into a single vector. Following a completely connected layer, which generates the best classification likelihoods, this vector is then passed through. As shown in Fig. 9.6, the specifics of the Dense-Net201 design, according to the variety of layers, the quantity of filters inside each level, the size of the filters, and the types of activation functions employed can change based on the job at hand and the database being utilized. However, the dense blocks and transition layers that make up the core of the DenseNet201 design are the same in all deployments.

4.2.3 AlexNet

AlexNet is a deep CNN architecture described in the article "ImageNet Classification with Deep CNN (DCNN)" by Alex Krizhevsky, Ilya Sutskever, and Geoffrey Hinton in 2012. It was the first deep CNN to take first place in the ImageNet Large Scale Visual Recognition Competition and is widely regarded as one of the deep learning's most significant accomplishments. AlexNet is made up of several convolutional and dense layers, coupled with pooling layers, which are used to lower the spatial resolution of feature maps and boost the number of channels. As it progresses, the network is supposed to learn progressively complicated characteristics, with the last layers learning the most speculative and high-level information. AlexNet's framework consists of several convolutional layers, all of which comprise a series of filters that learn to recognize certain characteristics in input pictures. To

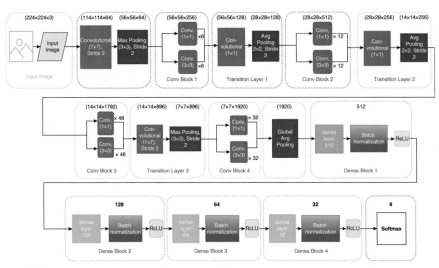

FIGURE 9.6

DenseNet201's optimized architecture.

generate feature maps, the screens are put on the input imagery, which are then pooled to lower spatial resolution and boost the number of channels. Following the creation of the feature maps, they were processed through one or more thick layers, which are intended to acquire additional abstract features and create the final classification predictions as depicted in Fig. 9.7. Numerous neurons inside the dense layers are linked to all neurons in the preceding layer and are utilized to learn complicated correlations between the characteristics in the input picture. AlexNet was given training on the ImageNet dataset, which comprises over 1 million photos from 1000 distinct categories, and it outperformed all previous condition systems on the collection at the time.

AlexNet has influenced the creation of many additional deep learning architectures since its inception, and its discoveries have been integrated into numerous cutting-edge models in computer vision and other disciplines. It is still used as a standard for deep learning in computer vision and is frequently used in picture classification, object recognition, and semantic segmentation. AlexNet's architecture comprised the underlying layers.

- Convolutional layers: AlexNet has five convolutional layers, each followed by an activated ReLU function. These layers learn to recognize low-level picture properties such as boundaries, patterns, and forms.
- Max pooling layers: A max pooling layer is added after each convolutional layer to lower the spatial resolution of the feature maps. The pooling layer takes the maximum value from each feature map subregion and discards the rest of the data.
- Normalization layers: AlexNet also incorporates normalization layers, which serve to attenuate the impacts of internal covariate shift and increase the network's capacity to learn.

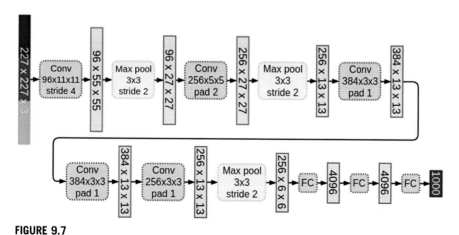

FIGURE 9.7

The streamlined architecture of AlexNet.

- Dense layers: Following the last pooling layer, there are three dense layers, all of which are completely coupled to all neurons in the preceding layer. These deep layers are responsible for learning increasingly abstract information and making final classification prognostications.
- AlexNet's last layer is an output layer, which yields the occurrences for every one of the 1000 classifications within ImageNet dataset.

It is worth noting that AlexNet was trained on the ImageNet dataset, which contains a significant number of color photographs, and that its framework was particularly built to deal with this sort of data. To get promising performance, AlexNet's design might have to be amended for usage with various types of data, such as grayscale photos or medical imaging data.

4.2.4 VGG16

Karen Simonyan and Andrew Zisserman proposed VGG16, a DCNN Method, in their publication "Very DCNN for Large-Scale Image Identification" in 2014. It is extensively employed in a range of computer vision applications, including image classification, object recognition, and semantic segmentation, and is acknowledged to be among the most notable and influential CNN structures in computer vision. VGG16 is distinguished by the use of several convolutional layers, each encompassing a pile of filters that train to distinguish specific geographical character traits in the input pictures. The filters are incorporated into the input picture to generate feature maps, which are then pooled to lower spatial resolution and increase the number of channels. VGG16's architecture is quite simple and uncomplicated, consisting of a succession of convolutional layers accompanied by pooling layers. As the network intensifies, the convolutional layers are supposed to learn progressively complicated characteristics, with the last levels acquiring the most abstract and high-level information. VGG16's use of extremely tiny filters (3×3) in the convolutional layers is one of its significant advances, allowing the network to accumulate fine-grained features and capture more local information.

VGG16 also employs many convolutional layers, allowing it can learn increasingly complicated features and improve its capacity to gather high-level information. VGG16 was capable of achieving exceptional performance by comparing several existing method, but on the ImageNet data set, which comprises over 1 million pictures from 1000 distinct methods. Since its debut, VGG16 has sparked the creation of several alternative deep learning approaches and its breakthroughs have been integrated into other cutting-edge models in computer vision and other fields. VGG16 is a well-liked option for many picture classification problems and is still in use and considered as a reliable model in computer vision despite its advanced age. The following outlines the architecture of VGG16.

- Input layer: The model's input is a picture with dimensions of $224 \times 224 \times 3$, wherein 3 refers to the number of color channels (red, green, and blue).
- Convolutional layers: The model is comprised of a succession of convolutional layers, each of which is accompanied by an activation function for ReLU and a

maximum pooling layer. The convolutional layers generate feature maps from the input data and gain knowledge of local patterns in it. These feature maps are then passed into the next layer.

- Dense layers: These layers are used to learn higher-level features and generate predictions. They are constructed by feeding the output of the final max pooling layer into a number of fully connected layers.
- Output layer: The output layer is the topmost layer of the model, and it is made up of a single neuron that generates the final prediction.

Because of its deep network design and usage of max pooling layers, one of VGG16's primary strengths is its ability to learn complicated features from picture data. The model has been pretrained on a large dataset of pictures and may be used as a basic model for a range of computer vision tasks such as image classification and object recognition. In summation, the VGG16 design is distinguished by the usage of an extremely deep network with 16 layers, which includes numerous convolutional layers, ReLU activation functions, max-pooling layers, and dense layers. The model is built to handle picture data and is useful for computer vision applications like image categorization (Fig. 9.8).

4.2.5 Base model with Keras Tuner

The base model in Keras Tuner corresponds to a preexisting deep learning model that is incorporated as a starting point for hyperparameter tweaking. Hyperparameters are values that influence the behavior of a model and are established before training, for instance, the attainment value, the number of layers, or the size of the filters. Keras Tuner is a package that allows you to tune hyperparameters in Keras, a prominent deep learning framework. It gives a quick and easy technique to find the optimum hyperparameters for a given model and dataset. The base model serves as the initial place for the hyperparameter tuning procedures, and its infrastructure and

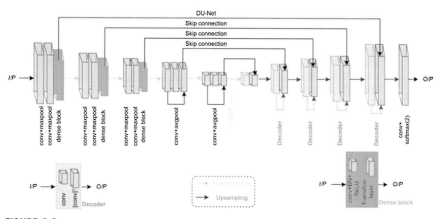

FIGURE 9.8

VGG16's systematic architecture.

parameters are often fixed during the tuning process. The Keras Tuner module then uses an optimization technique to find the ideal hyperparameters for the underlying model, often by training the model on the dataset and assessing its effectiveness. After determining the optimum hyperparameters, they may be added to the main model to fine-tune it for the given dataset and job. This procedure can aid in enhancing the model's performance and produce better outcomes on the specified job. In summation, the term "base model with Keras Tuner" refers to a deep learning model that has already been created and is used as the basis for tuning hyperparameters using the Keras Tuner library. Hyperparameter adjustment can help the model run more efficiently and produce better results for the given job (Fig. 9.9).

5. Findings and analysis

A sizable and varied collection of mammogram images is necessary for deep learning applications in mammography. To train the model to distinguish among these two types of tissue, the dataset should contain scans of both malignant

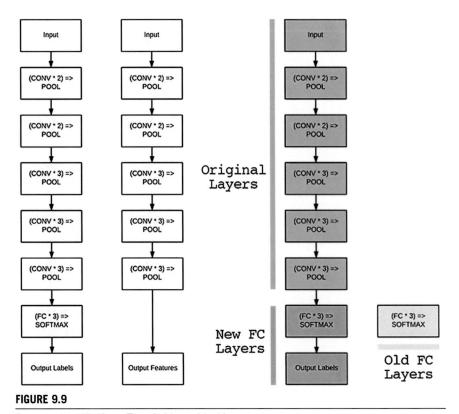

FIGURE 9.9

Base model with Keras Tuner's hierarchical layout.

(cancerous) and benign (noncancerous) breasts. These scans should be collected in accordance with suitable ethical standards, including getting patients' informed consent and protecting their privacy and medical data's security. The collection should include labels or annotations that specify whether each mammography scan is of a benign or carcinogenic breast in addition to the mammograms itself. The deep learning model is trained using these labels, and its effectiveness is assessed. Here Fig. 9.10 depicts scans of malignant and benign breasts from our data collection.

5.1 Discussions

5.1.1 Base model with augmentation

The term "base model with augmentation" refers to the process of utilizing data intensification methods to boost the effectiveness of a deep learning model. Data augmentation is the process of expanding the size of a training dataset deliberately by generating various specimens from existing ones using different transformations such as rotation, scaling, cropping, flipping, and so on. This improves the model's generalization and lowers overfitting, resulting in higher performance on unknown data. When a basic model is combined with data augmentation approaches, the model's performance on a particular job is frequently improved. Fig. 9.11 shows how the training and validation accuracy oscillates in the model accuracy graph, as well as how the training and validation loss graph is likewise inconsistent.

5.1.2 Base model without augmentation

A "base model without augmentation" deep learning model is one that has not been modified using data augmentation methodologies. In this case, the model is trained exclusively on the raw training dataset, with no false examples added via manipulations. This sort of model may still perform well, particularly if the original training

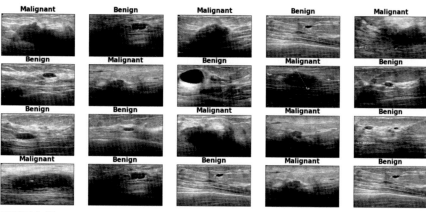

FIGURE 9.10

Breast radiographs that were employed during this research to show both benign and cancerous abnormalities.

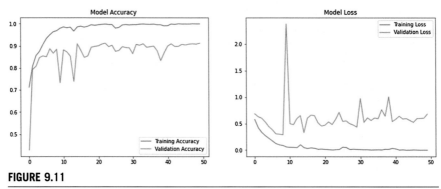

FIGURE 9.11

Comparison of the model accuracy and model loss data graphically in augmentation.

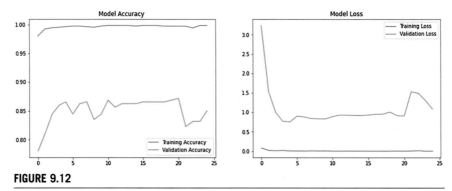

FIGURE 9.12

Comparison of the model accuracy and model loss data graphically without augmentation.

dataset is wide ranging and diverse, but it may also suffer from overfitting, particularly if the training data is restricted or does not match the real data distribution. As a result, using data augmentation approaches to increase the performance and durability of deep learning models is frequently advised. Fig. 9.12 shows how the training and validation accuracy oscillates in the model accuracy graph, as well as how the training and validation loss graph is likewise inconsistent.

5.2 **Model summary**

5.2.1 *Base model with Keras Tuner (Table 9.1) shows the summarization of base model*

Table 9.1 Summarization of the base model.

Method	Out turn	Parameters
Conv-2d	(−, 124, 124, 16)	416
max_pooling-2d	(−, 62, 62, 16)	0
Conv-2d-1	(−, 58, 58, 16)	6416
max_pooling-2d-1	(−, 29, 29, 16)	0
Flatten	(−, 13456)	0
Dense	(−, 128)	1722496
Dense-1	(−, 2)	258

Total parameters: 1,729,586.
Trainable parameters: 1,729,586.
Nontrainable parameters: 0.

5.2.2 *AlexNet (Table 9.2) shows the summarization of AlexNet model*

Table 9.2 Summarization of the AlexNet model.

Method	Out turn	Parameters
Conv-2D	(None, 30, 30, 96)	11712
batch_normalization	(None, 30, 30, 96)	384
Max-Pooling-2D	(None, 14, 14, 96)	0
Conv-2D	(None, 14, 14, 256)	614656
batch_normalization	(None, 14, 14, 256)	1024
Max_Pooling-2D	(None, 6, 6, 256)	0
Conv-2D	(None, 6, 6, 384)	885120
batch_normalization	(None, 6, 6, 384)	1536
Conv_2D	(None, 6, 6, 384)	1327488
batch_normalization	(None, 6, 6, 384)	1536
conv2d_6	(None, 6, 6, 256)	884992
batch_normalization	(None, 6, 6, 256)	1024
Max_Pooling-2D	(None, 6, 6, 256)	0
Flatten	(None, 1024)	0
Dense-2	(None, 4096)	4198400
Dropout	(None, 4096)	0
Dense-3	(None, 4096)	16781312
Dropout	(None, 4096)	0
Dense-4	(None, 2)	8194

Total parameters: 24,717,378.
Trainable parameters: 24,714,626.
Nontrainable parameters: 2752.

5.2.3 ResNet50 (Table 9.3) shows the summarization of the ResNet50 model

Table 9.3 Summarization of the ResNet50 model.

Layer (type)	Output shape	Parameters
Model (functional)	(−, 32768)	23581440
Dense_5 (dense)	(−, 2)	65538

Total parameters: 23,646,978.
Trainable parameters: 65,538.
Nontrainable parameters: 23,581,440.

5.2.4 DenseNet 201 (Table 9.4) shows the summarization of the DenseNet201 model

Table 9.4 Summarization of the DenseNet201 model.

Layer (type)	Output shape	Parameters
Model_1 (functional)	(None, 30720)	18315712
Dense_6 (dense)	(None, 2)	61442

Total parameters: 18,377,154.
Trainable parameters: 61,442.
Nontrainable parameters: 18,315,712.

5.2.5 VGG16 (Table 9.5) shows the summarization of the VGG16 model

Table 9.5 Summarization of the VGG16 model.

Method	Out turn	Parameters
Input_Layer	[(None, 299, 299, 1)]	0
block1conv1	(None, 299, 299, 64)	640
block1conv2	(None, 299, 299, 64)	36928
MaxPooling-2D	(None, 149, 149, 64)	0
blockllconv1	(None, 149, 149, 128)	73856
blockllconv2	(None, 149, 149, 128)	147584
blockllpool	(None, 74, 74, 128)	0
blocklllconv1	(None, 74, 74, 256)	295168
blocklllconv2	(None, 74, 74, 256)	590080

Table 9.5 Summarization of the VGG16 model.—*cont'd*

Method	Out turn	Parameters
blockIIIconv3	(None, 74, 74, 256)	590080
blockIIIpool	(None, 37, 37, 256)	0
blockIVconv1	(None, 37, 37, 512)	1180160
blockIVconv2	(None, 37, 37, 512)	2359808
blockIVconv3	(None, 37, 37, 512)	2359808
blockIVpool	(None, 18, 18, 512)	0
blockVconv1	(None, 18, 18, 512)	2359808
blockVconv2	(None, 18, 18, 512)	2359808
blockVconv3	(None, 18, 18, 512)	2359808
MaxPooling2D	(None, 9, 9, 512)	0
Flatten	(None, 41472)	0
Dense-7	(None, 2)	82946

Total parameters: 14,796,482.
Trainable parameters: 82,946.
Nontrainable parameters: 14,713,536.

5.3 Results

Implications of testing and training precision for all deep learning algorithms employed.

5.3.1 AlexNet

The recall, probability, and F1 scores in consonance with Fig. 9.13 are not 100% accurate. The testing accuracy has several alterations that may have shown an impact on the precision and F1 scores. Given the results of the research above, it is evident that the AlexNet Model is trying to perform to its full potential but only achieved 80% accuracy in Fig. 9.13.

5.3.2 ResNet50

The recall, probability, and F1 scores in consonance with Fig. 9.14 are not 100% accurate. The testing accuracy has several alterations that may have shown an impact

```
              precision   recall  f1-score   support

     benign       0.91      0.79      0.84       227
  malignant       0.63      0.82      0.72       101

   accuracy                          0.80       328
  macro avg       0.77      0.81      0.78       328
weighted avg      0.82      0.80      0.80       328
```

FIGURE 9.13

AlexNet's F1 score, precision, and recall.

	precision	recall	f1-score	support
benign	0.79	0.90	0.85	227
malignant	0.69	0.48	0.56	101
accuracy			0.77	328
macro avg	0.74	0.69	0.70	328
weighted avg	0.76	0.77	0.76	328

FIGURE 9.14

ResNet50's F1 score, precision, and recall.

on the precision and F1 scores. Given the results of the research above, it is evident that the ResNet50 Model is trying to perform to its full potential but only achieved 77% accuracy in Fig. 9.14.

5.3.3 DenseNet 201
The recall, probability, and F1 scores in consonance with Fig. 9.15 are not 100% accurate. The testing accuracy has several alterations that may have shown an impact on the precision and F1 scores. Given the results of the research above, it is evident that the DenseNet201 Model is trying to perform to its full potential but only achieved 75% accuracy in Fig. 9.15.

5.3.4 VGG16
The recall, probability, and F1 scores in consonance with Fig. 9.16 are not 100% accurate. The testing accuracy has several alterations that may have shown an impact on the precision and F1 scores. Given the results of the research above, it is evident that the VGG16 Model is trying to perform to its full potential but only achieved 73% accuracy in Fig. 9.16.

5.3.5 Base model with Keras Tuner and augmentation
The recall, probability, and F1 scores in consonance with Fig. 9.14 are not 100% accurate. The testing accuracy has several alterations that may have shown an impact on the precision and F1 scores. Given the results of the research above, it is evident that the Base Model is trying to perform to its full potential but only achieved 88% accuracy in Fig. 9.17. Comparison of accuracy is shown in Fig. 9.18.

	precision	recall	f1-score	support
benign	0.75	0.95	0.84	227
malignant	0.71	0.30	0.42	101
accuracy			0.75	328
macro avg	0.73	0.62	0.63	328
weighted avg	0.74	0.75	0.71	328

FIGURE 9.15

DenseNet's F1 score, precision, and recall.

	precision	recall	f1-score	support
benign	0.72	0.98	0.83	227
malignant	0.80	0.16	0.26	101
accuracy			0.73	328
macro avg	0.76	0.57	0.55	328
weighted avg	0.75	0.73	0.66	328

FIGURE 9.16

VGG16's F1 score, precision and recall.

	precision	recall	f1-score	support
benign	0.85	0.89	0.87	328
malignant	0.91	0.88	0.90	437
accuracy			0.88	765
macro avg	0.88	0.88	0.88	765
weighted avg	0.89	0.88	0.88	765

FIGURE 9.17

Base model's F1 score, precision, and recall.

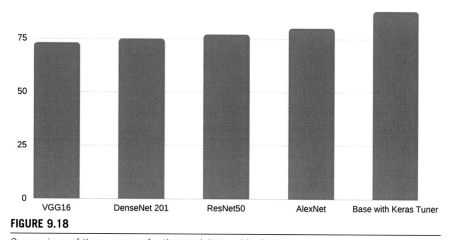

FIGURE 9.18

Comparison of the accuracy for the models graphically.

6. Conclusions

The categorization findings demonstrated that the suggested design had potential in terms of many established quality measures. Transfer learning and a bespoke baseline model are used in this technique and programmed detection of breast carcinoma from sonogram pictures. There seem to be 891 healthy and 421 cancerous cases in the sample. Following deployment and evaluation, the observation indicates that the best-performing architectures are Base Model, AlexNet, and ResNet50. Later, by tweaking the algorithms and expanding the amount of the dataset with additional ultrasound pictures obtained from various places and healthcare equipment, the model may be made more resilient and exact. Deep learning has demonstrated significance in mammography, notably in the early diagnosis of breast cancer. Many theses have shown that deep learning algorithms can reliably detect hazardous areas in mammography pictures and, in certain situations, outperform human radiologists. It is crucial to highlight, however, that these techniques are still in the process of development and require more validation and refining before they can be extensively used in clinical practice. Furthermore, the application of deep learning in mammography must be weighed against ethical and legal considerations like privacy and data security.

7. Future work

In the use of deep learning in mammography pictures, there is still much potential for improvement and enhanced study. Some potential future developments include:

- Improving generalization and robustness: To effectively categorize mammogram pictures from a range of demographics and imaging methods, deep learning techniques must be trained on vast and diverse datasets.
- Referencing bias and fairness: When implementing deep learning algorithms for mammography, prejudice and fairness must be carefully evaluated since they might effect the precision and reliability of findings.
- Clinical decision-making integration: Deep learning methodologies must be included in the clinical decision-making procedure in such a manner that enhances and integrates rather than replaces the knowledge of human radiologists.
- Melding with other imaging modalities: Deep learning algorithms can be used in conjunction with other imaging techniques to increase the precision of breast cancer diagnosis, such as magnetic resonance imaging (MRI) or ultrasound.
- Investigating comprehensibility and predictability: Deep learning algorithms must be made easily decipherable and transparent so that radiologists and other healthcare professionals may understand and justify their decisions and outputs.
  Overall, deep learning has tremendous potential in mammography, and further research in this field facilitates promise for enhancing the effectiveness and precision of breast cancer detection and therapy.

References

[1] Altan G. Deep learning-based mammogram classification for breast cancer. International Journal of Intelligent Systems and Applications in Engineering Dec. 2020; 8(4):171–6.

[2] Elmoufidi A, El Fahssi K, Jai-Andaloussi S, Sekkaki A. Automatically density-based breast segmentation for mammograms by using dynamic k-means algorithm and seed-based region growing. In: Proceedings of the I2MTC 2015 international instrumentation and measurement technology conference; May 2015. Pisa, Italy.

[3] Nemade V, Pathak S, Dubey AK, Barhate D. Breast mass classification using convolutional neural network. Intelligent System Design 2023:85–92. https://doi.org/10.1007/978-981-19-4863-3_8.

[4] Yemini M, Zigel Y, Lederman D. Detecting masses in mammograms using convolutional neural networks and transfer learning. Proceedings of the International Conference on the Science of Electrical Engineering, Eilat, Israel 2018:1–4.

[5] Agarwal R, Díaz O, Yap MH, Lladó X, Martí R. Deep learning for mass detection in full field digital mammograms. Computers in Biology and Medicine 2020;121(April): 1–10. https://doi.org/10.1016/j.compbiomed.2020.103774.

[6] Terumalasetti S. A comprehensive study on review of AI techniques to provide security in the digital world. In: 2022 third international conference on intelligent computing instrumentation and control technologies (ICICICT); 2022. p. 407–16. https://doi.org/10.1109/ICICICT54557.2022.9917931.

[7] Hoffman-Goetz L, Apter D, Demark-Wahnefried W, Goran MI, McTiernan A, Reichman ME. Possible mechanisms mediating an association between physical activity and breast cancer. Cancer: Interdisciplinary International Journal of the American Cancer Society 1998;83(S3):621–8.

[8] Reeja SR, Kavya NP. Noise reduction in video sequences-the state of art and the technique for motion detection. International Journal of Computer Application 2012;58(8).

[9] Bandi M. A review on thermal infrared semantic distribution for nightfall drive. In: 2022 third international conference on intelligent computing instrumentation and control technologies (ICICICT); 2022. p. 645–53. https://doi.org/10.1109/ICICICT54557.2022.9917651.

[10] Shen L, Margolies LR, Rothstein JH, Fluder E, McBride R, Sieh W. Deep learning to improve breast cancer detection on screening mammography. Scientific Reports 2019;9(1):12495.

[11] Becker AS, et al. Deep learning in mammography: diagnostic accuracy of a multipurpose image analysis software in the detection of breast cancer. Investigative Radiology 2017;52:434–40.

[12] Yala A, Lehman C, Schuster T, Portnoi T, Barzilay R. A deep learning mammography-based model for improved breast cancer risk prediction. Radiology 2019;292:60–6.

[13] Arevalo J, González FA, Ramos-Pollán R, Oliveira JL, Lopez MAG. Convolutional neural networks for mammography mass lesion classification. In: 2015 37th annual international conference of the IEEE engineering in medicine and biology society (EMBC); 2015. p. 797–800. https://doi.org/10.1109/EMBC.2015.7318482.

[14] Deepa N, Chokkalingam S. Deep convolutional neural networks (CNN) for medical image analysis. International Journal of Engineering and Advanced Technology 2019; 8(3S):607–10.

[15] Krizhevsky A, Sutskever I, Hinton G,E. ImageNet classification with deep convolutional neural networks. In: Advances in neural information processing systems; 2012. p. 1097−105.

[16] Dias N, Reeja SR. A quantitative report on the present strategies of Graphical authentication. International Journal of Computer Science and Engineering 2018;06: 64−73. https://doi.org/10.26438/ijcse/v6si10.6473.

[17] Reeja SR, Kavya NP. Real time video denoising. In: IEEE international conference on engineering education: innovative practices and future trends (AICERA). IEEE; 2012. 2012.

Computational techniques for analysis of breast cancer using molecular breast imaging

10

Revathi[1], Sharmila[1], Daniela Danciulescu[2], Anitha Jude[3], Dharani[1], Harini[1], Muthu Manoj[1]

[1]*Dr. N.G.P. Institute of Technology, Coimbatore, Tamil Nadu, India;* [2]*Department of Computer Science, University of Craiova, Craiova, Romania;* [3]*Karunya Institute of Science and Technology, Coimbatore, Tamil Nadu, India*

1. Introduction

1.1 Anatomy and physiology of the breast

The breasts are constituted of fatty, connective, and glandular tissue. A cluster of lobes presents in the tissue. Every lobe is made of lobules, which are tiny, tube-like structures that consist of milk glands. Dense breasts are those that have an abundance of glandular and fibrous structures. Breast tissue types include fatty breast tissues, scattered dense fibrous and glandular tissues, and heterogeneously dense and highly dense tissues. A group of ducts that converge on the nipple is encased and enclosed with subcutaneous fat [1]. Hence, these tissues determine the size and form of the breast. The combination of several hormones includes progesterone and estrogens. Fascia, a thin layer of connective tissue, encircles the breast tissue. The superficial layer of this fascia lies just below the skin, whereas the deep layer rests directly on top of the pectoralis muscle. Fig. 10.1 shows the anatomy of the breast.

2. Breast cancer

Mammary tissue can grow into cancer in cases of breast cancer. It refers to a condition where the breast's cells proliferate out of control. Males are much less likely to experience it than females. The most common form of cancer among females is breast cancer. Surgery can cure the majority of benign breast tumors, one-quarter have a dormant and sneaky nature that spreads quickly but grows slowly. Current treatments considerably slow tumor growth, but regrowth is unavoidable, leading to high fatality rates [2]. Through blood and lymphatic vessels, breast cancer cells can spread outside of the breast. Breast cancer is known to have metastasized

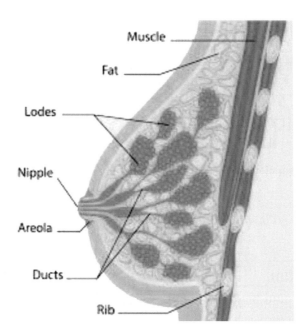

FIGURE 10.1

Anatomy of breast.

when it extends toward other body regions. Oestrogens have been linked to a variety of identified risk factors. Only a small percentage of occurrences of breast cancer is caused by mutations in specific genes [3].

3. Statistics

Breast cancer is the most prevalent type of cancer in women that affects 14% of Indian women. Recent research results show that Kerala has India's highest cancer rate. Some other states like Mizoram, Haryana, Delhi, and Karnataka are in India having high cancer incidence rates. The highest death rate in the nation was found in Mizoram, followed by Kerala and Haryana. Graphical representation describes the age-wise growth of breast cancer in X axis and a number of cases in percentage represented in Y axis.

Breast cancer was increasing gradually both year-wise and age-wise. This was due to lifestyle changes all over the country. According to statistics, breast cancer accounts for around 24.5% of all cancer cases worldwide. Figs. 10.2 and 10.3 shows the statistics of cancer rate.

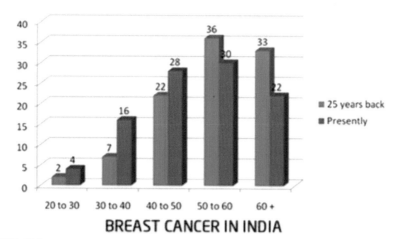

FIGURE 10.2

Graphical representation of the growth of breast cancer.

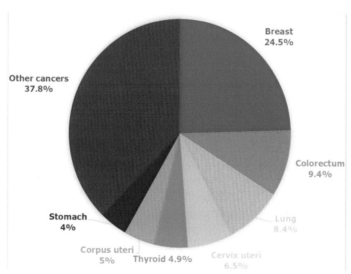

FIGURE 10.3

Statistics of cancer rate estimated by WHO in the year 2020.

4. Types of breast cancer

Typically tumors are the consequence of benign (noncancerous) alterations inside the breast. For instance, fibrocystic change, a noncancerous disorder, causes women to experience lumpiness, fibrosis (the creation of bruise-like connective tissue), cysts (cumulative packets of fluid), and areas of enlargement, discomfort, or breast pain [4]. Fig. 10.4 shows histopathological types of breast cancer.

4.1 Invasive ductal carcinoma

Cancer (carcinoma) known as invasive ductal carcinoma develops when irregular cells that proliferate in the lining of the duct. Invasive ductal carcinoma is a type of cancer (carcinoma) that develops when aberrant milk duct lining cells transform and spread into the nearby breast tissue. Breast cancer that has spread to other organs and parts of the body might develop from a cancer cell escaping through the lymph nodes or bloodstream. Intraductal cancer is commonly known as ductal carcinoma in situ of the breast [6].

4.2 Ductal carcinoma in situ

Ductal carcinoma in situ (DCIS) is the term for cancer that has developed and is still present (or "in situ") within the breast ducts. The cancer cells are not yet capable of penetrating out of these ducts into the breast tissue surrounding them or to other regions of the body [7]. DCIS shown in Fig. 10.5 is a controversial mammary gland disease.

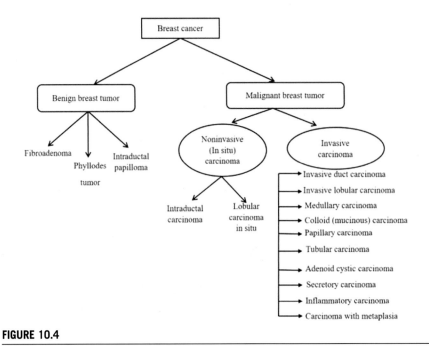

FIGURE 10.4

Histopathological types of breast cancer [5].

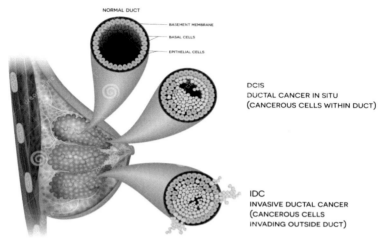

NORMAL DUCT
BASEMENT MEMBRANE
BASAL CELLS
EPITHELIAL CELLS

DCIS
DUCTAL CANCER IN SITU
(CANCEROUS CELLS WITHIN DUCT)

IDC
INVASIVE DUCTAL CANCER
(CANCEROUS CELLS
INVADING OUTSIDE DUCT)

FIGURE 10.5

Cancerous cell of DCIS.

4.3 Invasive lobular carcinoma

Breast cancer that starts in the lobules that produce milk is known as invasive lobular carcinoma. It has the potential to affect the lymph nodes as well as other parts of the body. Little of all breast malignancies are invasive lobular carcinomas. Invasive lobular cancer may show no symptoms at all. Invasive lobular carcinoma might result in the following as it spreads: a portion of the breast that has thickened; a fresh patch of breast heaviness or swelling. A modification in the skin's texture or appearance over the breast, such as thickening or dimpling. A newly flipped nipple. After invasive ductal carcinoma, invasive lobular carcinoma (ILC) is the most prevalent kind of invasive breast cancer (IDC) [8].

4.4 Lobular carcinoma in situ

In the milk glands (lobules) of the breast, abnormal cells can develop, a condition known as lobular carcinoma in situ (LCIS) which is shown in Fig. 10.6. It does not come under cancer. But receiving a diagnosis of LCIS suggests that your risk of getting breast cancer is higher. LCIS is a nonobligatory precursor of breast carcinoma as well as a risk factor for it [9].

4.5 Inflammatory breast cancer

Fig. 10.7 shows the diagnosis of inflammatory breast cancer. It primarily affects the breast skin, where the tumor develops along the skin's lymphatic vessels [10]. Breasts with inflammatory cancer have a recognizable red, swollen appearance because cancer cells obstruct the lymphatic veins in the skin surrounding the breast.

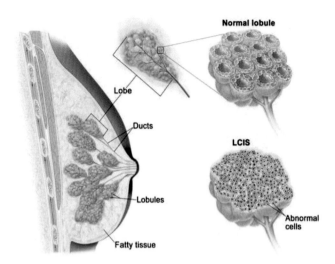

FIGURE 10.6

Lobular carcinoma in situ (LCIS).

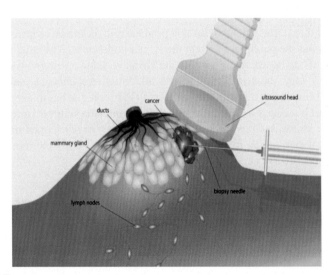

FIGURE 10.7

Diagnosis and staging of inflammatory breast cancer.

A locally progressed malignancy, such as inflammatory breast cancer, has migrated from its original site to adjacent tissue and may even have reached nearby lymph nodes.

Breast infections, which are a considerably more frequent cause of breast swelling and redness, are frequently confused with inflammatory breast cancer. Unlike other types of breast cancer, inflammatory breast cancer rarely develops a lump.

4.6 Pagets disease of the breast

An uncommon form of breast cancer called Paget's disease of the breast originates in the skin of the nipple. Fig. 10.8 shows Paget's disease of the breast. This particular type of cancer can have skin changes that affect the breast's nipple and may appear like eczema from the outside. A rare form of nipple—areola complex cancer called Paget's disease of the breast is frequently accompanied by an underlying in situ or invasive carcinoma [11].

4.7 Angiosarcoma

Blood vessel tumors known as angiosarcomas are uncommon, severe, and have low overall survival rates. Secondary neoplasia is caused by the increased use of radiation therapy for breast cancer. Angiosarcoma is the most typical secondary breast neoplasia. It happens 4—7 years following radiation therapy. In rarer cases, primary neoplasia can be seen. The preferred course of treatment for angiosarcoma is the radical surgical removal of the tumor with a substantial margin of safety [12].

4.8 Phyllodes tumors

A rare form of fibroepithelial lesion is phyllodes tumors [13]. The stroma, or connective tissue of the breast that contains ligaments and the fatty tissue that encircle the

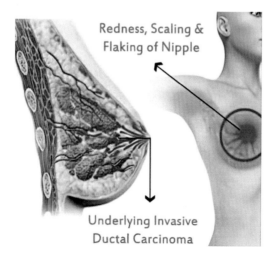

FIGURE 10.8

Paget's disease of the breast images.

ducts, lobules, blood arteries, and lymphatic vessels, is where phyllodes tumors begin. Phyllode tumors can also contain ductal and lobular cells along with stromal cells [14].

5. Screening methods

Breast cancer screening will differ greatly from existing standards in the coming decade due to the expected continuous, rapid advancements in technology. For ease of testing initial cancer specificity and sensitivity, new technology was introduced into clinical practice. This article investigates cutting-edge, novel methods for nuclear medicine, mammography, ultrasound, magnetic resonance imaging (MRI), and breast cancer screening. Fig. 10.9 shows various screening methods based on the sources.

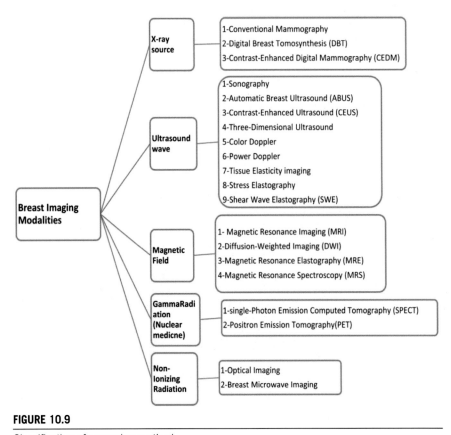

FIGURE 10.9

Classification of screening method.

5.1 **Mammography**

Mammography, which examines a person's breasts, uses X-rays as a diagnostic tool, just like other medical diagnostics systems. The results of these tests are saved as specialized pictures, which radiologists review for any potential anomalies [15]. The next paragraphs outline a few methods for detecting breast cancer early using mammography.

5.2 **Digital breast tomosynthesis**

Digital breast tomosynthesis (DBT), often utilized in conjunction with two-dimensional (2D) complete online digital mammography (DM), is now frequently used. The breast tissue was divided into thin slices. It reduces tissue overlap, resulting in a three-dimensional radiograph. It enables enhanced precision and sensitivity. Although DBT may improve the chances of cancer detection, certain studies demonstrate that the majority of cancers discovered with DBT tend to be less malignant [7,8], raising questions about overdiagnosis [16−18]. Tomosynthesis' key benefit is the ability to detect masses and lesions that may not be visible on a traditional mammogram because of overlap with dense breast tissue [19,20]. DBT is more effective than traditional mammography for detecting noncalcified lesions [21].

5.3 **Contrast mammography**

Contrast mammography is also known as contrast-enhanced spectral mammography shown in Fig. 10.10. It provides functional data regarding the anatomical and morphologic data that traditional (standard) mammography and DBT offer. Contrast-improved spectral mammography, similar to breast MRI, shows the (neo)vasculature connected to breast lesions and is practically independent of breast density.

   To perform contrast-enhanced spectral mammography as shown in Fig. 10.10, a dual-energy mammography system must be added to a number of existing mammography units. Iodinated contrast material must also be injected intravenously at a dose of 1.5 mL/kg, then the typical craniocaudal and mediolateral mammographic views must be performed. Each image is composed of a "low-energy" acquisition that resembles a standard mammogram and a "high-energy" image that was created by enhancing the contrast material signal with a kilo electron volt above the iodine K-edge. For the identification of masses and calcifications, the low-energy image is equivalent to typical mammography, and it takes the place of the traditional mammogram [22−24]. After postprocessing, the background signal is removed from the low- and high-energy images, and the combined iodine-only image aids in the detection of enhancing lesions. Contrast-enhanced spectral mammography is secure, and severe adverse responses to the contrast medium are uncommon [25]. Contrast-enhanced spectral mammography uses a larger radiation dosage than traditional mammography [26,27], but it is still below the US Food and Drug Administration's mammography safety standards [28]. It is a drawback that there is

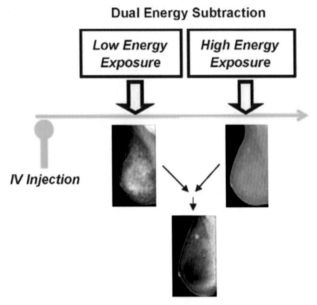

FIGURE 10.10

Contrast-enhanced dual-energy mammography [22].

currently no contrast-enhanced spectral mammography biopsy instrument for lesions detected only during contrast-enhanced spectral mammography, though prototypes are being evaluated.

According to research, contrast-enhanced spectral mammography works better than combined mammography and ultrasound for detecting lesions and determining the severity and range of the diseases [29,30]. The majority of research using contrast-enhanced spectral mammography provides greater precision while just slightly reducing sensitivity as compared to MRI [31,32].

5.4 Ultrasound approach

It has been found that ultrasound breast imaging is effective at finding cancers in dense breasts that mammography often missed [33]. A noninvasive and nonirradiating approach is ultrasound. A transducer sends high-frequency sound waves which are greater than 20 kHz to produce the breast images. Numerous developments in ultrasound imaging technology have been made, such as automated breast ultrasound (ABUS), power Doppler, color Doppler, 3D ultrasound, and sonoelastography.

5.4.1 Sonography

The wave sound reflected from the interior tissue is used to generate the image. Differentiating between the cystic or solid nature of palpable lesions was the primary indication for breast ultrasonography [34]. In contradiction to mammography, ultrasonography can produce a thorough description of the breast's anatomical structure. Sonography is capable of diagnosing many benign and malignant disorders of the breast on its own [35]. In dense breasts, ultrasound can assess the internal structures, shape, and orientation of lesions on several surfaces. Evaluation of these characteristics facilitates the distinction between benign and solid breast lesions [38].

5.4.2 Automatic breast ultrasound

The operator applies flat ultrasonic plates to each breast to get three different locations. A 3D reconstruction of the image is then obtained. Precise accuracy for diagnosis and good assessment of lesion size are two benefits of ABUS [34]. For females with dense breasts, this approach is effective. This technology uses a transducer 15.3 cm long that is automatically tuned between 6 and 15 MHz to produce 3D datasets with 340 images per capture.

5.4.3 Contrast-enhanced ultrasound

The intensity and extent of a tumor's vascularization are determined by its pathology. Malignant breast lesions have vascular distribution abnormalities. The clinical study employs contrast-enhanced ultrasound (CEUS) to view various vascular structures. To enhance the rebound of the waves during CEUS, gas microbubbles are injected into the veins. Some gases are trapped in various kinds of capsular shells to form microbubbles [36]. Their sizes range from 1 to 7 m in diameter. Microbubbles do not penetrate through the vascular wall and are much more echogenic than R's. The ultrasound contrast medium primarily functions as a nonlinear impurity because of the variations in acoustic impedance as well as compression between both the microbubbles and nearby tissues.

5.4.4 Ultrasonography in three dimensions

Two basic varieties of 3D ultrasonography exist. Utilizing instruments for double-dimensional imaging is the first type. Features of unique mechanics, which ultimately rebuild the three-dimensional ultrasonic volume. The second method makes use of an array matrix converter to scan a 3D area. The matrix array generates a beam and produces a pyramidal region [37].

5.4.5 Color Doppler

Tumor angiogenesis can be detected using a standard ultrasonography approach [34]. Applying mean Doppler shift coding in the area of interest, color Doppler calculates the local blood flow and colors it. To represent blood flow in color Doppler, the mean frequency shift can be calculated. The Doppler signal is typically increased by malignant breast lesions. Doppler imaging in color is heavily reliant on technical considerations. In Doppler, scale reduction utilizing power Doppler imaging and raising the color rate to improve vascular flow sensitivity. Reduced color box size results in increased color Doppler sensitivity. The decrease in transducer pressure on the breast has an impact on the color Doppler as well. The veins can easily collapse when pushed amid the chest wall and transducer due to the relatively superficial nature of the breast tumors.

5.4.6 Power Doppler

It refers to a method that shows the Doppler signal in a specific color while encoding pulse power through the signal. In contrast to the frequency shift, the parameter is distinct. Red blood cell (RBC) velocity determines the frequency. In other words, the signal's strength is dependent on how much blood is in the target location. In this technique, the direction of flow is typically not significant. Power Doppler's benefits over color. Doppler's advantages include its great sensitivity to blood flow, improved resolution, particularly at edges, and identification of continuous flow of blood. To diagnose firm breast lesions, a power Doppler is frequently utilized [38].

5.4.7 Tissue elasticity imaging (sonoelastography)

The use of ultrasound to image tumor elasticity is one of many recent initiatives to increase the specificity and sensitivity of the diagnosis of breast cancer via ultrasound. Solid tumors are more solid structurally than the surrounding tissue. Young's modulus determines the tissue's stiffness (often referred to as elasticity; the distortion of tissue brought on by pressure). From 1 to 100 kPa is the soft tissue elasticity range.

5.4.8 Stress elastography

This technique looks at how the texture alters in response to an external force. When exposed to an ultrasonic wave, soft tissue varies more than hard tissue. The required elastogram image is obtained and merged with the nonforce state image, and the necessary static force is given to the tissue. This pressure may be brought on by the probe's self-controlled movement, the patient's movements, or an outside cause. Elastogram noise results from excessive compression or propagation force, and an absence of an elastogram results from insufficient compression. However, user performance is heavily reliant on the sonoelastography approach, which does not require complicated software [39,40].

5.4.9 Shear wave elastography

Shear waves, which are transverse ultrasound waves, are created when the volume of a material changes as a result of ultrasonic pressure. More quickly than in soft tissues, these waves migrate through malignant mass-like hard tissues. The frequency is between 10 and 2000 Hz, whereas the wave speed is between 1 and 50 m/s. Shear wave velocity is measured in units of meters per second (m/s) or kilopascals (Young's modulus). In this technique, the ultrasonic waves themselves dynamically compress the tissue in the imaging area, and the wave velocity is concurrently manifested and transformed into a picture.

5.4.10 Diffusion-weighted imaging

Due to the existence of cell membranes as well as other physiological barriers, the pathway for the discharge of water molecules in the body environment is indeed not random. Water diffuses among inner cellular molecules, creating a vivid image that can contrast the reduction of normal tissue caused by tumors with the change in water diffusion patterns. Diffusion-weighted imaging (DWI) is a quick method (120−180 s scan time). It is less expensive than other contrast-enriched MRI methods and does not call for intravenous gadolinium infusion. Even for structures smaller than a millimeter, DWI is an extremely sensitive approach according to Brownian movements of molecules. Images are created by the water in biological tissue. The majority of DWI sequences employ (EPI) echo planar imaging. Quick imaging methods like EPI boost signal-to-noise ratio while minimizing scan duration and motion-induced deterioration. In tissue, cellularity and the integrity of cell membranes are inversely correlated with water permeability. However, in areas with slow propagation, like malignant tissues, the emission strength in DWI is strong. DW pictures show larger signal strengths for molecules with higher T2 relaxation durations [34,41,42].

5.5 Magnetic resonance elastography

To ascertain the biomechanical characteristics of each organ's tissue and any changes brought on by disease, magnetic resonance elastography (MRE) is performed. One of the crucial elements for assessing MRE is the organ of interest's ability to transmit shear waves adequately. Although the manual mass examination makes the operation easier for the breast, it can be challenging for deeper organs like the pancreas and heart. The tissue shows biomechanical changes during carcinogenesis. As a result, using MRE to measure biomechanical factors can yield useful diagnostic data. The MRE generates shear waves with the help of an engine coil that is placed close to the breast. These coils' magnetic fields induce visual artifacts that can be seen in the final product. A tensile-compression mechanism is used to put the coils away from the breast to solve this issue. This makes the patient more comfortable and enhances image quality.MRE offers data on tissue stiffness to ascertain lesion features, much to sonoelastography [42].

5.6 Magnetic resonance spectroscopy

To learn more about the chemical composition of breast lesions, magnetic resonance spectroscopy (MRS) can be used in conjunction with MRI. This method's results are utilized for a variety of purposes, such as tracking the effectiveness of cancer treatments and enhancing lesion diagnosis precision. Before any morphological alterations, breast MRS indicates changes in intracellular metabolism [43]. MRS is typically employed to evaluate the rate of tissue metabolism within the body. The apparatus of it is comparable to a traditional MRI and uses spectral signal acquisition sequences to determine the distribution of a certain metabolism in a volume.

PET is a metabolic and molecular imaging modality, whereas SPECT is a physiological imaging technique [44]. 99mTc is uptaken by inactive mitochondria in cancerous breast cells during SPECT after being injected. Gamma rays with an energy of 140 keV were released and captured in a sizable planar gamma detector underneath the compressed breast [45]. Due to the poor spatial resolution, the SPECT sensitivity for tiny lesions is minimal. Additionally, SPECT exposes patients to radiation at a rate that is 20–30 times greater than that of digital mammography.

5.7 Magnetic resonance imaging

A precise image of the breast is produced by breast MRI (BMRI), which employs a magnetic field and radio waves. Typically, radiographs of both breasts are produced using a 1.5 T magnet and a contrast material, most often gadolinium. The patient is placed in a posture during image acquisition to reduce breathing activity and enhance breast tissue expansion. The applied magnetic field temporarily realigns the water molecules. When radio waves are being used, highly designed coils placed at the breast positions emit the radiation and convert it into electrical signals [46,47]. Conversely, one drawback of BMRI is its tendency to diagnose benign lesions smaller than 5 mm in size.

5.8 Optical imaging

All imaging using ultraviolet, electromagnetic waves, and nonionized visible is referred to as optical imaging, which is a broad term. Currently, this noninvasive imaging technology uses visible light (400–700 nm) and near-infrared (NIR) wavelengths to examine the tissue's molecular, morphological, and functional characteristics as well as its fluorescence properties, scattering, and absorption. Currently, this noninvasive imaging technology uses visible light (400–700 nm) and NIR wavelengths to examine the tissue's molecular, morphological, and functional characteristics as well as its absorption, dispersion, and fluorescence characteristics. The capability of optical imaging to gather data at the metabolic and molecular scale, which may then be used to create images of microscopic structures and tissue function, is one of its benefits [48,49]. The optical technique has the features of using—antiionizing infrared light beams as its source. It is a cheap technology that produces images of the body's physiological and molecular processes. The depth to which light penetrates tissue when using optical techniques for clinical imaging is one of the main issues and challenges.

5.9 Nuclear imaging techniques

These procedures involve the intravenous injection of particular radiopharmaceuticals, and scintillation imaging is used to gather images showing drug uptake in the organs. Current technology of breast cancer to analyze the screening and staging methods generate a flawed diagnosis effect. Therefore, molecular imaging of tumor metabolism, proliferation, and other more tumor-specific targets may be contended to use therapy for breast cancer. The benefit of the use of molecular imaging is to create a focus on drug progress and the interpretation of breast cancer science.

Optical imaging, radionuclide imaging with positron emission tomography (PET), or single photon emission computed tomography are methods for molecular breast cancer imaging, these imaging techniques are sensitive serial noninvasive information of tumors.

5.10 **Molecular breast imaging**

According to the Society of Nuclear Medicine, molecular imaging involves the Image diagnosis, analysis, and measurement of imaging processes at the molecular and cellular functions of physical bodies of living systems. It can be carried out using a variety of imaging modalities. For magnetic resonance imaging, the ligands employed for this can be tagged with a contrast agent of magnetic resonance imaging and tomographic techniques.

A new nuclear medicine procedure called molecular breast imaging (MBI) uses miniature semiconductor-based gamma cameras in mammograms to produce high-resolution functional images of the breast.

Fig. 10.11 indicates the invasive ductal cancer image from mammography, MBI, and MRI. The existence of cancer is more clearly visible in MBI and MRI than in mammography. Presence of cleavage in the image from mammography and pectoral muscles in the image form MBI is identified in Fig. 10.12. The physiological lifespan of the isotope is chosen based on the proposed imaging technique, such as

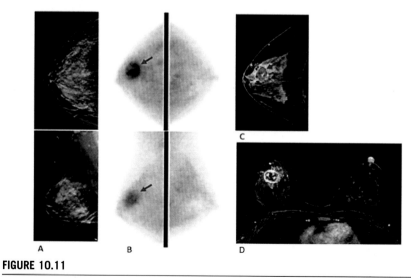

FIGURE 10.11

Right breast invasive ductal cancer is not detected by mammography (A); however is seen on MBI (B) as well as with breast MRI (C, D).

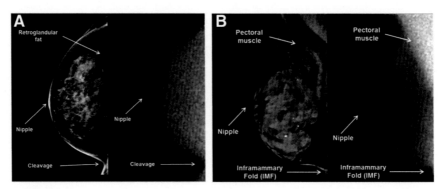

FIGURE 10.12

Proper breast positioning on mammography (*left*) and MBI (*right*). (A) Presence of cleavage; (B) presence of pectoral muscle in breast.

PET or SPECT, and it is ideal for it to agree with the biological life of the binding. After interacting with an electron, the positron and electron are annihilated, and their combined masses are split into two 511 keV gamma rays that are released 180 degrees apart. Generated photons are detectable by a PET camera, which has sets of two detectors that both capture a photon while simultaneously timing the usage of radioisotopes to emit single gamma photons in medical imaging.

In the MBI scan procedure, a single injection of 20 mCi Tc-99m sestamibi has been used in recent MBI research surveys, with breast cancer imaging starting about 5 min after injection. During the third of the mammography compression force ($\sim$ 15 lb force), the breast appears to be compressed between two detectors. Naturally, 10-min images are obtained for both the breast in the MLO Projection as well as in CC with the total time for imaging being about 40 min. Even though the technique takes longer than mammography, patients have reported that the lack of pain associated with significant compression overcomes the drawback of the lengthy imaging time. The MBI procedure is essentially similar to conventional mammography, except using a lighter compression force, and the images can be quickly compared. Nuclear medicine technologists with training in mammographic positioning procedures carry out the imaging.

5.11 Positron-emission mammography

PET using 2-deoxy-2-[18F] fluoro-D-glucose (FDG) seems to have inadequate pixel density for the diagnosis of small tumors (under 10 mm), with a sensitivity of about 57%, similar to conventional cancer screening, which lacks pixel density for the detection of tiny breast tumors with Tc-99m sestamibi [50]. An advanced technology called PEM has been created, using detectors composed of multicrystal arrays either bismuth germanate or lutetium orthosilicate [5,51]. Similar to mammography, PEM uses two high-resolution detecting heads that are positioned on either side of a squeezed breast and obtained the tomographic breast slices. In-plane resolution is 1.5 mm as compared to 4−6 mm for the normal PET scanner because detectors are close in proximity to the breast. Initial detection has demonstrated good specificity of greater than 90% and sensitivity equivalent to the MRI for the identification of breast cancer [52].

5.12 CZT-based molecular breast imaging system

CZT detectors can only attain energy resolutions of 3%−4%, while compact multicrystal Y- y-cameras used to scan Tc-99m typically obtain energy resolutions of 12% −20%. Reduced dispersion and better visual contrast result from this enhancement in energy resolution. Additionally, it is currently possible to build CZT detectors with pixels that are as low as 1.6 × 1.6 mm, with the potential for much lower pixels in the approach. Image resolution has become more reliant on the inherent resolving power of the detector since the breast can be placed in contact with its surface. Generally, multicrystal systems are restricted to a minimum pixel density of about 3 mm. Although these intrinsic quality differences have little relevance in nuclear imaging, they are crucial to the advancements in the accuracy and resolution of tiny detectors designed for breast imaging. Due to its superior inherent spatial and energy efficiency compared to multicrystal detectors, CZT allows better detection of lesions smaller than 10 mm, according to a comparison of three different detector technologies [53]. A drawback of the material is still quite new but more expensive than the extensively used detector materials.

5.13 Alternative radiopharmaceuticals

Tc-99m sestamibi is used in MBI studies conducted in the past. Other additional radiopharmaceuticals have great potential for breast imaging. Research is being conducted to determine the effectiveness of neoadjuvant therapy with a revolutionary Tc-99m-labeled angiogenesis antagonist [54]. In addition, several previous radiopharmaceuticals are now in use and need to be reviewed using advanced technologies. Comparatively, Tc-99m sestamibi has proven to be substantially more effective for breast imaging. Other substances, like Tc-99m DMSA (V) [55] as well as Tc-99m tetrofosmin [56], may also be more beneficial than previously seen clinical investigations using scintimammography.

Detecting tiny tumors in the breast, MBI could accurately detect small boobs lesions, or lesions less than 15 mm in diameter. The design is to select the patients with small lesions in mammography and analyze the capability of detection. Database of 150 patients' breast imaging and data system (BI-RADS), fewer lesions are detected in the size of 20 mm in estimated time [57].

6. Image processing techniques

6.1 Artificial intelligence technique

ML is a branch of AI, and it refers to algorithms that benefit from data to work better. There are two ways that data given into an ML program can be depicted: as features or as raw data. Features are quantifiable variables in data. DM/US/MRI scans represent the raw information in breast cancer screening, whereas lesion size. When raw data are used as the input, the algorithms must identify the features on their own. Learning features are challenging for these algorithms, even if they often perform better than utilizing hand-crafted features. The subtype of ML techniques known as DL can be used to overcome this issue. The characteristics in DL are expressed in terms of other, more basic features. As DL algorithms are made up of many (deep) layers of neuron layers, they are frequently described as deep neural networks (DNNs). CNNs are a specific kind of DNN which are frequently employed in breast cancer image recognition because they were created particularly to detect valuable features in images.

Medical research targeting breast cancer is not new and its under research since the 16th century. Medical image processing has proven to be quite helpful in the diagnosis of a variety of diseases, as shown by the data, and is not just restricted to cancer conditions. Early diagnosis and treatment can reduce the probability of the disease in the early stage. Before a diagnosis can be made, medical images must undergo particular processing due to their unique characteristics. Different categories and methods are implemented for the diagnosis of diseases In this study, we made an effort to give readers a better understanding of the various breast cancer detection methods with performance and accuracy. Each method has been discussed with the distinct features of breast cancer.

Four phases combined to form the breast cancer diagnosis method: acquiring the image, segmenting, and preprocessing it, extracting the features, and classifying it. In Fig. 10.13, it is observed that the internal tissue dynamics of the breast are acquired using a number of different technologies. The second step involves computational algorithms that carry out fundamental operations on the images allowing for segmentation to identify the region of interest (ROI). In the third stage, the classification technique is implemented to diagnose a disease. Numerous advancements have been made in all the above-mentioned areas as a result of the quick invention of new technologies that adequately capture the dynamic behavior of breast tissues

FIGURE 10.13

BC detection using image processing strategies.

[58,59]. Recent reviews of various approaches for feature classification and its interpretation have been discussed in various research articles [60–63].

6.2 Dataset

There are numerous publicly accessible databases for the diagnosis of breast cancer. Several datasets include viewpoint, malignant growth box, obstruction, and other features to help with cancer detection. The major objective of discussing these datasets is to assist researchers and enhance the functionality of the CAD system. The various datasets for breast cancer from MBI images are discussed in this section.

6.3 QIN-breast

These longitudinal PET/CT and statistical MR images were gathered for the investigation of treatment evaluation in preoperative breast cancer. Three time points were used to obtain the images: before the first cycle of therapy (t1), during the second cycle of treatment, or when all treatments had been completed (t), before surgery (t3). A support system that was created in-house allowed the patient to be in the supine position during the acquisition of the PET/CT images, which made it easier to register the MRI data. This collection is valuable because it offers clinical visual information for the creation and use of quantitative imaging techniques for treatment assessment early in the therapy process for breast cancer. The QIN-BREAST-02 set serves as an addition to the QIN-BREAST collection. This dataset was gathered at Vanderbilt University Medical Center.

6.4 Databases management of querying breast cancer image

In many parts of the world, breast cancer continues to be the primary cause of cancer mortality in women. Over 40,000 women every year in the United States alone pass away from the illness. Nowadays, mammography is the most economical way to find breast cancer early. When it comes to finding malignancies or evaluating malignancies in certain types of women, complementary health imaging techniques like ultrasound or MRI may be more helpful than mammography. Research on the efficiency and use of each approach for cancer screening and the determination of malignancy will be made possible by a collection of images from several technologies, including mammography, MRI, PET, as well as ultrasound. This database was built using SQL Server, but using it to construct a solution that will give doctors a web-based query tool to access the data via web services [64]. The reliability of the prognosis will also be improved because doctors would be able to locate cases similar to a current patient. By providing high-quality datasets, the preservation of datasets for comparison, a tool for the complex combination of attributes through spatial arrangement across images, and designed data analysis, this directory will be an excellent tool for the advancement in computer-aided detection methodologies. For this tool, a visual query description system will allow users to express questions by pointing out the desirable characteristics, structures, or qualities and describing the spatial information between them through direction and distance. Additionally, long-distance, electronic visual transmission for clinical consultations is made possible by secure data archiving and restoration. The database will allow for the comparison of images taken from the same patient over time, enhancing the ability to identify extremely early malignant signs and, hopefully, increasing the likelihood that a patient will survive their malignancy.

7. Image processing techniques

Fig. 10.14 shows various phases of image processing techniques. It includes preprocessing, segmentation, feature extraction, feature selection, and classification techniques. Preprocessing approach performs denoising and image enhancement process. Segmentation process highlights the interested area in the image. Feature extraction involves the various characteristic parameters obtained from the breast image. It is followed by the classification of an image to diagnose a disease.

7.1 Preprocessing techniques

The key limitations of mammography include noise, uneven lighting, and low contrast. As a result, AOI recognition and extraction of features are challenging in this situation. Before segmentation and feature extraction, denoising, cropping, and augmentation of images can be implemented at the preprocessing stage to minimize the consequences of these flaws. Cropping can be used to get rid of extra labels,

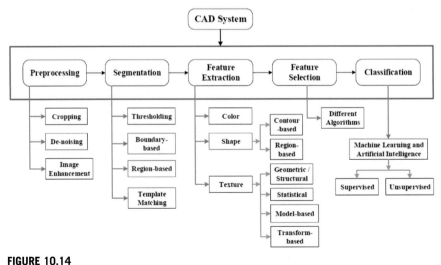

FIGURE 10.14

Different phases of a CAD system.

artifacts, and information-free areas of the image. Readout and shot noise are two types of noise that could be present when a digital image is being acquired [65]. When an image is denoised, the signals are smoothed as well as the noise is removed. The enhancement process raises the contrast level of the image, making the image details easier to view. The detection of masses is much more difficult than the detection of microcalcifications because the characteristics of masses are sometimes difficult to distinguish from normal breast tissues [66]. The microcalcifications can be easily spotted through denoising and enhancement of image using dyadic wavelet processing [67]. Because of the presence of high-frequency components, it produces a fine contrast image in the region of the interest. Masses with different densities, spiculated patterns, and low-frequency components can be detected with a sensitivity and specificity of 96.2% and 94.4%, respectively, due to the application of contrast limited adaptive histogram equalization (CLAHE) and median filtering [68].

7.1.1 Thresholding

This highly popular approach is image partitioning. In this approach, the background image with no clinical information is eliminated. The threshold value is chosen based on the gray-level histogram, and the image is divided into useful and background portions by the difference in pixel intensity values [69]. Although it is a quick and easy approach to implement, it does not ensure object consistency, for which certain other operators may need to perform postprocessing. A single thresholding value T, is set for the entire image $x(i, j)$ is called global. A local threshold is chosen for each subregion of a segmented picture based on both and a particular region image property $L(i, j)$. Bilevel and multilevel thresholding are the two types of thresholding. In bilevel thresholding, an image is split into two divisions, the main region is denoted by white, and the backdrop is denoted by black. Images with various surface properties require multilevel thresholding [70]. The traditional thresholding strategies are the least error method, the maximum entropy method, and Otsu's method. Otsu's thresholding is responsive to salt and pepper noise, thus the image must first be denoised to smooth it out [71]. Thresholding was employed by researchers to divide a mammogram at various levels, and from each segmented region, a set of characteristics were generated [72]. Another study [73] suggested a probabilistic adaptive thresholding method that depends on textural information and its likelihood to find the most practical threshold values for particular mammography regions. The threshold values in this adaptive thresholding method weren't determined by the shape of the region or by a histogram. This was done to solve the problem of nonuniform intensities in the background region of mammography, which may be a problem for global threshold-based approaches. Three kinds of threshold approaches and edge detection algorithms were applied for segmentation [74]. In Ref. [75], Otsu's thresholding and hybrid image segmentation were employed for the precise diagnosis of a breast tumor and the measurement of its size.

Even in real-time practical uses, implementing threshold is straightforward. It is affordable to compute and quick. Furthermore, no knowledge of the image is needed beforehand. However, it performs poorly for images that are chaotic and for images that lack peaks or have broad or plain troughs. Thresholding's primary flaw is that it skips an image's spatial data, leaving uninformed about whether the segmented portions are continuous. Additionally, only a proper threshold choice shall prevent poor nutrition or oversegmentation. As evidenced by the studies in Refs. [76−79], thresholding can produce superior results when combined with other methods.

7.1.2 Segmentation based on region

The essential tasks in the segmentation are the elimination of the image background and the AOI selection. Common techniques for segmenting images include boundary-based segmentation, thresholding, template matching, and region-based segmentation. Fig. 10.15 shows various classifications of segmentation.

By using region-based segmentation, distinct areas in an image that share attributes like color, structure, or grayscale can be distinguished. This procedure is referred to as the growth area or splitting process. At this stage, to identify tumor locations, the AOI is chosen using a specified criterion according to the prior acquired result with the image's intensity or edge details. Furthermore, extra functions like uniform blocking, merge, and split are required for this strategy [69,80]. This technique was employed to separate the pectoral tissues from the mammography and then it was used for categorization [81]. The implementation of mean-based region growing segmentation (MRGS) was established for segmentation [82]. Scientists used an artificial neural network (ANN) to determine the threshold for an automatic sector growth segmentation approach [83].

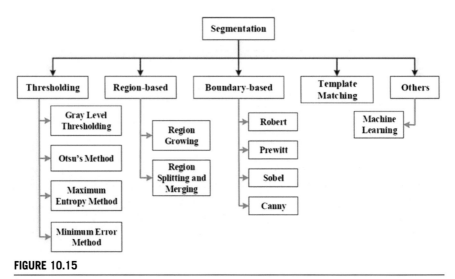

FIGURE 10.15

Classification of segmentation.

Automatic seed selection was carried out before the local thresholding approach was used in both [84,85] studies. Before the introduction of the region growth method, automatic seed selection was carried out. With region-based segmentation, an active and automated approach can be implemented. The movement from an inner point to the outer region results in the generation of borders in the interested area in the image. With the right seed, this method produces results that are superior to those of all other segmentation techniques. On the other hand, a flawed segmented region may result from noisy seed selection. As it is sequential by origin, it has less impact on minor locations. The major drawbacks of this approach are storage, longer computation times, and halting criteria.

7.1.3 Template matching

The challenge of identifying an object in an image is crucial. With prior knowledge of the recognized object or template, this issue can be handled. That can be used to pinpoint its exact position inside a scenario. The biggest disadvantage of this strategy is how challenging it is to apply this procedure. To find the questionable regions and improve them with thresholding, investigators in Ref. [26] used the Sech template. In Ref. [86], the contour was optimized using the template-matching method, a local cost operation, and dynamic programming.

7.1.4 Boundary-based segmentation

Robert operator:

Its 2D spatial gradient measuring of an image and 2×2 convolution mask enable it to be a gradient-based edge detection operator. The Roberts method's adaptive threshold sensitivity is provided. Edges that are weaker than the criterion are ignored [86].

Prewitt operator:

It is a gradient-based edge detector operator and a basic approach to determine the direction and intensity of an edge and estimate in the 3×3 vicinity for eight directions [86].

Laplacian of Gaussian (LoG):

The digital representation of the Laplacian function is accomplished by employing the mask, and also the log of an image is a second-order derivative. It initially blurs the image before computing the Laplacian [86].

Canny edge detection:

The canny operator limit consists of two components: low barrier and high barrier. Canny edge detection is a frequently used technique for processing images. It is a very reliable way to discover edges in the image by reducing noise from the input image and is less adaptable to noise. It is quicker than some other edge detection techniques and preserves image features while extracting the image's features [86].

Sobel operator:

This operator is used to identify the image's border direction and intensity. Using the Gaussian mask of the Gaussian function, gradient measurements of the original picture are obtained using Sobel. 3×3 convolution kernels are used [63].

8. Classification techniques

Convolutional neural networks (CNNs) and their derivatives are among the most well-known DL techniques that have been frequently used in molecular imaging. The translation-invariant property of CNNs is a crucial attribute and a key factor in their prominence in computer vision and molecular imaging applications. Fig. 10.16 indicates the CNN model to diagnose breast cancer. The convolution operator is typically used in the convolution layer to identify and extract spatial characteristics from the image. The feature map output from this layer may contain negative values, which could lead to numerical instability during training [87]. Hence, it is processed by the pooling layer to prevent negative values. The output of the pooling layer is then sent to the fully connected layer to be classified after the feature map has been processed, where it is reduced in size to get rid of redundant information. In this regard, CNN has been used in various works [88] to identify benign and cancerous regions in mammography or MRI images. As they point out, the depth of Lecun et al. introduced the first CNN model. The growing interest in medical imaging applications was sparked by the success of AlexNet. AlexNet won an ImageNet competition. convolutional and pooled layers are two examples of the many specialized layers that make up a standard CNN [88].

The network's central node is the convolutional layer and it filters the picture field effectively using a kernel (e.g., PET image voxels). In comparison, the completely associated layer's computing process is fairly time consuming. To generate an accurate description of the molecular image that is more clearly relevant to the goal, such as categorization or lesion identification in a PET image, the pooling layers undertake data reduction with functions like greatest or average pooling. Other neural networks, besides CNNs and their variants like the aforementioned LeNet and AlexNet, have shown enhanced classification and object recognition capabilities over the original CNN in a number of medical applications, including PET imaging. These neural networks include PyramidalNet, Inception, ResNet, VGG, DenseNet, and U-Net. These algorithms differ in terms of their attention processes, channel boosting, depth and width, and spatial invariant features [89—92]. The recurrent neural network is another kind of DL method that works with temporal information and data sequences. RNNs have achieved outstanding results in a variety

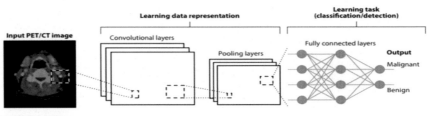

FIGURE 10.16

Classification techniques.

of applications, including dynamic image analysis, audio processing, genome sequencing, and natural language processing (NLP). Initial RNNs had problems with optimization and could not handle data that was spread out. Current RNNs are built using an architecture known as long short-term memory (LSTM) (e.g., transformer networks applied in NLP). With the help of a data sequence, LSTM's fundamental components, known as gates, can learn to memorize and then forget information stored in internal memory and establish long-term dependencies [93,93a]]. As a result, researchers suggested a more straightforward architecture termed gated recurrent units, which only has two gates (update and reset) [94].

The generative adversarial network (GAN) is another form of DL method that has been extensively used in medical and molecular imaging [95]. In GANs, two neural networks—one acting as a generator and the other as a discriminator compete against one another in a zero-sum game. While the discriminator tries to tell any bogus data from real data, the generator tries to create additional data from noisy data to mimic the original data. This design has been used to generate fake data (like synthetic brain PET) and change one diagnostic modality into another (e.g., MRI into synthetic CT or PET). The role of transfer learning and data preprocessing techniques in enhancing and supplanting the limited clinical and molecular imaging data available is responsible for many of the successful uses of DL in medical imaging [96–99].

9. Conclusions and future directions

The use of advanced data analytics for quantitative molecular imaging has attracted a great deal of attention in the last 10 years. This area of research has concentrated primarily on the function of ML/DL methodologies and their capability to change the molecular imaging landscape and address many of its issues. In addition to discussing instrumentation layout and optimization, reconstruction, classification and registration, screen resolution, CAD, treatment planning, and result prediction, this paper has shown examples of ML/DL in PET images. In every instance, ML/DL algorithms have shown performance on par with or better than cutting-edge methods. Despite their initial success, it is anticipated that these algorithms will still have more to contribute, particularly in bridging the gap between technological development and clinical care implementation. The availability of superior interpreted data, intensive testing, and validation to ensure consistency, and novel tools to increase data comprehensibility in research and clinical practice are all obstacles facing AI-guided PET imaging.

A new generation of quantitative PET imaging technologies will most likely expand the scope of radiological diagnosis, treatment administration, and prognosis that may be accomplished with PET. However, to fully realize the potential of AI and specifically, DL algorithms for better molecular imaging, massive curated and annotated data points for training and validation will be required, which may necessitate top scientific societies. Although the focus of this analysis has been on PET imaging,

SPECT, other molecular imaging, and theranostic techniques are also applicable [100].

The current limitations and challenges with deep learning in molecular imaging are that the use of ML/DL in medical imaging generally and molecular imaging specifically is not without difficulties or disagreements. These include the dearth of high-quality interpreted data and incompatibilities between the development dataset and the intended application environment [101], in addition to the relationship between imaging variables and clinical or biological outcomes.

References

[1] Bazira PJ, Ellis H, Mahadevan V. Anatomy and physiology of the breast. Surgery 2022;40(2):79–83.

[2] Key TJ, Verkasalo PK, Banks E. Epidemiology of breast cancer. The Lancet Oncology 2001;2(3):133–40.

[3] Cowin P, Rowlands TM, Hatsell SJ. Cadherins and catenins in breast cancer. Current Opinion in Cell Biology 2005;17(5):499–508.

[4] Sharma GN, Dave R, Sanadya J, Sharma P, Sharma KK. Various types and management of breast cancer: an overview. Journal of Advanced Pharmaceutical Technology & Research 2010;1(2):109.

[5] Schilling K, Conti P, Adler L, Tafra L. The role of positron emission mammography in breast cancer imaging and management. Applied Radiology 2008;37:26–36.

[6] Burstein HJ, Polyak K, Wong JS, Lester SC, Kaelin CM. Ductal carcinoma in situ of the breast. New England Journal of Medicine 2004;350(14):1430–41.

[7] Grimm LJ, Rahbar H, Abdelmalak M, Hall AH, Ryser MD. Ductal carcinoma in situ: state-of-the-art review. Radiology 2022;302(2):246–55.

[8] Cristofanilli M, Gonzalez-Angulo A, Sneige N, Kau S-W, Broglio K, Theriault RL, et al. Invasive lobular carcinoma classic type: response to primary chemotherapy and survival outcomes. Journal of Clinical Oncology 2005;23(1):41–8.

[9] Wen HY, Brogi E. Lobular carcinoma in situ. Surgical Pathology Clinics 2018;11(1): 123–45.

[10] Irakleidis F, Tan PH. Inflammatory breast cancer; diagnostic and therapeutic challenges. Therapy 2019;2(1):126.

[11] Ashikari R, Park K, Huvos AG, Urban JA. Paget's disease of the breast. Cancer 1970; 26(3):680–5.

[12] Ayadi M, Berrazaga Y, Adouni O, Meddeb K, Gamoudi A, Mezlini A. Breast angiosarcoma: a case report. International Journal of Case Reports and Images 2017; 8(10):643–7.

[13] Rowell MD, Perry RR, Hsiu JG, Barranco SC. Phyllodes tumors. The American Journal of Surgery 1993;165(3):376–9.

[14] Mishra SP, Tiwary SK, Mishra M, Khanna AK. Phyllodes tumour of breast: a review article. International Scholarly Research Notices 2013;2013.

[15] Yasmin M, Sharif M, Mohsin S. Survey paper on diagnosis of breast cancer using image processing techniques. Research Journal of Recent Sciences 2013. ISSN 2277-2502.

[16] Lång K. The coming of age of breast tomosynthesis in screening. Radiology 2019; 291(1):31–3.

[17] Hovda T, Holen ÅS, Lång K, et al. Interval and consecutive round breast cancer after digital breast tomosynthesis and synthetic 2D mammography versus standard 2D digital mammography in BreastScreen Norway. Radiology 2020;294(2):256–64.

[18] Conant EF, Zuckerman SP, McDonald ES, et al. Five consecutive years of screening with digital breast tomosynthesis: outcomes by screening year and round. Radiology 2020;295(2):285–93.

[19] Svahn TM, Houssami N, Sechopoulos I, Mattsson S. Review of radiation dose estimates in digital breast tomosynthesis relative to those in two-view full-field digital mammography. Breast 2015;24(2):93–9.

[20] Bae MS, Moon WK. Is synthetic mammography comparable to digital mammography for detection of microcalcifications in screening? Radiology 2018;289(3):639–40.

[21] TMIST/EA1151 Study: Tomosynthesis Mammographic Imaging Screening Trial. American College of Radiology. https://www.acr.org/-/media/ACR/Files/Research/TMIST-Site-Process-Summary.pdf?la=en. Accessed January 7, 2020.

[22] Fallenberg EM, Dromain C, Diekmann F, et al. Contrast-enhanced spectral mammography: does mammography provide additional clinical benefits or can some radiation exposure be avoided? Breast Cancer Research and Treatment 2014;146(2):371–81.

[23] Lalji UC, Jeukens CR, Houben I, et al. Evaluation of low-energy contrastenhanced spectral mammography images by comparing them to full-field digital mammography using EUREF image quality criteria. European Radiology 2015;25(10):2813–20.

[24] Francescone MA, Jochelson MS, Dershaw DD, et al. Low energy mammogram obtained in contrast-enhanced digital mammography (CEDM) is comparable to routine full-field digital mammography (FFDM). European Journal of Radiology 2014; 83(8):1350–5.

[25] James JR, Pavlicek W, Hanson JA, Boltz TF, Patel BK. Breast radiation dose with CESM compared with 2D FFDM and 3D tomosynthesis mammography. American Journal of Roentgenology 2017;208(2):362–72.

[26] Phillips J, Mihai G, Hassonjee SE, et al. Comparative dose of contrast-enhanced spectral mammography (CESM), digital mammography, and digital breast tomosynthesis. American Journal of Roentgenology 2018;211(4):839–46.

[27] Jeukens CR, Lalji UC, Meijer E, Bakija B, Theunissen R, Wildberger JE, et al. Radiation exposure of contrast-enhanced spectral mammography compared with full-field digital mammography. Investigative Radiology 2014;49(10):659–65.

[28] Sorin V, Yagil Y, Yosepovich A, et al. Contrast-enhanced spectral mammography in women with intermediate breast cancer risk and dense breasts. American Journal of Roentgenology 2018;211(5):W267–74.

[29] Patel BK, Garza SA, Eversman S, Lopez-Alvarez Y, Kosiorek H, Pockaj BA. Assessing tumor extent on contrast-enhanced spectral mammography versus full-field digital mammography and ultrasound. Clinical Imaging 2017;46:78–84.

[30] Lee-Felker SA, Tekchandani L, Thomas M, et al. Newly diagnosed breast cancer: comparison of contrast-enhanced spectral mammography and breast MR imaging in the evaluation of extent of disease. Radiology 2017;285(2):389–400.

[31] Fallenberg EM, Schmitzberger FF, Amer H, et al. Contrast-enhanced spectral mammography vs. mammography and MRI - clinical performance in a multi-reader evaluation. European Radiology 2017;27(7):2752–64.

[32] Sung JS, Lebron L, Keating D, et al. Performance of dual-energy contrastenhanced digital mammography for screening women at increased risk of breast cancer. Radiology 2019;293(1):81–8.

[33] Gordon PB, Goldenberg SL. Malignant breast masses detected only by ultrasound. a retrospective review. Cancer 1995;76(4):626–30.

[34] Mortezazadeh T, Gholibegloo E, Riyahi Alam N, Haghgoo S, Musa AE, Khoobi M. Glucosamine conjugated gadolinium (III) oxide nanoparticles as a novel targeted contrast agent for cancer diagnosis in MRI. Journal of Biomedical Physics and Engineering 2020;10(1):25–38.

[35] Hille H. Advances in breast ultrasound. 2015.

[36] Vourtsis A. Three-dimensional automated breast ultrasound: technical aspects and first results. Diagnostic and Interventional Imaging 2019;100(10):579–92.

[37] Guo R, Lu G, Qin B, Fei B. Ultrasound imaging technologies for breast cancer detection and management: a review. Ultrasound in Medicine and Biology 2018;44(1):37–70.

[38] Carpentier B, Hayward J, Strachowski L. Enhancing your acoustics: ultrasound image optimization of breast lesions. Journal of Ultrasound in Medicine 2017;36(7):1479–85.

[39] Kim HJ, Kim SM, Kim B, La Yun B, Jang M, Ko Y, et al. Comparison of strain and shear wave elastography for qualitative and quantitative assessment of breast masses in the same population. Scientific Reports 2018;8(1):6197.

[40] Barr RG. The role of sonoelastography in breast lesions. Seminars in Ultrasound, CT and MRI 2018;39(1):98–105.

[41] Greenwood HI, Dodelzon K, Katzen JT. Impact of advancing technology on diagnosis and treatment of breast cancer. Surgical Clinics of North America 2018;98(4):703–24.

[42] Partridge SC, Nissan N, Rahbar H, Kitsch AE, Sigmund EE. Diffusion weighted breast MRI: clinical applications and emerging techniques. Journal of Magnetic Resonance Imaging 2017;45(2):337–55.

[43] Patharkar A, Raval N, Kalyane D, Tambe V, Anup N, More N, Tekade RK. Glucosamine-conjugated nanoseeds for chemo-magneto hyperthermia therapy of cancer. Journal of Drug Delivery Science and Technology 2021;61:102295.

[44] Bolan PJ, Nelson MT, Yee D, Garwood M. Imaging in breast cancer: magnetic resonance spectroscopy. Breast Cancer Research 2005;7(4):149.

[45] Drukteinis JS, Mooney BP, Flowers CI, Gatenby RA. Beyond mammography: new frontiers in breast cancer screening. Americas Journal of Medicine 2013;126(6):472–9.

[46] Jochelson M. Advanced imaging techniques for the detection of breast cancer. Alexandria, VA, USA: American Society of Clinical Oncology Educational Book; 2012. p. 65–9.

[47] Mumin NA, Hamid MTR, Ding Wong JH, Rahmat K, Hoong K. Magnetic resonance imaging phenotypes of breast cancer molecular subtypes: a systematic review. Academic Radiology 2022;29:S89–106.

[48] Godavarty A, Rodriguez S, Jung Y-J, Gonzalez S. Optical imaging for breast cancer prescreening. Breast Cancer 2015;7:193 (Dove Med Press).

[49] Tromberg BJ, Pogue BW, Paulsen KD, Yodh AG, Boas DA, Cerussi AE. Assessing the future of diffuse optical imaging technologies for breast cancer management. Medical Physics 2008;35(6Part1):2443–51.

[50] Avril N, Rosé CA, Schelling M, et al. Breast imaging with positron emission tomography and fluorine-18 fluorodeoxyglucose: use and limitations. Journal of Clinical Oncology 2000;18:3495−502 [PubMed: 11032590].

[51] Raylman RR, Majewski S, Smith MF, et al. The positron emission mammography/tomography breast imaging and biopsy system (PEM/PET): design, construction and phantom-based measurements. Physics in Medicine and Biology 2008;53:637−53 [PubMed: 18199907].

[52] Berg WA, Weinberg IN, Narayanan D, et al. High-resolution fluorodeoxyglucose positron emission tomography with compression ("positron emission mammography") is highly accurate in depicting primary breast cancer. Breast Journal 2006;12:309−23 [PubMed: 16848840].

[53] Hruska CB, O'Connor MK, Collins DA. Comparison of small field of view γ camera systems for scintimammography. Nuclear Medicine Communications 2005;26:441−5 [PubMed: 15838427] of considerable interest Head-to-head comparison of the various technologies available for small field of view γ cameras.

[54] Bach-Gansmo T, Danielsson R, Saracco A, et al. Integrin receptor imaging of breast cancer: a proof-of-concept study to evaluate 99mTc-NC100692. Journal of Nuclear Medicine 2006;47(9):1434−9 [PubMed: 16954550].

[55] Papantoniou V, Tsiouris S, Koutsikos J, et al. Scintimammographic detection of usual ductal hyperplasia with increased proliferation rate at risk for malignancy. Nuclear Medicine Communications 2006;27:911−7 [PubMed: 17021432].

[56] Mansi L, Rambaldi PF, Procaccini E, et al. Scintimammography with technetium-99m tetrofosmin in the diagnosis of breast cancer and lymph node metastases. European Journal of Nuclear Medicine 1996;23:932−9 [PubMed: 8753682].

[57] Hruska CB, Phillips SW, Whaley DH, Rhodes DJ, O'Connor MK. Molecular breast imaging: use of a dual-head dedicated γ camera for detection of small breast tumors. American Journal of Roentgenology 2008;191:1805−15 [PubMed: 19020253] of interestShows advantages of a dual-head system compared with a single-head system in detecting small breast lesions.

[58] Mann RM, Cho N, Moy L. Breast MRI: state of the art. Radiology 2019;292:520−36.

[59] Vobugari N, Raja V, Sethi U, Gandhi K, Raja K, Surani SR. Advancements in oncology with artificial intelligence—a review article. Cancers 2022;14:1349.

[60] Chougrad H, Zouaki H, Alheyane O. Multi-label transfer learning for the early diagnosis of breast cancer. Neurocomputing 2020;392:168−80.

[61] Le EPV, Wang Y, Huang Y, Hickman S, Gilbert F. Artificial intelligence in breast imaging. Clinical Radiology 2019;74:357−66.

[62] Yassin NIR, Omran S, Houby EMF, Allam H. Machine learning techniques for breast cancer computer aided diagnosis using different image modalities: a systematic review. Computer Methods and Programs in Biomedicine 2018;156:25−45.

[63] Basurto-Hurtado JA, Cruz-Albarran IA, Toledano-Ayala M, Ibarra-Manzano MA, Morales-Hernandez LA, Perez-Ramirez CA. Diagnostic strategies for breast cancer detection: from image generation to classification strategies using artificial intelligence algorithms. Cancers 2022;14(14):3442.

[64] Poornima B, Ramadevi Y, Sridevi T. Threshold based edge detection algorithm. International Journal of Engineering and Technology 2011;3(4).

[65] Bray F, Ferlay J, Soerjomataram I, et al. Global cancer statistics 2018: GLOBOCAN estimates of incidence and mortality worldwide for 36 cancers in 185 countries. CA: A Cancer Journal for Clinicians 2018;68(6):394−424.

[66] Hung M, Liu C, Teng C, Hu Y, Yeh C. Risk of second non-breast primary cancer in male and female breast cancer patients : a population-based cohort study. PLoS One 2016;11(2):1–12.

[67] Bagchi S, Huong A. Signal processing techniques and computer-aided detection systems for diagnosis of breast cancer – a review paper. Indian Journal of Science & Technology 2017;10:1–6.

[68] Ong MS, Mandl KD. National expenditure for false-positive mammograms and breast cancer overdiagnoses estimated at $ 4 billion a year. Health Affairs 2015;34(4): 576–83.

[69] Lehman CD, Wellman RD, Buist DSM, Kerlikowske K, Tosteson ANA, Miglioretti DL. Diagnostic accuracy of digital screening mammography with and without computer-aided detection. JAMA Internal Medicine 2015;175(11):1828–37.

[70] Lee H, Chen YPP. Image based computer aided diagnosis system for cancer detection. Expert Systems with Applications 2015;42(12):5356–65.

[71] Mousa R, Munib Q, Moussa A. Breast cancer diagnosis system based on wavelet analysis and fuzzy-neural. Expert Systems with Applications 2005;28(4):713–23.

[72] Suckling J, et al. The mammographic image analysis society digital mammogram database. Excerpta Medica International Congress Series 1994;1069:375–8.

[73] Islam MS, Kaabouch N, Hu WC. A survey of medical imaging techniques used for breast cancer detection. In: IEEE International Conference on Electro/Information Technology; 2013. p. 10–4.

[74] Andreea GI, et al. The role of imaging techniques in diagnosis of breast cancer. Expert Medica, International Congress Series 2013;1069(12):5356–65.

[75] Uematsu T, Sano M, Homma K, Shiina M, Kobayashi S. Three-dimensional helical CT of the breast: accuracy for measuring extent of breast cancer candidates for breast conserving surgery. Breast Cancer Research and Treatment 2001;65(3):249–57.

[76] Tang X, Zhang L, Zhang W, Huang X, Iosifidis V, Liu Z, Zhang J. Using machine learning to automate mammogram images analysis. In: 2020 IEEE international conference on Bioinformatics and biomedicine (BIBM). IEEE; December 2020. p. 757–64.

[77] Du G, Cao X, Liang J, Chen X, Zhan Y. Medical image segmentation based on u-net: a review. Journal of Imaging Science and Technology 2020.

[78] Pal UM, Saxena M, Anil Vishnu GK, Parsana D, Sarvani BSR, Varma M, Pandya HJ. Optical spectroscopy-based imaging techniques for the diagnosis of breast cancer: a novel approach. Applied Spectroscopy Reviews 2020;55(8):778–804.

[79] Zheng X, Yao Z, Huang Y, Yu Y, Wang Y, Liu Y, Zhou J. Deep learning radiomics can predict axillary lymph node status in early-stage breast cancer. Nature Communications 2020;11(1):1236.

[80] O'Loughlin D, O'Halloran M, Moloney BM, Glavin M, Jones E, Elahi MA. Microwave breast imaging: clinical advances and remaining challenges. IEEE Transactions on Biomedical Engineering 2018;65(11):2580–90.

[81] Wang L. Early diagnosis of breast cancer. Sensors 2017;17(7).

[82] Medina Y, Augusto M, Paz AV. Microwave imaging for breast cancer detection: experimental comparison of confocal and holography algorithms. In: Proceedings of the 2016 IEEE ANDESCON, ANDESCON 2016; 2017. p. 0–3.

[83] Bagchi S, Huong A, Tay KG. Investigation of different spatial filters performance toward mammogram de-noising. International Journal of Integrated Engineering 2017; 9(3):49–53.

[84] Tang J, Rangayyan RM, Xu J, El Naqa IE, Yang Y. Computer-aided detection and diagnosis of breast cancer with mammography: recent advances. IEEE Transactions on Information Technology in Biomedicine 2009;13(2):236–51.

[85] Ha SM, Cheun JH, Lee SH, Kim SY, Park AR, Kim YS, Chang JM. Ipsilateral lymphadenopathy after COVID-19 vaccination in patients with newly diagnosed breast cancer. Journal of Breast Cancer 2022;25(2):131.

[86] Nigam RK, Waghmare NP, Gupta AK. Evaluation and analysis of different type of edge detection techniques on cartridge case image. International Journal on Recent and Innovation Trends in Computing and Communication 2014;2(10).

[87] Gong K, Berg E, Cherry SR, Qi J. Machine learning in PET: from photon detection to quantitative image reconstruction. Proceedings of the IEEE 2020;108:51–68.

[88] Zaharchuk G. Next generation research applications for hybrid PET/MR and PET/CT imaging using deep learning. European Journal of Nuclear Medicine and Molecular Imaging 2019;46:2700–7.

[89] Ravishankar H, Sudhakar P, Venkataramani R, Thiruvenkadam S, Annangi P, et al. Understanding the mechanisms of deep transfer learning for medical images. In: Carneiro G, Mateus D, Loïc P, Bradley A, Tavares JMRS, et al., editors. Deep learning and data labeling for medical applications. Berlin: Springer; 2016. p. 188–96.

[90] El Naqa I, Haider MA, Giger ML, Ten Haken RK. Artificial intelligence: reshaping the practice of radiological sciences in the 21st century. British Journal of Radiology 2020;93:20190855.

[91] Hochreiter S, Schmidhuber J. Long short-term memory. Neural Computation 1997;9:1735–80.

[92] Cho K, van Merrienboer B, Gulcehre C, Bahdanau D, Bougares F, et al. Learning phrase representations using RNN encoder-decoder for statistical machine translation. arXiv:1406.1078 2014 [cs.CL].

[93] Goodfellow IJ, Pouget-Abadie J, Mirza M, Xu B, Warde-Farley D, et al. Generative adversarial networks. arXiv:1406.2661 2014 [stat.ML].
[93a] Liu X, Faes L, Kale AU, Wagner SK, Fu DJ, et al. A comparison of deep learning performance against health-care professionals in detecting diseases from medical imaging: a systematic review and meta-analysis. Lancet Digit. Health 2019;1:e271–97.

[94] Sari H, Teimoorisichani M, Mingels C, Alberts I, Panin V, Bharkhada D, Rominger A. Quantitative evaluation of a deep learning-based framework to generate whole-body attenuation maps using LSO background radiation in long axial FOV PET scanners. European Journal of Nuclear Medicine and Molecular Imaging 2022;49(13):4490–502.

[95] Yaakub SN, McGinnity CJ, Kerfoot E, Mérida I, Beck K, Dunston E, Hammers A. Brain PET-MR attenuation correction with deep learning: method validation in adult and clinical paediatric data. arXiv preprint arXiv:2212.01294 2022.

[96] Abdou MA. Literature review: efficient deep neural networks techniques for medical image analysis. Neural Computing & Applications 2022;34(8):5791–812.

[97] Hosny A, Parmar C, Quackenbush J, Schwartz LH, Aerts HJ. Artificial intelligence in radiology. Nature Reviews Cancer 2018;18(8):500–10.

[98] Gong K, Catana C, Qi J, Li Q. PET image reconstruction using deep image prior. IEEE Transactions on Medical Imaging 2018;38(7):1655–65.

[99] Hong Y, Commandeur F, Cadet S, Goeller M, Doris M, Chen X, Dey D. Deep learning-based stenosis quantification from coronary CT angiography. In: Medical imaging 2019: image processing. 10949; March 2019. p. 643–51 [SPIE].

[100] Stoykow C, Erbes T, Maecke HR, Bulla S, Bartholomä M, Mayer S, Meyer PT. Gastrin-releasing peptide receptor imaging in breast cancer using the receptor antagonist 68Ga-RM2 and PET. Theranostics 2016;6(10):1641.

[101] Khan KS, Wojdyla D, Say L, Gülmezoglu AM, Van Look PF. WHO analysis of causes of maternal death: a systematic review. The Lancet 2006;367(9516):1066–74.

Machine learning and deep learning techniques for breast cancer detection using ultrasound imaging

11

Prerna Ajmani[1], Vandana Sharma[2], Richa Hirendra Rai[3] and Sheetal Kalra[3]

[1]*Vivekananda Institute of Professional Studies-TC, GGSIPU, New Delhi, India;* [2]*Computer Science Department, CHRIST (Deemed to be University), Delhi-NCR, India;* [3]*School of Physiotherapy, Delhi Pharmaceutical Sciences and Research University (DPSRU), New Delhi, India*

1. Introduction

Metastatic breast tumor is a category of illness where the tissue cells of the breast change as well as multiply uncontrollably, typically resulting in a tumor or lump. Majority of breast cancer problems are found in the mammary glands or the ducts that link the lobules to the nipple [1]. The world's second biggest cause of mortality is breast cancer, which is one of the most prevalent malignancies in females. Worldwide, the prevalence of breast cancer has increased over time and each year [2]. Compared to different malignancies, this carcinoma affects women more frequently. If this condition is not detected in time, it might be fatal [3]. This cancer is split into two categories: benign (relatively harmless) and malignant (progressive and has the capacity to spread to different parts of the body). Benign tumors are relatively slow-growing, do not infiltrate nearby tissues, and do not disseminate to other body regions [4]. Usually, breast cancer is discovered after a woman discovers a lump or during screening before symptoms show. Most breast lumps are the largest tumor visible on mammography and are thought to be benign (not malignant). If cancer is found, tissue is normally withdrawn using a fine or large core needle for microscopic inspection. Reviewing past diagnostic information and gathering useful information from it are the main goals in the early and accurate detection of this disease. Furthermore, medical imaging and machine learning methods will facilitate the detection of cancer in the breast.

1.1 Medical imaging

Image processing approach of machine learning can help in quick tumor detection and diagnosis. It can now significantly enhance the differential diagnosis and can precisely detect cancer in breast. In the past, medical visualization has greatly benefitted in the diagnosis of severe clinical illness, treatment evaluation, and spotting of imperfections in various organs of the body, including the eye retina, lungs, nervous system, breast, and abdomen [5—8]. Medical visualization or imaging is the umbrella term for a number of specialized procedures used to examine the organism, particularly to identify, monitor, or treat diseases. Every technique of imaging provides precise information on the examined area of the body [9]. These visualization techniques categorize the size, position, and features of the body organ in query and are thought to be an efficient way to sort through the vast amount of data to get the information that is actually valuable. To identify the majority of disorders, various researchers concentrated heavily on the creation and analysis of medical images. Thus, medical images aid in the diagnosis of sickness, the identification of pathological abnormalities, the therapeutic treatment of the patient, and the assumption of a variety of illnesses. Recent advances in machine learning and artificial intelligence have had a significant impact on the medical industry, including the processing of medical pictures. The best technique to find breast cancer is by medical imaging, which generally employs several techniques such as magnetic resonance imaging (MRI), mammography, thermography imaging, radiography or duplex ultrasound, and historical examination [10—12] to detect the multiplying cells in a particular tissue. An overview of the most popular forms of medical photographs (such as MRI, ultrasound, mammography, histological, and thermography images) is given in this review to discuss the characteristics and features of that related to ultrasound imaging technique.

1.1.1 Mammography

A mammogram screening aids in the early detection and diagnosis of breast lumps in females. This X-ray technique scans the female breast that creates a picture of the breast using low-dosage X-rays [13]. Although diagnostic mammography imaging is carried out on patients with erratic breast nodules, mammogram screening is also beneficial for determining the risk of cancer even in women without obvious symptoms and also those at high risk for developing the same. This results in an image that displays the fibro-gland region, pectoral muscle, dense tissue, and soft tissue. Professional radiologists can review these mammograms to see if the breast contains any irregularities. A slight change in characteristics in two or more mammograms taken after a year or two may indicate early malignancy. Until before the person herself or doctor notices the cancer symptoms, a mammography can detect breast alterations. Early detection of substantial alterations in cancer allows for the prevention of more rigorous treatments and an increase in the likelihood that a breast cancer patient will survive.

American Cancer Society suggests that all women over 40 should get a mammogram once a year. In mammography test, thick tissues in the breast seem either white or light gray in color. Mammograms of younger women who appear to have bigger breasts may be easier to view as a result however difficult to analyze due to the density. The majority of disorders in the breast are indications of cancer, which necessitates testing and frequent biopsy to detect cancer so that it can be treated timely.

Mammography continues to be the gold standard for community breast screening because it is less aggressive than MRI and ultrasound [14]. It is a procedure in which an X-ray of the breast is taken after compressing the tissue between two metal plates. Full-field digital mammograms are considered to be almost accurate in screening the carcinogenic lesion in the breast and the ductal carcinoma in situ. However, there are chances of some false negative results especially with women having dense breast tissue and also some chances (around 20%) of overdiagnosis, false positives, of not being able to differentiate those tissues which would not be mitotic in activity and would not cause a problem in future. Overdiagnosis and overtreatment and minimal exposure to radiation during the process lead to deterioration of quality of life [15]. Patients who have undergone partial mastectomy or breast conservation surgery still have to continue to get their mammography images done as they are still at risk of developing a new lesion. The changes observed in the skin and breast tissues postsurgery and the radiotherapy sessions make it all the more a challenge for the mammographic images to be precise.

It is alarming to see that the rates of radical mastectomy have not come down even though we have good provisions for screening; thus, there is a dire need to increase the efficacy of medical imaging techniques, and there is how artificial intelligence and deep learning can come as a savior. It can help to better locate, recognize, segment, and distinguish the features of the mass of breast cancer from malignant versus mild cancer. Moreover, this strong machine learning with AI technology can be used to encapsulate the physiological presentations of the patients and their other pathological diagnostic tests including the genetic study variations. Finally, the deeply convolutional neural networks (CNN) finetuned to conserve the information between the pixels of the image and that with pretrained weights along with the AI-based platform would help to diagnose the different categories of tumors by reducing the noise between the images. Authors have also used superimposed convolutional autoencoder (SCAE) using several mammograms to eliminate errors considering the variability in the human data and also thus self-learning spiking neural networks was incorporated. Faster R−CNN models have been generated and tested over several databases like DDSM, private (GURO), and public (INbreast) datasets. You only Look once (YOLO) deep learning method was also adapted to give a better efficiency in the speed of detection of the mass. Several AI-based softwares have been introduced, which propose unique algorithms for tumor detection including deep learning with random forest classifiers and AI-based platforms for deep learning strategies for comparison between well-recognized source annotations to target annotations and self-paced sampling algorithms for better predictions and classification of the breast cancer [16−19]. Even after this, the basic limitation of mammography still prevails as the dense breast tissue comes as a hindrance and thus people at high risk may go back undiagnosed and then might have to undergo the prolonged treatment of breast cancer and palliative care.

1.1.2 Ultrasound imaging and manifold advantages

It is a common modality used for screening subjects under 40 years old, for subjects with a high density of breast tissue and those at lower risk for breast cancer. Thus, mammography is not recommended for subjects below 40 years old due to its ionization limitations across fatty, dense, and glandular structures. Ultrasonography having high resolution and the potential to discriminate between the cyst and solid masses is recommended. It is thus a modality of choice and is additionally used along with MRI in a few cases. Moreover, the flexibility of the process to assess the structures from the chest cage, neck, and pectoral region to the axilla and lymph nodes provides a comprehensive detailed assessment and thus can develop a better fusion with the histopathological examinations, contrasting methods and molecular and metabolic activity in situ. However, the manual process is a tedious and lengthy task (sensitivity of 84% and specificity of 91%) and requires the intervention of different ultrasound equipment at different setups and a team of skilled and professional sonographers, breast surgeons, and histopathologists. This imaging technique has advanced from the extraction of features and their classification depending on the type of breast tissue from the (shear wave elastography) SWE images using 2-layer deep learning (DL) model with neural networks [20] to automate analyzed data generated from large datasets and from inputs by an experienced team of specialists including histopathologists included as annotations for deep learning [21]. This is a robust multiplaner clinical method to adjudge the location, dimension, and texture of tissue under consideration along with an additional advantage of being noninvasive in nature [22]. Authors in 2019 demonstrated a similar technique but a low-cost technique, to differentiate malignant breast lesions from benign ones by noninvasive method. They utilized the data of 150 subjects and achieved 93.6% AUC on the receiver operating curve, by utilizing the full potential of CNN, fine-tuning, and transfer learning strategies to train and thus obtain color conversation through back-propagation and echo patterns. In the same year, authors [23] suggested use of CNN to study the cutaneous and subcutaneous adipose and lymph tissue images on the ultrasound to conclude that differentiation with respect to the type of tissue. Possibly this technique helped to reduce the ambiguity in the traditional method which was due to intra and interobserver variability and their experience and as well as skill determined. U-net was introduced in 2018 [24] for mass detection based on real-time image segmentation. They used ultrasound with a deep convolutional neural network system to identify and categorize the nodule size. In 2016, studies also supported such advanced methods, as proposed by certain authors [25]. Ultrasound is mostly employed for its convenience, volume, noninvasiveness, and affordability, and it can detect and locate breast mass nodes.

1.1.3 MRI

Along with ultrasound and mammography, MRI is a method for early cancer cell detection. Instead of using X-rays, MRI creates extremely precise three-dimensional (3D) transverse pictures using magnetic fields. Therefore, when MRI is done, the variations in the affected area are prominently noticeable. Contrarily, as MRI technology is costlier than mammography, this alone is insufficient for screening. By altering the data collecting parameters, distinct differences between soft tissues can be observed with high spatial precision.

Due to the high sensitivity of MRI, a persistent problem with it remains overdiagnosis by diagnosing the clinical irrelevant breast tumors, which grow slowly and actually are not alarming. Thus, MRI alone is not considered a good screening method other than for preoperative staging in people with high risk or those requiring radical mastectomy. Once there is an established diagnosis, it is used to differentiate multifocal to multicentric conditions and also invasion deeper into the chest cage or the bony and soft tissue around the target region or away from it. Extensive research done on 700 women having different categories of breast cancer stated that MRI does not offer any significant screening benefit over other radiological imaging techniques [26]. Gibson et al. [27] proposed an open-source NiftyNet system using TensorFlow platform and libraries, which offered classification and regression functions analysis for breast cancer through 2D and 3D images and graphs through the better user interface that helps to remove the variability across radiologists and images uploaded by different hospitals. In this software, the Image quality of the MRI is assessed by studying the density of pixels in the designated target area. Researchers have also devised the categorization of breast cancer lesions on neural network and image quality assessment (IQA) algorithms [28]. Featured learning and integration by using the multisequence MRI method has demonstrated one of the best clinical methods with a sensitivity of 84.6%, specificity of 85.7%, and an accuracy of 85.0% in differentiating benign lesions from malignant ones in comparison to other algorithms [29]. But these deep learning neural networks are being fed with continuous data from ultrasound images which are considered to be more efficient if classifying the tumor growth by analyzing the features noninvasively and wherever available an added advantage of annotations from histological image analysis was also added.

1.1.4 Histopathological images

In the case of breast cancer, the fluid/tissue around the lump is generally either aspirated by fine needle aspiration (FNA) or some amount of tissue from the suspected area is removed by core needle biopsy under local anesthesia generally on an outpatient basis. There are times when vacuum may be used in the later procedure for vacuum-assisted core biopsy. Also, in highly suspected cases, where either of the tests is nonconclusive, the surgeon may prefer going for an open biopsy where the localized tissue may be extracted using incisional biopsy or the entire abnormal area may be removed using excisional biopsy. In both cases, the suspected area is first demarcated in the preoperative preparation phase by using local devices like radiofrequency reflectors or radioactive pellets. Histopathological examination of such tissue using hematoxylin–eosin (H&E) stained microscopy images is considered to be a gold standard for the diagnosis of breast cancer, including the genomic pattern expression and methylation profiling, and the histopathology of the tissue extracted was proposed to enhance the process of learning. This examination is a tedious task and skill based and is an invasive procedure. It requires expert skills and experience to develop the right report and avoid operator bias. High resemblance in the features of the microscopic structures leads to a lot of intraclass and interclass variations. Thus, machine learning techniques including support vector and compressed sensing were recruited for better results. Deep learning-based transfer learning and fine-tuning based on large data available promises to make this complex technique simple, efficient, and accurate by use of a deep cascade network for histological images depicting cell proliferation. FCN model, CaffeNet, and ImageNet were used to first identify the cell demonstrating mitotic activity and in the next stage establish the mitotic activity by studying a series of histological slides and assembling the information from the data from cross-domain using deep cascade network and results from other modalities such as ultrasound images [30]. Staining reactions of different tissues were studied using ensemble of multiscale convolutional neural networks (EMS-Net) and algorithms to demonstrate high accuracy. A combination of several fine-tuned models and thus generated algorithms were preferred by the researchers in 2019 [31]. Breast cancer and its proliferation may involve several tissues and thus pathological samples from different regions were studied using CNN which suggested the nature of and the invaded tissue around the lesion. The pathological samples taken from different regions were compared with reference samples of histopathological and other modality images, like those from ultrasound images, and also compared pixel by pixel. Authors have proposed a convolutional network and a recurrent deep neural network, transfer learning, pseudocodes, and mathematical equations for enhancing the accuracy of classification of the breast tissue to locate the mitotic activity by studying the cell proliferation and identifying the mitotic and nonmitotic nuclei by studying several samples and slides and generating multisequential images and using an established dataset of several patients integrating them from other modalities also [32−36]. To be more precise, the authors suggested to enhance and use large data and results from image-guided biopsy procedures and integrate them with other tests like mammography, ultrasound, MRI, thermography, and PET/CT scans (positron emission tomography/computerized tomography) so that deep learning can make the preventive, diagnostic, and follow-up procedures more accurate without being invasive in nature. Survival of the patient depends on the stage in which the cancer is detected as well as the morphology of the cancer. Thus, despite its limitations and its being invasive in nature, this test is still indicated in patients with high-risk and suspected lumps or growths in addition to Ultrasound/Mammography.

1.1.5 Thermography images

Thermography is a different imaging technique for breast cancer that first appeared in 1982. Breast thermography, also known as thermal imaging, is a painless, noninvasive technique that is frequently used to find breast alterations that could be signs of breast cancer. In this technique, infrared radiation from the object placed in front of the cameras is captured, and a thermal image is created to study the heat distribution in the target area in comparison to the surrounding structures. With the mitotic activity in the cells, the area under consideration invites more blood flow and thus records higher temperatures than their immediate surroundings. It is being used widely as it is a noninvasive procedure, which does not require to touch or compress the breast tissue, does not have radiation exposure, also it can detect changes in the dense breast tissue, those with implants or reconstructed breast. Menstrual cycle and the hormonal milieu of the body do not change its results. However, major limitations are, as we understand from the basic discussion of the technique itself that it does not detect carcinoma, as it majorly is dependent on vascular features. It has always thus been recommended in combination with other imaging modalities like mammography, ultrasound, and MRI. Recent studies on benchmarking datasets are being conducted to study the accuracy of this method and have shown positive results. Advancements in AI and ML-based technology have shown high accuracy (>94%) in combining thermography with deep learning techniques for the recognition and classification of breast cancer lesions [37–39].

2. Ultrasound and imaging techniques for staging of breast tumor

The American Joint Committee on Cancer (AJCC) developed the TNM staging based on the physical assessment, clinical presentation, histological examination of the tissue, and imaging techniques. Here "T" stands for the size and extent of the tumor, "N" stands for how many nearby lymph nodes it has affected, and "M" stands for metastasis to the distant sites. Depending on the location and the area of invasion by the tumor, its stage is determined. Based on the protein molecular expression example that human epidermal growth factor receptor 2, estrogen and progesterone receptor, etc. or genetic features, cancer with the integration of the hormonal tests values can be categorized further with more precision. However, the presentation symptoms are quite similar for each but on examination they can be further classified according to their stages based on the permutation and combination of the above features making it a complicated task. By using the breast ultrasound images already present, several researchers have tried to segment and classify the same using AI and deep learning techniques. The annotation and transfer learning techniques using the power of matching layers and rescaling pixels along with color conversion can help to better differentiate the tissue. Using advanced mathematical operations and segmentation techniques, region of interest is being developed and is being fine-tuned further as more datasets are being integrated into the system by utilizing the power of information technology. Box classifiers and bounding techniques are also being used to extract features especially for the texture, shape, and extent. In TNM classification, T followed by "is" stands for carcinoma in situ, whereas T followed by a number 1–4 indicates the size of the tumor starting from 2 cm to any size of the tumor and thus numbered accordingly, where N, which represents the involvement of the lymph nodes or cells involved around or with the tumor cells on or away from the site but in the breast region and connected tissue, thus defined by a number between 1 and 4, also define whether they are isolated or having micro or smaller areas of spread to larger or macro areas of spread and thus accordingly numbered and described in abbreviation. M stands for affection of other organs/tissues invaded by breast cancer, which will require collaboration with other test measures to investigate and confirm. Moreover, once the treatment starts, the tissue density changes further to complicate the further viewing and classification process, and to reconcile the old images with the new images, the deep convolutional network and several machine learning techniques may be incorporated [40].

3. Machine learning techniques incorporated with ultrasound imaging

Machine learning algorithms are being utilized in healthcare to identify breast cancer. Based on data from PubMed, Fig. 11.1 shows the statistics for publications of usage of machine learning techniques for predicting breast cancer from 2000 to 2022.

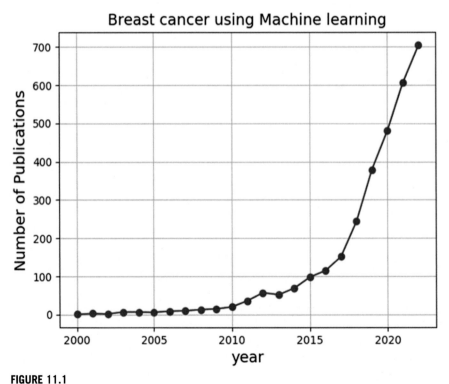

FIGURE 11.1

Breast cancer prediction using machine learning publications from 2000 to 2022.

According to Ref. [41], machine learning (a subset of AI) connects given problems from datasets to the basic principle of implication. It employs logical, mathematical, and statistical techniques to enable the machine to comprehend data [42]. In 1959, Arthur Samuel conceived the phrase "machine learning" to make the computer learn from its own experience; since then, artificial intelligence is used in several applications such as gaming, predictions, pattern recognition, and classifications. The science of machine learning techniques has enjoyed unmatched success over the years.

Breast cancer prediction has been done using various machine learning techniques such as Naive Bayes, random forest, KNN, logistic regression (LR), decision tree, AdaBoost, support vector machine (SVM), and multilayer perceptron algorithms. SVM, LR, and a C5.0 Decision Tree model were employed in Ref. [43] to forecast British Columbia's survival. It has been found that the SVM categorization of breast cancer is the most popular technique.

3.1 Support vector machine

The support vector machine (SVM) is a popular machine-learning technique for problem regression and categorization. It was first popularized in the last decade of the 20th century. SVM was applied in a variety of applications, including chemoinformatics [44], biometrics [45], and bioinformatics [46]. The SVM classifier builds a model for the classification using training data. A later step is the classification of an unidentified sample. The main idea behind SVM is the use of hyperplanes to separate diverse groups. In cases where data can be separated linearly, SVM has attained great levels of precision. However, the output of SVM cannot nonlinearly distinguish separable data. The data can be split linearly after being mapped to a new, high-dimensional space utilizing kernel functions to tackle this problem. The correct kernel function selection and its parameters are two of the main issues with SVM [47].

3.2 Decision tree

Decision trees (DT) are frequently used "divide and conquer" data classification strategies. With this method, the data can be shown as a tree, with the leaf nodes serving as the labels for the sample data labels and the inside nodes serving as the different attributes. The appropriate dataset is located by moving up the tree from root to leaf. The final grade is stored in the leaf node [48]. The most popular DT algorithm for classification is C4.5 [49]. Lim et al. contrasted C4.5 with other DT algorithms [50]. When compared to C4.5, EC4.5 offers five times the efficiency for the identical decision tree [51], making it a more reliable alternative to the standard algorithm. The same decision tree as C4.5 is present.

3.3 *k*-Nearest neighbor

A dataset is compared to other sample data using a distance metric in the *k*-nearest neighbor (*k*-NN) algorithm. The distance between two identical data samples can be reduced or increased using a distance metric [52]. Euclidean distance is typically used to calculate the separation between two different samples. Convenience, volume, noninvasiveness, and affordability finally help in detecting the position of breast mass nodules. Euclidian distance between any two set of points (x_1, y_1) and (x_2, y_2) can be calculated as follows:

$$d = \sqrt{[(x_2 - x_1)^2 + (y_2 - y_1)^2]} \qquad (11.1)$$

3.4 Naïve Bayesian network

It is a directed acyclic graph that uses probability to represent the relationship between numerous variables (features). The variables are represented by the graph nodes, while the relationships between the variables are shown by the graph arcs. In a particular instance of BN, the directed acyclic graph contains a single parent and numerous numbers of children. Within their parent node, child nodes continue to be autonomous. Instead of making predictions, BN classifiers estimate the chances of the likelihood of cancer cells in a particular area [53].

3.5 Artificial neural network

Artificial neural network (ANN) is strikingly similar to interlinked neurobiological networking in a person's mind [54]. In a direct-acting ANN, a single neuron's basic structure is depicted in Fig. 11.2. In an ANN, a single neuron gets input from various other neurons and then multiplies it by the appropriate weight Wij and finally uses a bias and activation function to create a weighted output ý.

In this kind of network, the neurons are arranged in the shape of strata. Single input and output layer but numerous hidden layers are common for direct action ANNs, as shown in Fig. 11.3. In Feedforward ANN, during preparation, the technique for weight adjustments is provided by the training algorithm for backpropagation. In this type of network, the desired output values are compared with the actual output generated by the ANN. The difference between the actual (determined) and anticipated (target) outputs is then used to calculate the error. The errors in updating weights are then sent back to the network during the following round [55].

4. Deep learning techniques and ultrasound imaging

In contrast to conventional ML extraction techniques, DL is a subclass of machine learning and AI that concentrates on a complicated hierarchy of picture properties. To make significant advancements in the way computers extract information from images, deep learning techniques are applied to various recently utilized computer models. These algorithms have been used to complete jobs in many different

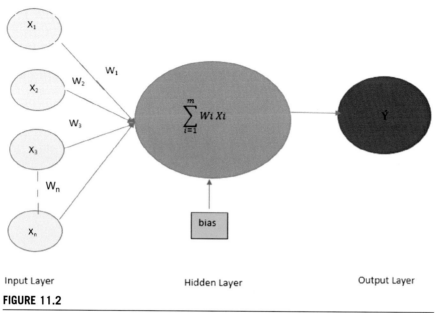

Input Layer   Hidden Layer   Output Layer

FIGURE 11.2

Basic structure of single neuron in artificial neural network (ANN).

medical fields, particularly radiology and pathology, and in certain instances, their performance has been on par with that of human professionals. Aside from providing information on molecular status, prognosis, and treatment sensitivity, DL can also be utilized to extract information from medical images that would be indistinguishable from human analysis [56]. Multilevel neural networks make up DL, which uses raw input images to derive a hierarchical structure of features. The rapid development of powerful DL algorithms that can practice with millions of photos and are insensitive to image differences has been made possible by the rapid improvement in graphics processor processing power. Due to its most recent successes, particularly in applications for image segmentation and classification, DL has gained more recognition. For a variety of uses, such as the identification and segmentation of objects in images, speech recognition, the identification of genotypes and phenotypes, and illness categorization, numerous types of DL algorithms have been created. CNN, deep Boltzmann machines, stacked autoencoders, and deep neural networks are a few popular DL algorithms [57]. Fig. 11.4 shows the usage of deep learning in breast cancer classification.

4.1 Convolution neural network and ultrasound imaging

CNNs have emerged as a crucial tool for image processing, especially when it comes to identifying or detecting features, language, human body parts, and biomedical pictures [57]. One of the most popular image techniques is CNNs. The CNNs

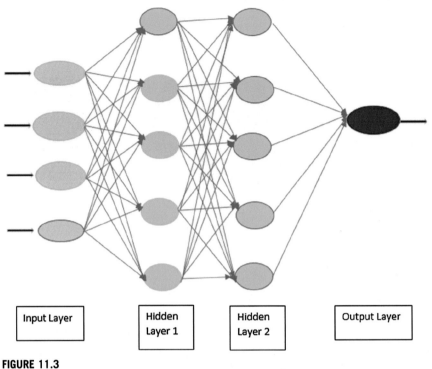

FIGURE 11.3

Artificial neural network (ANN).

have been successfully employed for the categorization and segmentation of pictures since its inception in 1989 [58–60]. The most common use for CNN is the classification of visual pictures. CNN is a feed-forward network that can extract the topological characteristics of a picture. CNNs are models powered by Multilayer Perceptron. The multilayer perceptron typically refers to networks that are entirely connected, with each neuron. Convolutional, pooling, and fully linked layers make up CNNs, which resemble neural networks more than anything else. Each layer's function is distinct. An extractor of features has been implemented in the convolutional layer. To determine which category the current entry belongs to, it employs the extracted function. Minimizing the size of the map feature and network parameters is the responsibility of the pooling layer. Fig. 11.5 displays an illustration of CNN.

a. Convolutional layers: Using the connection-oriented concept and weight distribution theory, convolutional layers are arranged into feature maps. Local connection describes the fact that each neuron in a feature map is only connected to local patches of the feature map at the previous stage through a weight group called a filter bank. A filter row is shared by all of the units on a feature map. This is the weight distribution model. Different feature maps also employ different filter banks. The aim of local connectivity and weight distribution is to

Breast cancer prediction using deep learning

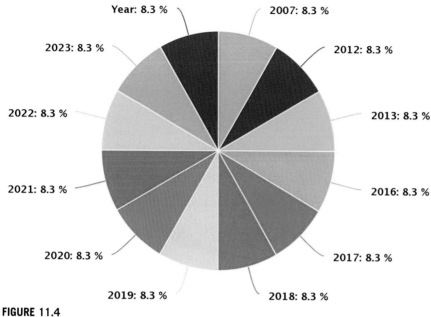

FIGURE 11.4

Publication statistics showing prediction of breast cancer using deep learning from 2007 to 2023.

curtail the number of parameters while maximizing the advantages of the strongly correlated local pixel neighborhood and location-independent local picture statistics. The activation function is a nonlinear transformation function that receives the weighted sum of each unit. The activation function allows the transmitted information to be transformed nonlinearly for use in later processing stages [61].

b. Pooling layer: It uses a subsampling process to combine convolutional layer characteristics that are similar into one (semantically). A unit within a pooling layer computes the maximum or average patch value at the output using a local patch as input from a prior entity map (convolutional layer). By lowering the number of parameters required at later phases, it reduces the size of the representation and improves the resilience of the representation by offering an invariance against tiny displacements and distortions.

c. Fully connected layer: The units in this layer are entirely connected to all the units in the previous layer, as can be seen in a typical neural network (i.e., a multilayer perceptron).

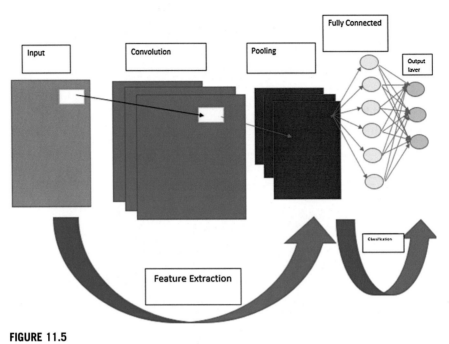

FIGURE 11.5

An example of convolution neural network (CNN).

5. Comparison of popular AI methods employed for various image modalities

Based on the characteristics gleaned from medical imaging, a variety of machine-learning approaches are employed to identify, categorize, and diagnose breast cancer. The popularity of various ML and CNN methods for finding breast cancer using mammogram, ultrasound, MRI, histological, and thermography pictures is illustrated in Figs. 11.6–11.8.

According to above statistics, it has been clearly found that SVM and CNN are both popular methods and are mostly used in mammogram imaging techniques. ANN scores second in finding breast cancer using mammogram, but LR and RF scores the least popularity in breast cancer prediction.

6. Limitations of ML and DL in imaging techniques

a. Quality data set: Human interface is an unavoidable factor for medical imaging, and it is known to enhance the knowledge of the deep learning process. However, it may act as a barrier as well as an avenue to enhance the quality of the dataset. They can decrease the quality if errors or artifacts interfere with the

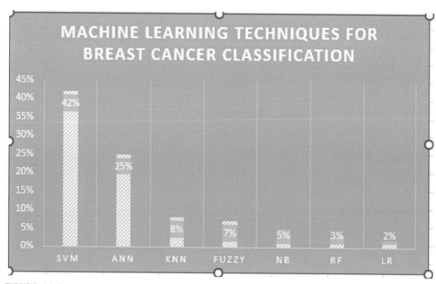

FIGURE 11.6

Statistics showing popularity of ML techniques used for breast cancer classification.

system or invariably enter the system, thus decreasing the efficiency of the imaging modality. Due to the paucity of quality datasets, it may be difficult to overcome the limitations of individual imaging technologies, which lead to under or overdiagnosis of the condition.

b. Physiological variation presurgery and postsurgery: As there is a lot of physiological variations in human and also the variation in the morphology of the tissues present in and around the breast, thus to build a stronger deep learning model for imaging techniques, Large and accurate, as well as good quality and large datasets are required. This technology will thus be helpful in establishing new epidemiological general data across different scenarios and autonomous learning paradigm and self-learning spiking neural networks as well as the transfer learning algorithms for different kinds of treated patients including those with reconstructed breasts and subcutaneous mastectomy as well.

c. Data manipulation: A quest for actual and large datasets is itself a deterrent for the process as it may lead to several data overfitting and underfitting rather than transfer learning. Moreover, cross-validation with other modalities and then knowledge-based deep learning will be a need in the near future.

d. Environmental variations: It is important to have the least confounding factor variability in data collection, to cover up for the nonmodifiable factors like genetic variability.

e. Image modality limitation: Robust data can be generated by integration data of all imaging modalities in ML and DL systems. Distance of the camera from the object being imaged or captured, surrounding temperature, and humidity during

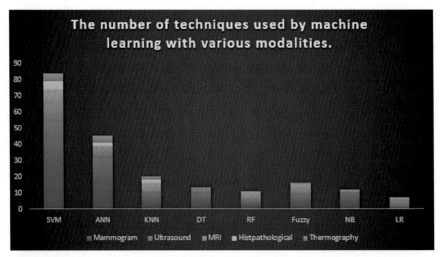

FIGURE 11.7

The percentage of any particular ML technique used by different modalities.

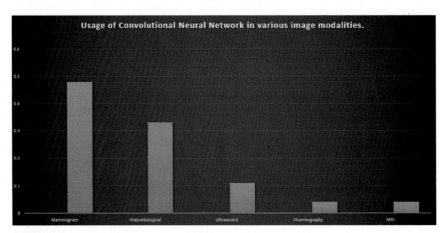

FIGURE 11.8

Usage of CNN in various image modalities.

thermography can create variation in data. Moreover, quality cameras are required to capture images in thermography and biopsy techniques; the precision and procedure of biopsy to extract the right tissue need to be maintained. This can reduce the accuracy of the tests and missing out of the detection of this life-threating cancer in the early stages.

7. Open research problems and future trends

Significant efforts are needed to improve the efficacy of breast cancer classification, and this section offers future directions for research that can be employed in 695 breast cancer diagnoses and classifications. ML and DL approaches for breast cancer diagnosis and classification still have some limits and 697 problems to be solved, despite the excellent results of the examined literature. The review discovered a number of significant problems, and the underlying patterns, future research paths, and problems are outlined below:

a. The absence of extensive training datasets has been a significant barrier in developing AI models to facilitate pathological imaging. This is primarily because the effectiveness of the AI classifier greatly depends on the size and caliber of the data set. Deep learning techniques require a vast amount of training data. But the biggest barrier to DL's effectiveness in medical scans is a paucity of data. Making vast amounts of medical imaging data is also challenging because, to eliminate human error, interpreting the data consumes plenty of time and expects a lot of hard work from a number of professionals as well as any amateur. Furthermore, any large medical imaging datasets are extremely challenging to build.

b. The majority of the examined studies evaluated and assessed these using various datasets that were privately acquired by cancer research organizations or clinics. The execution of such models across several research is difficult to compare, which is one of this argument's main flaws.

c. A problem and a lack of flexibility were perceived in the absence of benchmarks.

d. Most of the primary papers that were chosen classified breast cancer using the supervised learning method. These methods have produced superior outcomes when labeled images are used throughout the training process. However, in reality, it can be challenging to find breast cancer photos with accurate, medically labeled indications. Most of the time, there are many unidentified medical photographs available. Many blank labels serve as important knowledge sources but cannot be used for supervised learning. As a result, there is a great demand for the model of classification for breast cancer, which may be developed using a variety of grouping strategies without supervision.

e. Methodology of reinforcement learning for breast cancer classification. It is quite difficult to make an ML model capable of learning from its environment at the same time. The main issue is supplying enough breast cancer image examples to illustrate all types of breast cancer. Thus, employing a learning-based reinforcement model can effectively boost the effectiveness and performance of methods for identifying breast cancer from medical photographs.

f. Robustness in comparison to data gathering techniques: To gradually incorporate new datasets, the robustness issue of various clinical and technical scenarios must be addressed. Different image acquisition scanners, diverse lighting environments, varying sizes and views in multiple image modalities, and various presentation features of the coloring and enlargement factors are a few examples of such variations.

g. A deep learning classifier is another recent advancement that is fascinating. The use of computer-aided diagnosis tools has drawn more attention in recent years, creating a computer-aided diagnosis system with a hybrid architecture that includes many picture modalities.

h. Another significant advance is the emergence of 3D mammography-based CAD systems, a recent development that has the potential to improve CAD performance. These concerns must be taken into consideration for CAD system production in the future.

i. Instead of relying solely on imaging techniques such as ultrasound, MRI, mammography, and histopathology, other forms of imaging techniques such as computed tomography (CT) images or thermal imaging can be employed to improvize the accuracy of classification models for the disease. The identical patient must receive MRI or CT scans. Additionally, pictures of all different breast cancer cases will be collected. Boundary pictures should be labeled for the classification of multiclass breast cancer as they enable researchers to evaluate the efficacy of the recently created multiclass breast cancer classification model.

j. Building a reliable and computer-efficient CAD system is essential for assisting physicians in making early detection of cancer.

k. Cross-validation, a technique for model validation, can be used to assess the generalizability of the model findings. The goal is to fetch samples to test a model during the initial phase of training, to address issues like underfitting and overfitting, and to demonstrate how the learned model generalizes to a different dataset.

8. Conclusion

We assess the most current studies that have focused on the recognition and categorization of malignancy in the breast using major ML and DL algorithms in several picture modalities. The review's strengths include presenting SVM, nearest neighbor, DT, BN, and ANN, widely used machine learning algorithms. A deep learning technique known as CNN is also reviewed as this technique is popularly employed to detect breast malignancy. Breast imaging techniques such as ultrasounds, MRIs, mammograms, histopathological tests, and thermography are also described in this chapter. Furthermore, open research challenges are also being discussed.

References

[1] American Cancer Society. Breast cancer facts & figures 2019—2020. Atlanta: American Cancer Society, Inc.; 2019.

[2] Hamidinekoo A, Denton E, Rampun A, Honnor K, Zwiggelaar R. Deep learning in mammography and breast histology, an overview and future trends. Medical Image Analysis 2018;47:45—67. https://doi.org/10.1016/J.MEDIA.2018.03.006.

[3] Arevalo J, González FA, Ramos-Pollán R, Oliveira JL, Guevara Lopez MA. Representation learning for mammography mass lesion classification with convolutional neural networks. Computer Methods and Programs in Biomedicine 2016;127:248—57.

[4] Mohammed MA, Al-Khateeb B, Rashid AN, Ibrahim DA, Abd Ghani MK, Mostafa SA. Neural network and multi-fractal dimension features for breast cancer classification from ultrasound images. Computers & Electrical Engineering 2018;70:871—82. https://doi.org/10.1016/J.COMPELECENG.2018.01.033.

[5] Akbar S, Akram MU, Sharif M, Tariq A, Khan SA. Decision support system for detection of hypertensive retinopathy using arteriovenous ratio. Artificial Intelligence in Medicine 2018;90:15—24. https://doi.org/10.1016/J.ARTMED.2018.06.004.

[6] Rajinikanth V, Satapathy SC, Fernandes SL, Nachiappan S. Entropy based segmentation of tumor from brain MR images - a study with teaching learning based optimization. Pattern Recognition Letters 2017;94:87—95. https://doi.org/10.1016/J.PATREC.2017.05.028.

[7] Fonseca P, Mendoza J, Wainer J, Ferrer J, Pinto J, Guerrero J, et al. Automatic breast density classification using a convolutional neural network architecture search procedure. Medical Imaging 2015: Computer-Aided Diagnosis 2015;9414:941428. https://doi.org/10.1117/12.2081576.

[8] Khan MA, Sharif M, Akram T, Yasmin M, Nayak RS. Stomach deformities recognition using rank-based deep features selection. Journal of Medical Systems 2019;43(12): 1—15.

[9] Ashour AS, Dey N, Mohamed WS. Abdominal imaging in clinical applications: computer aided diagnosis approaches. Studies in Computational Intelligence 2016;651: 3—17. https://doi.org/10.1007/978-3-319-33793-7_1.

[10] Dhawan AP. Medical image analysis. 2nd ed. John Wiley and Sons; 2010. https://doi.org/10.1002/9780470918548.

[11] Deserno TM. Fundamentals of biomedical image processing. In: Deserno TM, editor. Biomedical image processing. Berlin, Heidelberg: Springer Berlin Heidelberg; 2011. p. 1—51. https://doi.org/10.1007/978-3-642-15816-2_1.

[12] Pluim JPW, Maintz JBAA, Viergever MA. Mutual-information-based registration of medical images: a survey. IEEE Transactions on Medical Imaging 2003;22: 986—1004. https://doi.org/10.1109/TMI.2003.815867.

[13] Dheeba J, Albert Singh N, Tamil Selvi S. Computer-aided detection of breast cancer on mammograms: a swarm intelligence optimized wavelet neural network approach.

Journal of Biomedical Informatics 2014;49:45—52. https://doi.org/10.1016/J.JBI.2014.01.010.

[14] Shen R, Yan K, Tian K, Jiang C, Zhou K. Breast mass detection from the digitized X-ray mammograms based on the combination of deep active learning and self-paced learning. Future Generation Computer Systems 2019;101:668—79. https://doi.org/10.1016/j.future.2019.07.013.

[15] Autier P, Boniol M. Mammography screening: a major issue in medicine. European Journal of Cancer 2018;90:34—62. https://doi.org/10.1016/J.EJCA.2017.11.002.

[16] Agarwal R, Díaz O, Yap MH, Lladó X, Martí R. Deep learning for mass detection in full field digital mammograms. Computers in Biology and Medicine 2020;121. https://doi.org/10.1016/J.COMPBIOMED.2020.103774.

[17] Kallenberg M, Petersen K, Nielsen M, Ng AY, Diao P, Igel C, et al. Unsupervised deep learning applied to breast density segmentation and mammographic risk scoring. IEEE Transactions on Medical Imaging 2016;35:1322—31. https://doi.org/10.1109/TMI.2016.2532122.

[18] Ionescu Gv, Fergie M, Berks M, Harkness EF, Hulleman J, Brentnall AR, et al. Prediction of reader estimates of mammographic density using convolutional neural networks. Journal of Medical Imaging 2019;6:1. https://doi.org/10.1117/1.jmi.6.3.031405.

[19] Shen R, Yao J, Yan K, Tian K, Jiang C, Zhou K. Unsupervised domain adaptation with adversarial learning for mass detection in mammogram. Neurocomputing 2020;393:27—37. https://doi.org/10.1016/j.neucom.2020.01.099.

[20] Zhang Q, Xiao Y, Dai W, Suo J, Wang C, Shi J, et al. Deep learning based classification of breast tumors with shear-wave elastography. Ultrasonics 2016;72:150—7. https://doi.org/10.1016/J.ULTRAS.2016.08.004.

[21] Qi X, Zhang L, Chen Y, Pi Y, Chen Y, Lv Q, et al. Automated diagnosis of breast ultrasonography images using deep neural networks. Medical Image Analysis 2019;52:185—98. https://doi.org/10.1016/J.MEDIA.2018.12.006.

[22] Byra M, Galperin M, Ojeda-Fournier H, Olson L, O'Boyle M, Comstock C, et al. Breast mass classification in sonography with transfer learning using a deep convolutional neural network and color conversion. Medical Physics 2019;46:746—55. https://doi.org/10.1002/mp.13361.

[23] Xu Y, Wang Y, Yuan J, Cheng Q, Wang X, Carson PL. Medical breast ultrasound image segmentation by machine learning. Ultrasonics 2019;91:1—9. https://doi.org/10.1016/J.ULTRAS.2018.07.006.

[24] Kumar V, Webb JM, Gregory A, Denis M, Meixner DD, Bayat M, et al. Automated and real-time segmentation of suspicious breast masses using convolutional neural network. PLoS One 2018;13:e0195816. https://doi.org/10.1371/JOURNAL.PONE.0195816.

[25] Chen Y, Huang Q. An approach based on biclustering and neural network for classification of lesions in breast ultrasound. In: ICARM 2016 - 2016 international conference on advanced robotics and mechatronics; 2016. p. 597—601. https://doi.org/10.1109/ICARM.2016.7606988.

[26] Laws A, Katlin F, Hans M, Graichen M, Kantor O, Minami C, et al. ASO visual abstract: screening MRI does not increase cancer detection or result in earlier stage at diagnosis in patients with high-risk breast lesions—A propensity score analysis. Annals of Surgical Oncology 2022;30:78—9. https://doi.org/10.1245/S10434-022-12656-0.

[27] Gibson E, Li W, Sudre C, Fidon L, Shakir DI, Wang G, et al. NiftyNet: a deep-learning platform for medical imaging. Computer Methods and Programs in Biomedicine 2018;158:113—22. https://doi.org/10.1016/J.CMPB.2018.01.025.

[28] Fang Y, Zhao J, Hu L, Ying X, Pan Y, Wang X. Image classification toward breast cancer using deeply-learned quality features. Journal of Visual Communication and Image Representation 2019;64:102609. https://doi.org/10.1016/J.JVCIR.2019.102609.

[29] Feng H, Cao J, Wang H, Xie Y, Yang D, Feng J, et al. A knowledge-driven feature learning and integration method for breast cancer diagnosis on multi-sequence MRI. Magnetic Resonance in Imaging 2020;69:40−8. https://doi.org/10.1016/J.MRI.2020.03.001.

[30] Chen H, Dou Q, Wang X, Qin J, Heng P-A. Mitosis detection in breast cancer histology images via deep cascaded networks. AAAI Conference on Artificial Intelligence; 2016.

[31] Yang Z, Ran L, Zhang S, Xia Y, Zhang Y. EMS-net: Ensemble of multiscale convolutional neural networks for classification of breast cancer histology images. Neurocomputing 2019;366:46−53. https://doi.org/10.1016/j.neucom.2019.07.080.

[32] Shallu MR. Breast cancer histology images classification: training from scratch or transfer learning? ICT Express 2018;4:247−54. https://doi.org/10.1016/J.ICTE.2018.10.007.

[33] Reddy A, Soni B, K. S. Breast cancer detection by leveraging Machine Learning. ICT Express 2020;6. https://doi.org/10.1016/j.icte.2020.04.009.

[34] Jiang Y, Chen L, Zhang H, Xiao X. Breast cancer histopathological image classification using convolutional neural networks with small SE-ResNet module. PLoS One 2019; 14. https://doi.org/10.1371/JOURNAL.PONE.0214587.

[35] Yan R, Ren F, Wang Z, Wang L, Zhang T, Liu Y, et al. Breast cancer histopathological image classification using a hybrid deep neural network. Methods 2020;173:52−60. https://doi.org/10.1016/J.YMETH.2019.06.014.

[36] Toğaçar M, Özkurt KB, Ergen B, Cömert Z. BreastNet: a novel convolutional neural network model through histopathological images for the diagnosis of breast cancer. Physica A: Statistical Mechanics and its Applications 2020;545:123592. https://doi.org/10.1016/J.PHYSA.2019.123592.

[37] Cabıoğlu Ç, Oğul H. Computer-aided breast cancer diagnosis from thermal images using transfer learning. In: Lecture notes in computer science (including subseries lecture notes in artificial intelligence and lecture notes in bioinformatics), 12108; 2020. https://doi.org/10.1007/978-3-030-45385-5_64. LNBI:716−26.

[38] de Freitas Barbosa VA, de Santana MA, Andrade MKS, de Lima R de CF, dos Santos WP. Deep-wavelet neural networks for breast cancer early diagnosis using mammary termographies. Deep Learning for Data Analytics 2020:99−124. https://doi.org/10.1016/B978-0-12-819764-6.00007-7.

[39] Ekici S, Jawzal H. Breast cancer diagnosis using thermography and convolutional neural networks. Medical Hypotheses 2020;137. https://doi.org/10.1016/J.MEHY.2019.109542.

[40] Edge SB, Compton CC. The American Joint Committee on cancer: the 7th edition of the AJCC cancer staging manual and the future of TNM. Annals of Surgical Oncology 2010;17:1471−4. https://doi.org/10.1245/S10434-010-0985-4.

[41] Tapak L, Shirmohammadi-Khorram N, Amini P, Alafchi B, Hamidi O, Poorolajal J. Prediction of survival and metastasis in breast cancer patients using machine learning classifiers. Clinical Epidemiology and Global Health 2019;7:293−9. https://doi.org/10.1016/J.CEGH.2018.10.003.

[42] Montazeri M, Naji HR, Montazeri M, Faraahi A. A novel memetic feature selection algorithm. In: IKT 2013 - 2013 5th conference on information and knowledge technology; 2013. p. 295−300. https://doi.org/10.1109/IKT.2013.6620082.

[43] Chao CM, Yu YW, Cheng BW, Kuo YL. Construction the model on the breast cancer survival analysis use support vector machine, logistic regression and decision tree. Journal of Medical Systems 2014;38. https://doi.org/10.1007/S10916-014-0106-1.

[44] Byvatov E, Schneider G. Support vector machine applications in bioinformatics. Applied Bioinformatics 2003;2:67−77.

[45] Vatsa M, Singh R, Noore A. Improving biometric recognition accuracy and robustness using DWT and SVM watermarking. IEICE Electronics Express 2005;2:362−7. https://doi.org/10.1587/ELEX.2.362.

[46] Doucet J-P, Barbault F, Xia H, Panaye A, Fan B. Nonlinear SVM approaches to QSPR/QSAR studies and drug design. Current Computer-Aided Drug Design 2007;3:263−89. https://doi.org/10.2174/157340907782799372.

[47] Tharwat A, Hassanien AE, Elnaghi BE. A BA-based algorithm for parameter optimization of Support Vector Machine. Pattern Recognition Letters 2017;93:13−22. https://doi.org/10.1016/J.PATREC.2016.10.007.

[48] Witten IH, Frank E, Geller J. Data mining. ACM SIGMOD Record 2002;31:76−7. https://doi.org/10.1145/507338.507355.

[49] Amin RK, Indwiarti, Sibaroni Y. Implementation of decision tree using C4.5 algorithm in decision making of loan application by debtor (Case study: bank pasar of Yogyakarta Special Region). In: 2015 3rd international conference on information and communication technology (ICoICT); 2015. p. 75−80. https://doi.org/10.1109/ICOICT.2015.7231400.

[50] Lim TS, Loh WY, Shih YS. A comparison of prediction accuracy, complexity, and training time of thirty-three old and new classification algorithms. Machine Learning 2000;40(3):203−28. https://doi.org/10.1023/A:1007608224229.

[51] Ruggieri S. Efficient C4.5. IEEE Transactions on Knowledge and Data Engineering 2002;14:438−44.

[52] Kotsiantis SB. Supervised machine learning: a review of classification techniques. In: Proceedings of the 2007 conference on emerging artificial intelligence applications in computer engineering: real word AI systems with applications in EHealth, HCI, information retrieval and pervasive technologies. NLD: IOS Press; 2007. p. 3−24.

[53] Kourou K, Exarchos TP, Exarchos KP, Karamouzis Mv, Fotiadis DI. Machine learning applications in cancer prognosis and prediction. Computational and Structural Biotechnology Journal 2014;13:8−17. https://doi.org/10.1016/J.CSBJ.2014.11.005.

[54] Ripley RM, Harris AL, Tarassenko L. Neural network models for breast cancer prognosis. Neural Computing & Applications 1998;7:367−75. https://doi.org/10.1007/BF01428127/METRICS.

[55] Saxena S, Gyanchandani M. Machine learning methods for computer-aided breast cancer diagnosis using histopathology: a narrative review. Journal of Medical Imaging and Radiation Sciences 2019;51:182−93. https://doi.org/10.1016/J.JMIR.2019.11.001.

[56] Akkus Z, Galimzianova A, Hoogi A, Rubin DL, Erickson BJ. Deep learning for brain MRI segmentation: state of the art and future directions. Journal of Digital Imaging 2017;30:449−59. https://doi.org/10.1007/S10278-017-9983-4.

[57] Yap MH, Pons G, Marti J, Ganau S, Sentis M, Zwiggelaar R, et al. Automated breast ultrasound lesions detection using convolutional neural networks. IEEE Journal of Biomedical and Health Informatics 2018;22:1218−26. https://doi.org/10.1109/JBHI.2017.2731873.

[58] Krizhevsky A, Sutskever I, Hinton GE. ImageNet classification with deep convolutional neural networks. Communications of the ACM 2012;60:84—90. https://doi.org/10.1145/3065386.

[59] Russakovsky O, Deng J, Su H, Krause J, Satheesh S, Ma S, et al. ImageNet large scale visual recognition challenge. International Journal of Computer Vision 2014;115:211—52. https://doi.org/10.1007/S11263-015-0816-Y.

[60] Deng J, Dong W, Socher R, Li L-J, Kai L, Fei-Fei L. ImageNet: a large-scale hierarchical image database. 2010. p. 248—55. https://doi.org/10.1109/CVPR.2009.5206848.

[61] Guo Y, Liu Y, Oerlemans A, Lao S, Wu S, Lew MS. Deep learning for visual understanding: a review. Neurocomputing 2016;187:27—48. https://doi.org/10.1016/J.NEUCOM.2015.09.116.

Efficient transfer learning techniques for breast cancer histopathological image classification

12

R. Deiva Nayagam[1], D. Selvathi[2]

[1]*Department of Electronics and Communication Engineering, Ramco Institute of Technology, Rajapalayaam, Tamil Nadu, India;* [2]*Department of Bio Medical Engineering, Mepco Schlenk Engineering College, Sivakasi, Tamil Nadu, India*

1. Introduction

One of the leading causes of illness and mortality in the modern period is cancer across the globe. By 2030, it is predicted that the number of cancer-related deaths will have increased to more than 28 million from the current figure of about 14.5 million. Males can also be impacted by breast cancer (BC), which is a prevalent condition that several women deal with during one's lifetimes [1]. According to the Breast Cancer Institute (BCI), it ranks among the most severe diseases that impact women globally. In a woman's lifetime, breast cancer will be detected in one out of every seven of them, according to statistics, making it one among the most prevalent types of tumors among females in the United Kingdom (UK). Actually, in the UK, there are 55,200 new instances of breast cancer detected each year [2], and a depressing estimate of 11,400 of those cases result in death. One of the worst diseases is breast cancer, which has a risk of mortality around 20% on average.

The best and most effective strategy for properly controlling cancer is early diagnosis. Contrarily, delaying a diagnosis may cause cancer to spread all through the body and make it more challenging to treat and control. Furthermore, the likelihood of effective treatments is decreased with a late diagnosis. The breast is made up of a variety of structures, including fibrous tissue, lipids, hormones, and duct, all of which are susceptible to cancer. Screening test, self-examination at home, contacting a doctor, and other methods are available for timely detection of BC. These techniques will increase the likelihood of successful therapy while lowering the rate of death. Breast ultrasonography, magnetic resonance imaging (MRI), computed tomography (CT), thermography, mammography, cytopathological, and histopathological imaging are the greatest popular imaging methods for diagnosing the BC.

Images from mammography are frequently used for the clinical settings for tumor screening, especially to detect the breast tumors, but regrettably their research cannot conclusively determine whether a tumor is benign or malignant [3]. The most accurate test to diagnose cancer in this situation is a biopsy. A biopsy intervention is considered as a foremost suggested procedure on the human body for diagnosis. This is a laborious analysis, as samples of the detected strata are taken with a special needle that is inserted into the tumor while being monitored by ultrasound or mammography. The samples collected are used for histological examination to confirm the diagnosis and special immunohistochemically analyses that can identify a number of conclusive factors such as tumor sensitivity and targeted therapies [4,5]. Histopathologists employ visual examination of tissue samples under a microscope to determine the grade and stage of BC [6]. Open biopsies allow for the collection of histopathological images, which can be used to accurately assess the condition of breast tissue. The technical process of histopathological analysis takes a lot of time, and it is made more difficult by things like exhaustion and drowsiness. The typical manual diagnosis, however, demands a heavy workload from qualified professionals. Pathologists with insufficient clinical expertise are more inclined to make diagnostic errors. Early intervention of BC is the primary motivating factor for this study. It has been demonstrated that using computer-aided diagnosis (CAD) [7] to categorize histopathological images automatically might increase diagnostic effectiveness and give doctors more objective and accurate diagnosis results.

The potential of CAD is a promising method to fully exploit the data obtained through digital whole-slide imaging (WSI). CAD systems and digital histopathological images can be used together for screening a lot of images and grading the cases of cancer for review by a pathologist. CAD and digital pathology may be used to fill the shortage of pathologists in resource-constrained environments. One of the essential tasks in CAD systems is the autonomous categorization of BC from histopathology images, and the artificial intelligence (AI) algorithms play a noteworthy role to identify, classify, and segment from a BC histology data. The development of reliable CAD tools to classify breast cancer as benign or malignant from the images of histopathological data using machine learning (ML) and deep learning (DL) algorithms has received significant attention from numerous researchers across the globe [8]. The usage of many steps like preprocessing, postprocessing, and data augmentation is a challenge in CAD since inaccuracy occurs in every stage of the model and results in an accumulated error that grows with each stage. The accumulated inaccuracy of classifying and segmenting and classification processes can be seen, for instance, in a features extraction method where characteristics like volume of the tumors can be obtained by evaluating the fragmented region of the image. The tumors are classified using machine learning. To get more precise findings, it is crucial to reduce errors at each phase. Additionally, despite the fact that there are now more online image datasets available than ever before, the medical image data are still very limited when compared to other online datasets like ImageNet. The accuracy of CAD systems is significantly influenced by the quality of the data used to train artificial intelligence (AI) algorithms [9].

As machines gather information and facts through experience, it becomes less and less necessary for human—computer operators to define all the knowledge a computer needs. A type of AI-based neural network called deep learning architecture known as a convolutional neural network (CNN) holds promise for solving computer vision problems, sometimes surpassing human experts. The multilayer architecture of CNN technology helps computers learn by storing properties such as shapes, colors, edges, and more complex concepts in feature maps. Using the Breast Cancer dataset BreakHis, we will develop and adapt pretrained CNN architectures in this work. Further, we will use data augmentation methods on low-resolution tissue samples to improve the models by modifying their parameters. The proposed model leverages an efficient and time consumed approach on transfer learning technique of ResNet-based CNN architecture that has been pretrained on ImageNet to learn and classify the BreakHis dataset into benign or malignant.

2. **Related works**

Many studies have been conducted recently on the evaluation of health data, particularly in the areas of tumor diagnostics and early cancer cell identification. Accurate nuclei detection is difficult due to the diversity of histology images. Hematoxylin and Eosin (HE), a common stain, highlight cytoplasm in pink and nuclei in blue/purple to assist visualize the feature of significance in the tissue [10]. Histopathology imaging analysis often employs the methods used to evaluate cytology imaging. Particularly, certain characteristics of nuclei serve as warning signs for cancerous conditions. Thus, to appropriately incorporate the broad perspectives of an experienced pathologist, quantitative assessments for malignant nuclei were developed and assessed utilizing cytology images. As long as the histology characteristics have been properly separated, the same metrics can be utilized to interpret histopathology images.

Recent publications have employed machine learning methodologies and its algorithms to analyze medical imaging data. With BC datasets, numerous algorithms and methods have yielded outstanding segmentation and classification outcomes. The BreakHis dataset was trained using the CNN architecture in Ref. [11], and they demonstrate the evaluation of various classifiers and mixtures of various image features extracted. They claim an accuracy range of between 80% and 85%, based on the image enlargement ratio. The random-patches method is believed to have boosted the models' accuracy by 4−6 percentage points. This method necessitates recovering subimages both during the training and testing phases.

Sliding window support vector machine (SVM) has been employed to categorize the BC images into benign and malignant states depending on what sliding windows experience the most frequently. A sliding window of 80×80 pixels was subjected to local binary pattern recognition on the BreakHis dataset to extract features that were later used to train the SVM. The BreaKHis BC datasets were used to evaluate on the SVM model to categorize them into benign and malignant [12]. The accuracy of the fold-dependent categorization ranged from 81% to 96%, whereas the accuracy of the fold-independent classification was 91%, with accuracy rising with increasing magnification.

The author [13], used a hybrid method of CNN and random forest model to categorize size, shape, and thickness to identify tumors from BC histological images. When compared to the random forest classifier model, the CNN-based classifier model performed better while using less computing power. For classifying aggressive adenocarcinomas in breast cancer as positive or negative from the BreaKHis dataset, a comparison of traditional and deep learning algorithms was conducted [14]. A data set of WSI from 279 patients with IDC positive and negative spots was used in this analysis. The typical machine learning techniques selected for this study are logistic regression, AdaBoost, and SVM classifiers. The deep learning architecture was used to implement the convolutional model of the Xception architecture, and it outperformed all other conventional machine learning models.

Now-a-days, most of the researchers/experts are paying attention to deep learning systems, a developing technique in the machine learning space. CNN has been very effective at recognizing images and videos on a big scale. In Ref. [15], the author classified the BC histopathology data into benign and malignant using AlexNet. Their categorization outcomes outperform conventional machine learning techniques by 6%. The authors [16] suggested that hybrid CNN units can fully exploit both local and global aspects of images to produce more accurate predictions. To improve classifier performance, the authors also introduce bagging strategies and hierarchical voting tactics. Finally, 87.5% of the various breast cancer classifications were correctly classified.

The author [17] used a unique transition module of the CNN regularization method to facilitate the layer transition from convolution to a fully connected layer. This method collects filters at different scales and uses global mean pooling to compress the filters. A modest dataset could be successfully fitted with 91.9% accuracy by the transition module. In Ref. [18], the author suggested using breast cancer class and subclass designations as prior information to constrain the various characteristic distances of BC histopathology images. The accuracy of binary classification was 97% and a strategy for data augmentation was presented at the same time. A multiple-instance learning (MIL) approach for CNN was presented by the author [19]. To aggregate the most beneficial features of the patches that make up the entire slide, we developed a new layer of pooling that does not require overlap between patches or global slide coverage. Nearly 88% accuracy is achieved for images of breast cancer cases.

Large amounts of data are required to implement deep learning models with acceptable performance. Creating huge datasets with millions of images, such as ImageNet, contains millions of images from 15,000 distinct taxonomies and requires a lot of effort and computational power [20]. Health care Datasets are not always widely available. Accessing the computer resources necessary to analyze the data and run the model, as well as overcoming the challenge of small quantities of information while preventing overfitting, are some of the challenges in setting up a deep learning CAD system. Mammogram databases barely reach 10,000 images. Overfitting happens when the system learns the sources very well, and because it only recognizes situations it has already seen, it is unable to generalize successfully to new cases.

Transfer learning, which uses CNN models pretrained on large general datasets, is a deep learning technique that is regularly deployed when data are scarce. High-performance CNNs can transfer insights from other general areas with larger datasets to related areas such as medical imaging. Using datasets like CBISDDSM [21], the author showed how transfer learning transferred the learned knowledge from ImageNet weights could be applied to the field of mammograms, achieving 78.4% accuracy with the ResNet-50 model. The author [22] demonstrated how employing popular CNN architectures like VGG and ResNet that had been pretrained on ImageNet led to improvements in accuracy. Depending on how many patches are applied, accuracy increases. The two different ResNet50 models were evaluated using alternative weight initializations, one with weights of ImageNet and the other with random on BC dataset, as the author illustrated in Ref. [23]. By achieving 84% accuracy as opposed to 75% accuracy, the model that used transfer learning demonstrated the benefit of utilizing CNNs with pretrained weights.

Deep learning-based image classification tasks have, in fact, attracted a great deal of attention recently. Researchers have presented innovative categorization techniques using the ImageNet Large Scale Visual Recognition Challenge (ILSVRC) as a legitimate framework. CNNs have at least one convolutional layer that uses convolution operations on the input matrix instead of matrix multiplication to identify the image's high-level and low-level features. Deepening the network allows deep CNN to learn additional features. However, vanishing and deteriorating gradients become a problem as the depth of the mesh increases. Since then, advances in deep neural networks have enriched the field of AI, and in 2015 residual learning frameworks were introduced to facilitate training of CNN deep networks. This approach simplified network optimization and improved accuracy. This network, later known as residual neural networks (ResNet), served as the foundation for participation in the ILSVRC competition, in which it won the tasks for ImageNet localization and identification [24]. These problems are addressed by residual neural networks (ResNet), which include a "residual block" with a "skip connection" which transfers the data of the preceding layer to the preceding layer. In this study, we propose to classify histopathological images using ResNet-50 (50 slices deep) to aid early detection of BC. The BC dataset catalogs various histology images of breast and categorizes them into cancer and noncancerous categories. Noncancerous one is considered as benign tumor, unlike a cancerous one is considered as a malignant tumor that moves to other parts of the body or invades nearby tissues. The ResNet-50 model is used to train with the approach of transfer learning using the ImageNet weights, which indicates that the model has good test accuracy.

Nawaz et al. [25] described how to use CNN in combination with the transfer learning techniques of DenseNet to categorize the BC into multiple categories. DenseNet-based CNN has 201 layers which were pretrained on the ImageNet. They implemented patient and image classification using the public database Break-His. For breast cancer transfer learning classification, the model's classification accuracy for images was 95.4, and its classification accuracy for patients was 96.48. Using a transfer learning approach, they adapted a DenseNet model to process images of histological breast cancer cases. Transfer learning is described as the optimization of CNN models previously trained on original image data sets for medical imaging applications. A pretrained deep neural network called DenseNet which effectively classifies cancers based on medical images.

Furthermore [26] demonstrated classifying histopathological images using deep mindful feature learning and created a new CNN architecture based on deep learning. In addition, the base model adopts VGG19 and adds several attention modules. The CNN-based pretrained VGG-19 contains 19 layers and has an input size of 224×224. The average patient categorization accuracy for the BreakHis dataset is 98.05%, while the average image classification accuracy is 91.75%. In Ref. [27], the author used a simple SE-ResNet module to classify BC histopathology images using CNN techniques. They used the open-source dataset BreakHis in their study to create a CNN model with a tiny squeeze and excitation ResNet module (SE-ResNet) for histopathological imaging of BC. Layer squeeze and excitation blocks are fully convolved in the SE-ResNet module. The authors achieved the same efficiency with limited training data by automatically classifying binary and multiclass data from the BreakHis database. In the end, they were able to attain multiclass classification performance ranges between 90.66% and 93.81%.

The authors [28] tested the performance of ResNet-50 on the BACH dataset to detect BC and found an area under the curve (AUC) of 0.99 and approximately 93% patchwise classification accuracy and 96% image classification accuracy. Break-His datasets are separately categorized into four cancerous and noncancerous subtypes using a multicategory deep residual network (MuDeRN), a variant of ResNet152 [29]. ResNet152 CNN classified the eight classes as benign or malignant. BreaKHis benign and malignant images were divided into subtypes using four iterations of the starting network and four iterations of the residual network [30]. The accuracy of ResNet-101 is maximal when trained with 6000 iterations. In this study, a mini-batch size of 32 was chosen, and one iteration corresponds to processing one mini-batch.

The remainder of this article is divided into the following sections: Section 3 discusses the methodology used in the model such as transfer learning techniques and the concept of ResNet networks. Section 4 discusses the implementation details along with BreakHis dataset and the steps required for preprocessing techniques for feature extraction. Experimental results and performance comparisons are described in Section 5. In Section 6, we conclude our discussion.

3. Methodology

This chapter outlines the concept of the proposed model that has been suggested as well as the experimental protocol used in this study. The suggested model will be put through the following procedures to see if it has been evaluated.

- To use the open-ended BC dataset—BreakHis dataset, a database of pathologists' labeled microscopic breast tumor histopathology images.

- To develop a thorough, effective classification model using deep learning algorithms that can distinguish the breast cancer in the BreakHis dataset, including benign and malignant.
- Additionally, before using deep learning algorithms on the gathered medical images, we describe thoroughly various preprocessing procedures.
- ResNet50 CNN, which was previously learned on the larger dataset—ImageNet, is used to acquire new weights for the BreakHis dataset using the transfer learning method. By using accuracy as a metric to calculate the transfer learning-based classification system's performance on testing unseen data.

Fig. 12.1 shows the proposed methodology used for the binary classification of benign or malignant breast cancer from the histopathology BreaKHis dataset.

3.1 Transfer learning approach on pretrained network

The CNN deep learning model is used to learn how to represent images. Without human aid, CNN models are capable of learning from basic to complex features. This model learns neighborhood information from images using a method called convolution. These features help you classify your photos. However, deep learning still requires a lot of data. Transfer learning is a concept that can be used as compensation. Employing a model that has been trained on a wider and more varied dataset for a problem is an approach called transfer learning which benefits the model training by fine tuning our proposed model and training it on BreakHis dataset. In transfer learning, the computer makes better predictions for new tasks using data from earlier tasks. Transfer learning is the practice of processing a model which was previously trained to new tasks or issues [31]. Other models need to be

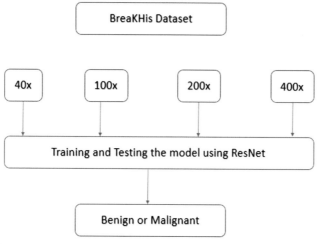

FIGURE 12.1

Proposed model for binary classification of breast cancer.

completely retrained. It takes a lot of calculation and data to make this operate properly. Transfer learning, however, reduces resource and computational costs. Additionally, it improves the effectiveness of categorization, grading, and natural language processing jobs. Deep neural networks can be trained using little to no data utilizing transfer learning, which is becoming more and more popular in deep learning. It expedites learning and cuts training time in half. Transfer learning models perform at their peak more quickly. This is because the networks which employ various information like weights and features from previously trained models already have an understanding of the properties. It is quicker than building the model from the starting phase. Transfer learning with CNNs aims to improve the performance of new or subsequent tasks by preserving previous information. This solved the problem of missing data in medical image analysis. It reduces training time and computational complexity. The task of classifying medical images benefits greatly from transfer learning techniques. Deep models such as ResNet, ShuffleNet, and InceptionNet are recommended in research studies for use as feature extractors because they reduce processing cost and time without sacrificing predictive power [32].

3.2 Residual network (ResNet)

The "vanishing gradient" problem is solved with a CNN design called ResNet. This makes networks with more convolutional layers perform better than networks with fewer layers. Networks with multiple layers (potentially thousands of layers) can be easily trained without increasing the training error rate. Skip connections are used in ResNet to skip some layers. By utilizing identity mapping, ResNets aid in the vanishing gradients task. A ResNet model is made up of these residual blocks designed to simplify the training of incredibly deep networks. The first thing to notice is the direct linking of the ResNet model bypassing some layers of the model. The so-called "skip connections" are the central component of the rest of the blocks. The output is different because of the jump connection. The source "X" is compounded by the input weights, and a biased term is included whenever there is no skip connection. The result of the activation function is mentioned in the following Eq. (12.1),

$$H(x) = f(x) \tag{12.1}$$

The output of $H(x)$ has changed due to the addition of a new skip connection mechanism, which was shown in Eq. (12.2),

$$H(x) = f(x) + x \tag{12.2}$$

However, the input dimensions and the output dimensions can vary when convolutional or pooling layers are used. Therefore, there are two ways to approach this issue: one is padding the skip connection with zeros to increase size, and the other is adding a 1 * 1 convolution layer to the input to adjust for the dimensions. The output of $H(x)$ has changed in this situation as mentioned in Eq. (12.3).

$$H(x) = f(x) + w1.x \tag{12.3}$$

In contrast to the first technique, this situation introduces an additional parameter called $w1$. ResNet's skip connections technique can be used to move the gradient through an additional shortcut channel to address the problem of gradient fading in deep CNNs [33]. Skip connections are useful even if layers affect the performance of your architecture. The layers of weight, activation, and normalization that make up a ResNet organized into a series of blocks called residual blocks. Combine the inputs and residuals using a shortcut combination using element-wise addition. Residuals are the output of weights, activations, and normalization layers. After a 3×3 max pooling layer with batch normalization, ReLU activation, and stride 2 of downsampling, the first residual module has 64 output channels and 7×7 convolutions with stride 2. A 1×1 convolutional input is added first, followed by the rest of the following blocks. Each remaining block has half the height and width and double the number of channels. After being stacked to create the desired neural network depth, the max pooling layer and the output convolution layers are contained in the remaining blocks. The source image for the ResNet model is 224×224 pixels. ResNet architectures come in many forms, each with a different number of layers. These include ResNet-18, ResNet-34, and ResNet-50. The number indicates the number of layers in ResNet. ResNet50 is used in our study.

For the ImageNet dataset, a deeper neural network, ResNet50, outperforms ResNet18 and ResNet34 in terms of accuracy [34]. The remaining blocks therefore deal with depth-induced degradation. This extremely deep neural network's training time is shortened via a bottleneck design. Three layers of convolution—1×1, 3×3, and 1×1 layers of convolution—replace the two layers of convolution present in each residue block. The first 1×1 convolutional layer reduces dimensionality, the third 1×1 layer restores dimensionality, and the bottlenecking is caused by the 3×3 convolutional layer's small inputs and outputs. These 3-layer blocks are stacked into modules of various sizes to create ResNet50 networks.

4. Implementation

4.1 Dataset used

We tested an effective ResNet50 architecture on histopathological imaging data of breast cancer (BreaKHis). Moreover, 7909 microscopic images (consisting of 5429 images of cancerous breast tumors and 2480 images of benign breast tumors) are included in the dataset [35]. The different magnification ranges are $40\times$, $100\times$, $200\times$, and $400\times$. Every picture has a dimension of 700×460 pixels, a 3-channel RGB color scheme, and an 8-bit per channel depth. Adenopathy (A), fibroadenoma (F), tubular adenoma (TA), and phyllodes tumor (PT) are four subtypes of benign tumor. In addition, Papillary carcinoma (PC), lobular carcinoma (LC), ductal carcinoma (DC), and mucinous carcinoma (MC) are the four subtypes of malignant cancer. Table 12.1 describes the cancer categories of benign and malignant and its subimages on BreaKHis dataset. Fig. 12.2 represents the images of subtypes of $400\times$ magnification factor of benign and malignant tumors.

Table 12.1 Details of BreaKHis dataset by cancer categories.

Classes	Subclasses	Number of patients	Magnification factors				Total
			40×	100×	200×	400×	
Benign	A	4	114	113	111	106	444
	F	10	253	260	264	237	1014
	TA	3	109	121	108	115	453
	PT	7	149	150	140	130	569
Malignant	DC	38	864	903	896	788	3451
	LC	5	156	170	163	137	626
	MC	9	205	222	196	169	792
	PC	6	145	142	135	138	560
Total		82	1995	2081	2013	1820	7909

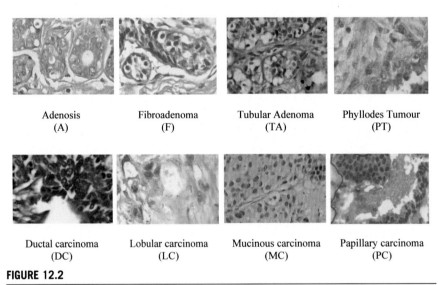

Adenosis (A)   Fibroadenoma (F)   Tubular Adenoma (TA)   Phyllodes Tumour (PT)

Ductal carcinoma (DC)   Lobular carcinoma (LC)   Mucinous carcinoma (MC)   Papillary carcinoma (PC)

FIGURE 12.2

Samples of 400× magnification factors of benign tumors (first row) and malignant tumors (second row).

4.2 Preprocessing steps

Preprocessing, which is the first stage in data preparation for deep learning models, is crucial. We removed speckle noise from the images in our proposed study to achieve excellent classification rates. Finally, we transformed the label string data into numeric data after conducting the compression of data, transformation of data, and extraction of features. By minimizing unneeded properties such as data separation, validity, and variation additions, updates, and deletions, data normalization is a dataset design technique. Z-score normalization, decimal scale normalization, and min−max normalization are examples of existing normalizing methods. Z-score normalization was used to normalize the dataset. Methods for extracting features from images are crucial for image processing because they separate images into groups that can be handled. By extracting a large number of attributes, our research recognizes patterns in multiple datasets. Extracting the features is the process of turning a collection of characteristics from a collection of inputs. During the feature extraction stage of machine learning, borrowed values, also known as features, are generated from a consistent dataset. These borrowed values aid the subsequent stage of learning and observation. These standards are intended to be non-verbose and descriptive. The dimensionality reduction process is most closely connected to this. The image collection was examined using principal component analysis (PCA) [36]. PCA enables the largest number of important components to be extracted.

Numeric input is the most prevalent type that machine learning algorithms need to process for claims. We keep going through the steps for each attribute to arrive at the figures on various scales. Additionally, this data must be organized and maintained to enhance model development and model validation for different deep learning systems. The label encoder tool from the experiment uses the Python functionality to interpret the two labels as malignant and benign into the numeric values 1 and 0, respectively. Without actually gathering new data, a method known as data augmentation introduces a drastically altered version of the existing accessible training data. The training set's size can be deliberately increased or delay model overfitting immediately through oversampling or data movement. By simply leveraging information already present in the training data, a technique called data augmentation can increase the number of training data and reduce the overfitting of the network. After splitting the dataset into training and validation sets, we used several enhancement settings at the right levels using rotation, random clipping, and mirroring [37].

4.3 Training the ResNet model on pretrained model

Transfer learning approaches can help with deep learning network. This enables the application of information learned from addressing on particular kind of problem to other problems of a similar type. The first layer of a deep network is typically trained using a larger image dataset- ImageNet includes approximately several millions of images from thousands of distinct categories. The network weights starting with randomly initialized weights for training are much slower and less effective than fine-tuning a deep network in conjunction with transfer learning. It becomes feasible to learn. Many applications, including speech identification and classifying the medical image and recognizing the image, have found a role for transfer learning. A deep learning model that has been pretrained using the ImageNet repository makes up the first section of the deep CNN. A fully convolved layer, a nonlinear activation layer, and an output classifier layer make up the last three output layers of the classifier. The final slice is learned with the BreakHis dataset of tissue images. In those other terms, everything from the pretrained network is extracted, with the exception of the final three layers. A fully integrated convolutional layer has the same number of outputs as classes in the dataset. One should raise the parameters of the weight learn rate factor (WLRF) and the bias learn rate factor (BLRF) to accelerate the weighted information gain of the output units. Deep network models that have already been trained using transfer learning learn common characteristics from the datasets which are not present in the target dataset.

Recent research has suggested ResNet configurations that are more effective and efficient. Deep convolutional neural networks can be effectively taught using this design. ResNet constructs the design using the remaining building elements. Between the first and last convoluted layer, there are a total of 50 layers. In this study, we develop a binary classifier for the detection of breast cancer using ResNet50. The next few fully convoluted layers are designed to be trainable. The input consists of an image with a resolution of 228 × 228 pixels, and it is transformed into an image with a convolutional layer that is half the size. The input images should be resized based on the network to be trained after implementing transfer learning, and the values of WLRF and BLRF are both set to 20. The adaptive moment estimation technique (Adam) was utilized to enhance the network and set the learning rate value as 0.0001, number of epochs as 20, decay factor as 0.99, and mini-batch size as 128. Throughout the investigation using the BreakHis dataset, we divided the dataset into training and testing sets using an image-based method by arbitrarily choosing 70% of each sample without replacement. Instead of cross-validation, we employed ensemble learning on the training set to equalize the categories and improve the classification accuracy. We also added dropout to each to prevent overfitting the pretrained neural network. We evaluated the performance of an efficient time-consuming ResNet model based on a transfer learning approach using standard metrics such as classification accuracy, precision, sensitivity, F1 score, and Roc AUC values.

5. Results and discussions

The software used in this work is Python Anaconda. We are using a Spyder notebook in it. To compare our results with those of previous studies, we accurately assessed the effectiveness of our network. Below are the steps in the training process. First, import the ResNet50 network model and its configuration optimized for the ImageNet dataset. Then, before the final layer of the network, we froze all parameters. Then adjust the fully connected layer and change that of the dense layer to nine to allow binary classification. Unfrozen layers of the network are then trained until the best results are archived. The entire classification method is designed and implemented using the deep learning framework TensorFlow. We choose the Adam algorithm because it is a fast and reliable optimizer. We decided to use Adam's learning rate as 0.0001 for the learning rate because it can explain the learning rate optimization. For the output, we chose Softmax as the output function. We chose to train on the raw data when it came to data augmentation to save time.

5.1 **Results for breast cancer classification**

We use the BreaKHis dataset to train our fine-tuned ResNet50 using transfer learning, and we then test the classification performance on a different set of photos from untrained data. On the BC histopathological images, we conducted classification diagnosis research using ResNet50, transfer learning, and fine-tuning techniques. The training accuracy reached 0.9292%, and the validation accuracy was 0.98324%. During the training test, the loss function's value reaches 0.1890, which drops quickly and smoothly before converging to a small value. The validating loss value reaches 0.1402. Fig. 12.3 represents the validation accuracy of 0.9834 after reaching its epoch value of 20.

Figs. 12.4 and 12.5 represent the accuracy, and the loss obtained is plotted as shown in Figs. 12.4 and 12.5, respectively.

Table 12.2 depicts the evaluation metrics of precision, sensitivity, F1 score, and ROC AUC values.

5.2 **Performance comparison**

This subsection compares the test results for classifying histology images of BC as benign and malignant using ResNet50 with transfer learning and other fine-tuning techniques. The classifier evaluation will be analyzed in terms of accuracy and loss because the vast number of currently available studies simply used this evaluation criterion. The results indicate that, among all the studies we could discover in the literature, our fine-tuned network had the best classification accuracy for the BreaKHis dataset's histological images. ResNet50's fine-tuned neural network with transfer learning is a fantastic option for categorizing the histological images of BC as a consequence. Assuming that the features in the new application are classified differently from those in the original application on which the network was trained but that the extracted characteristics are the same, we can reinforce these

```
Epoch 00015: val_acc did not improve from 0.97207
Epoch 16/20
45/44 [==============================] - 40s 884ms/step - loss: 0.1890 - acc: 0.9292 - val_loss: 0.1402 - val_acc: 0.9832

Epoch 00016: val_acc improved from 0.97207 to 0.98324, saving model to weights.best.hdf5
Epoch 17/20
45/44 [==============================] - 40s 887ms/step - loss: 0.2015 - acc: 0.9338 - val_loss: 0.2041 - val_acc: 0.8994

Epoch 00017: val_acc did not improve from 0.98324
Epoch 18/20
45/44 [==============================] - 40s 886ms/step - loss: 0.1891 - acc: 0.9079 - val_loss: 0.0932 - val_acc: 0.9609

Epoch 00018: val_acc did not improve from 0.98324
Epoch 19/20
45/44 [==============================] - 40s 883ms/step - loss: 0.1565 - acc: 0.9287 - val_loss: 0.0982 - val_acc: 0.9721

Epoch 00019: val_acc did not improve from 0.98324
Epoch 20/20
45/44 [==============================] - 40s 886ms/step - loss: 0.1736 - acc: 0.9329 - val_loss: 0.1259 - val_acc: 0.9385

Epoch 00020: val_acc did not improve from 0.98324
```

FIGURE 12.3

Training and validation accuracy level.

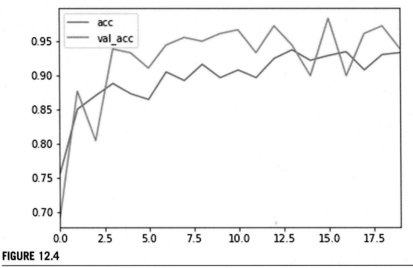

FIGURE 12.4

Accuracy plot.

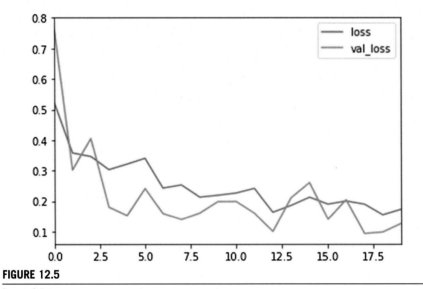

FIGURE 12.5

Loss plot.

Table 12.2 Metric evaluation scores.

Precision	Sensitivity	F1 score	ROC-AUC
0.65	0.95	0.77	0.69

Table 12.3 Performance comparison of different studies on breast cancer data for binary classification.

References	Classifier	Overall accuracy
[12]	SVM	91%
[16]	Hybrid CNN	87.5%
[17]	CNN	91.9%
[23]	ResNet	84%
[25]	DenseNet	96.48%
[26]	VGG	91.75%
[27]	SE-ResNet	93.81%
Our suggested model	ResNet50	98.32%

last layers of transfer learning models. The proposed ResNet50 resultant model is better than those of various earlier attempts and are shown in Table 12.3. The challenges to achieving lesser performance include fewer input data, appropriate data segmentation, noise and artifacts in images, ineffective preparation techniques, and improper hyperparameter model selections.

6. Conclusion

The ultimate testing accuracy rate for the proposed system is 98.32%, and the testing loss is 0.1402. In this study, deep learning-based efficient transfer learning techniques are used for classifying the image. Various parameters like accuracy and loss of classification are analyzed. The model training process requires very little computational time. As we employ a pretrained framework in our research, both the learning curve and accuracy are significantly reduced. A comparison with them shows that the suggested strategy outperforms most existing similar cancer diagnostic techniques. If pathologists employ this computer-based identification method, they will be able to diagnose more instances of breast cancer in hospitals with less time, money, and effort. We plan to enhance the classification model's design and engineer new sets of features from more histopathological images to increase its efficacy.

References

[1] WHO. Data of breast cancer sufferers in the world. World Health Organization (WHO); 2018. Jan. 3. 2019.
[2] Cancer Research UK. Breast cancer statistics. 2020. https://www.cancerresearchuk.org/health-professional/cancer-statistics/statistics-by-cancer-type/breast-cancer.

[3] Kuhl CK, Schrading S, Leutner CC, Morakkabati- Spitz N, Wardelmann E, Fimmers R, et al. Mammography, breast ultrasound, and magnetic resonance imaging for surveillance of women at high familial risk for breast cancer. Journal of Clinical Oncology 2005;23(33):8469–76.

[4] Lyman GH, E Giuliano A, Somerfield MR, Benson AB, Bodurka DC, Burstein HJ, et al. American society of clinical oncology guideline recommendations for sentinel lymph node biopsy in early-stage breast cancer. Journal of Clinical Oncology 2005;23(30): 7703–20.

[5] Elston DM, Stratman EJ, Miller SJ. Skin biopsy: biopsy issues in specific diseases. Journal of the American Academy of Dermatology 2016;74(1):1–16.

[6] Lakhani SR, Ellis IO, Schnitt S, Tan PH, van de Vijver M. Who classification of tumours of the breast. 2012.

[7] Araujo T, Aresta G, Castro E, Rouco J, Aguiar P, Eloy C, et al. Classification of breast cancer histology images using convolutional neural networks. PLoS One 2017;12(6): e0177544.

[8] Pallua JD, Brunner A, Zelger B, Schirmer M, Haybaeck J. The future of pathology is digital. Pathology, Research & Practice Sep. 2020;216(9):153040.

[9] da Costa Vieira RA, Biller G, Uemura G, Ruiz CA, Curado MP. Breast cancer screening in developing countries. Clinics Apr. 2017;72(4):244–53.

[10] Veta M, Pluim JPW, Van Diest PJ, Viergever MA. Breast cancer histopathology image analysis: a review. IEEE Transactions on Biomedical Engineering 2014;61(5): 1400–11.

[11] Spanhol FA, Oliveira LS, Petitjean C, Heutte L. Breast cancer histopathological image classification using convolutional neural networks. In: 2016 international joint conference on neural networks (IJCNN). IEEE; 2016. p. 2560–7.

[12] Alqudah A, Alqudah AM. Sliding window based support vector machine system for classification of breast cancer using histopathological microscopic images. IETE Journal of Research Mar. 2019;68(4):1–9.

[13] McKinney SM, et al. International evaluation of an AI system for breast cancer screening. Nature Jan. 2020;577(7788):89–94.

[14] Ibrahim A, et al. Artificial intelligence in digital breast pathology: techniques and applications. Breast Feb. 2020;49:267–73.

[15] Krizhevsky A, Sutskever I, Hinton GE. Imagenet classification with deep convolutional neural networks. In: Advances in neural information processing systems; 2012. p. 1097–105.

[16] Guo Y, Dong H, Song F, Zhu C, Liu J. Breast cancer histology image classification based on deep neural networks. In: International conference image analysis and recognition. Springer; 2018. p. 827–36.

[17] Akbar S, Peikari M, Salama S, Nofech-Mozes S, Martel A. The transition module: a method for preventing overfitting in convolutional neural networks. Computer Methods in Biomechanics and Biomedical Engineering: Imaging & Visualization 2018:1–6.

[18] Wei B, Han Z, He X, Yin Y. Deep learning model based breast cancer histopathological image classification. In: 2017 IEEE 2nd international conference on cloud computing and big data analysis (ICCCBDA). IEEE; 2017. p. 348–53.

[19] Das K, Conjeti S, Roy AG, Chatterjee J, Sheet D. Multiple instance learning of deep convolutional neural networks for breast histopathology whole slide classification. In: 2018 IEEE 15th international symposium on biomedical imaging (ISBI 2018). IEEE; 2018. p. 578—81.

[20] Krizhevsky A, Sutskever I, Hinton GE. ImageNet classification with deep convolutional neural networks. 2012.

[21] Falconi LG, Perez M, Aguilar WG. Transfer learning in breast mammogram abnormalities classification with Mobilenet and Nasnet. In: International conference on systems, signals, and image processing; 2019. p. 109—14.

[22] Shen L, Margolies LR, Rothstein JH, Fluder E, McBride RB, Sieh W. Deep learning to improve breast cancer early detection on screening mammography. Scientific Reports 2017;9(1).

[23] Diaz O, Marti R, Llado X, Agarwal R. Mass detection in mammograms using pretrained deep learning models. In: Krupinski EA, editor. 14th international workshop on breast imaging (IWBI 2018), 10718. SPIE; 2018. p. 12.

[24] He K, Zhang X, Ren S, Sun J. Deep residual learning for image recognition. Microsoft Research 2015.

[25] Nawaz M, Sewissy AA, Soliman THA. Multi-class breast cancer classification using deep learning convolutional neural network. International Journal of Advanced Computer Science and Applications 2018;9(6):316—22.

[26] Wu P, Qu H, Yi J, Huang Q, Chen C, Metaxas D. Deep attentive feature learning for histopathology image classification. In: IEEE 16th international symposium on biomedical imaging (ISBI 2019); 2019. p. 1865—8.

[27] Jiang Y, Chen L, Zhang H, Xiao X. Breast cancer histopathological image classification using convolutional neural networks with small SE-ResNet module. PLoS One 2019; 14(3).

[28] Vesal S, Ravikumar N, Davari A, Ellmann S, Maier A. Classification of breast cancer histology images using transfer learning. In: Image analysis and recognition; 2018. p. 812—9.

[29] Gandomkar Z, Brennan PC, Mello-Thoms C. MuDeRN: multi-category classification of breast histopathological image using deep residual networks. Artificial Intelligence in Medicine Jun. 2018;88:14—24.

[30] Motlagh MH, et al. Breast cancer histopathological image classification: a deep learning approach. Cold Spring Harbor Laboratory 2018:242818.

[31] Alruwaili M, Gouda W. Automated breast cancer detection models based on transfer learning. Sensors 2022;22(3):876.

[32] Aljuaid H, Alturki N, Alsubaie N, Cavallaro L, Liotta A. Computer-aided diagnosis for breast cancer classification using deep neural networks and transfer learning. Computer Methods and Programs in Biomedicine 2022;223:106951. https://doi.org/10.1016/j.cmpb.2022.106951.

[33] Ferreira CA, et al. Classification of breast cancer histology images through transfer learning using a pre-trained Inception Resnet V2. In: Image Analysis and recognition; 2018. p. 763—70.

[34] Al-Haija QA, Adebanjo A. Breast cancer diagnosis in histopathological images using ResNet-50 convolutional neural network. In: 2020 IEEE international IOT, electronics and mechatronics conference (IEMTRONICS), Vancouver, BC, Canada; 2020.

[35] Spanhol FA, Oliveira LS, Petitjean C, Heutte L. A dataset for breast cancer histopathological image classification. IEEE Transactions on Biomedical Engineering 2016; 63(7):1455−62.

[36] Chiu H-J, Li T-H, Kuo P-H. Breast cancer−detection system using PCA, multilayer perceptron, transfer learning, and support vector machine. IEEE Access 2020;8: 204309−24.

[37] Zhang C, Nan B, Hang S, Hong L, Jing L, Qian W, et al. A deep learning image data augmentation method for single tumor segmentation. Frontiers in Oncology 2022;12.

Classification of breast cancer histopathological images based on shape and texture attributes with ensemble machine learning methods

13

Vijayalakshmi G.V. Mahesh

Department of Electronics and Communication Engineering, BMS Institute of Technology and Management, Bangalore, Karnataka, India

1. Introduction

Breast cancer is one of the most prevalent diseases in women and is the result of uncontrolled division of breast cells. Breast cancer can be categorized depending upon the cells of which parts of the breast, such as lobules, ducts, or connective tissues, have resulted in cancer. Some of the common types are: (1) Ductal carcinoma (DC): The cancer in this condition starts in the ducts and does not spread to other parts. This is referred as precancer. (2) Invasive ductal carcinoma (IDC): The cells start dividing in an uncontrolled way in the ducts and start growing out and invade other parts of the breast and body. (3) Invasive lobular carcinoma: Here, the cancer cells begin in the lobules and then invade other parts of the breast and other parts of the body. Detecting the disease at the earliest is very crucial to follow appropriate procedures for controlling and improving the survival rate. The presence of breast cancer can be identified by analyzing the histological, biological, and clinical characteristics of the disease. Noninvasive imaging methods such as breast ultrasound (BUS), mammography, and magnetic resonance imaging (MRI) are widely utilized for diagnosing cancer [1–4]. The analysis of breast cancer biomarkers [5,6] is also found to be encouraging in providing clues for detecting cancer. Though these methods are successful, they also have some shortcomings related to (1) detecting the disease in the case of dense breast tissue, (2) identifying the cancer subtypes, and (3) reducing the false positives and negatives. So, histopathological image analysis of breast cancer, which is referred to be the gold standard [7,8], can improve the identification and diagnosis accuracy and also distinguish the cancer subtypes [9,10].

Computational Intelligence and Modelling Techniques for Disease Detection in Mammogram Images
https://doi.org/10.1016/B978-0-443-13999-4.00012-2

Providentially, the development of image processing algorithms, machine learning models, and deep learning architectures are offering consistent solutions and methodologies for analyzing histopathological images for acceptable recognition and detection of cancer. Histopathology is the microscopic observation of tissue. Thus, histopathological images provide tissue images that help in the detailed analysis of the cancer. Here it is very important to observe the changes in the cell characteristics as cancer cells have irregular/abnormal shapes, angular edges, and rigid texture as compared to normal cells. One of the challenges in using machine learning models and deep learning architectures for identifying breast cancer is the effective representation of the characteristics of the cells, which form the features/attributes that can discriminate the cancerous tissue from the normal one. A significant effort has been put by many researchers for the same.

The changes in shape and texture characteristics of the breast cancer cells/tissues are to be represented efficiently as they are the key factors that provide precise discrimination between cancerous and noncancerous tissues. This chapter explores the abilities of shape descriptors: local energy-based shape histogram (LESH), Hu moments, and texture descriptors based on Law's energy measures (LEM) and Gray level cooccurrence matrix (GLCM) in describing the details captured in histopathological images. The objectives of this work include (1) efficient representation of the breast cancer tissues/cells combining the shape and texture descriptors, (2) build appropriate classifier models to recognize the IDC cancerous tissue for classification, and (3) quantitative analysis of classification rate. The chapter is further organized into four sections. Section 2 presents a literature review on the different methodologies presented for classifying breast cancer histopathological images. The formulation of methodology is explained in Section 3. Simulation results and analysis is presented in Section 4. Finally, Section 5 concludes the chapter.

2. Literature review

Several works on utilizing histopathological images for breast cancer identification/recognition integrated with image processing algorithms for feature extraction and machine learning/deep learning models developed for histopathological image classification are presented in this section. A review on the analysis of histopathological images based on digital image processing techniques to detect breast cancer, the present status, and future directions are presented in Refs. [8,11,12]. A dataset of 7909 breast cancer histopathological images was introduced by Ref. [13] and 3771 images by Ref. [14], which is publicly available to work on the automatic classification of benign and malign images to aid in computer-aided diagnostic procedures.

The identification of breast cancer and classifying the histopathological images can be achieved by deriving the handcrafted features from the images and presenting them to machine learning algorithms to build classifier models or use deep learning methods that combine feature extraction and classification in an architecture. Both approaches have contributed significantly toward breast cancer histopathological image analysis.

A method based on texture features and machine learning algorithms was proposed by Ref. [15] to classify benign and malign breast tissues. The work utilized GLCM, Graph Run length (GRL) matrix, and Euler number features to train linear discriminant analysis, k nearest neighbors (kNN), and support vector machine (SVM) classifiers to analyze the histopathological images. The results and analysis showed the outperformance of LDA classifiers over SVM and kNN. Ref. [16] utilized a combination of color and texture features for histopathological image classification. The authors represented the images using Gabor, complex wavelet and chromatic, dual-tree complex wavelet transform, and opponent color local binary pattern features. The features were provided to the classifiers, and the classifier models were quantitatively analyzed using AUC. The authors conclude that a precise combination of features and classifiers can present a more reliable performance in classification. Ref. [17] derived four texture features based on local binary patterns, local-oriented statistic information booster, and fractal texture analysis to represent the global and local characteristics of the tissues in histopathological images and experimented on the datasets: Kather, BreakHis, Epistroma, and Warwick-QU with SVM and decision tree bagger classifiers. The results indicated a high accuracy of 92.5%, 91.73%, 98.04%, and 96.29% for the datasets mentioned.

A concept of data sampling was presented by Ref. [18] to classify breast cancer histological images. The authors used a method to create samples dependent on the data distribution of binary patterns. The idea here was to classify the region of interest in a histology image as benign or malign. The analysis of the experiment conducted on the images acquired from the MITOS 2012 dataset provided good accuracy and F1 score. Random subspace classifier ensemble was also found to be effective when combined with multiple features. Ref. [19] reported to have obtained a classification rate of 95.22% for a combination of GLCM, curvelet transform, and completed local binary pattern features on a publicly available dataset.

Nuclei segmentation methods were proposed by Refs. [20,21] that used watershed algorithm and wavelet decomposition with multiscale region growing methods. Ref. [20] performed a two-stage segmentation process to identify the nuclei and then extracted color autocorrelation and local binary pattern features for representation. The experiments on ICIAR 2018 dataset provided a recognition accuracy of 92.5%. Ref. [21] derived texture and shape features based on color spaces and tested the proposed method BCH images that provided an accuracy of 96.19% with SVM. Further, a neural networks-based approach was implemented by Ref. [22]. The work focused on combining the attributes derived using invariant moments, intensity histogram, GLRLM, and GLCM to train a probabilistic neural network (PNN). The approach was evaluated on the histopathological images from the BreakHis database, which provided a performance of 82.5% accuracy.

As mentioned earlier, identifying the right features for representing the images is essential as they describe the characteristics of the cells and aid in proper discrimination between cancerous and noncancerous tissues. Deep learning methods have the advantage of extracting the features in a hierarchical form by itself making the learning faster. Ref. [23] proposed two deep learning architectures: single task convolutional neural network (CNN) to identify malign tissues and multitask CNN to identify malignity and image magnification level. The method was implemented on the BreakHis dataset with 2480 benign and 5429 malign images acquired under different magnification levels 40×, 100×, 200×, and 400×. The results indicated better performance irrespective of magnification level. Ref. [24] tried to improve the problem of intraclass classification and high false negative rates employing deep CNN combined with ensemble learning methods. The implementation proved to be successful, attaining a classification rate of 91.5% on the BreakHis dataset.

Ref. [14] developed a hybrid deep learning architecture combining CNN and recurrent neural network that was able to classify four classes of breast cancer with an average accuracy of 91.3% on the 3771 images available at http://ear.ict.ac.cn/?page_id=1616. Multiple compact CNNs were assembled based on the squeeze excitation pruning and bagging approach with different data partitions and compositions by Ref. [25]. The assembled models voted for the final result. The method was validated on BreakHis and Breast Cancer histology (BACH) datasets and found to have an effective generalization ability. Incremental boosting CNN was presented in Ref. [26] for improving breast cancer diagnosis. Transfer learning-based method using a pretrained CNN model was proved to be effective as compared to other models by Ref. [27]. Ref. [28] utilized pretrained CNN for deep feature extraction and provided these features to another classifier to perform the classification task. The works justify that this can be an alternate solution toward developing precise breast cancer recognition systems. On the same line, Ref. [29] utilized transfer learning approach to extract features from Inception V3 CNN to train SVM classifier for histopathological image classification and was able to improve the accuracy by 3.7%.

An ensemble of deep learning methods for effective diagnostic procedures was implemented by Ref. [30]. The method intended to distinguish carcinoma and noncarcinoma histopathological images based on pretrained VGG 19 and VGG 16 architectures. Four models that include fully trained VGG 16, fine-tuned VGG 16, fully trained VGG 19, and fine-tuned VGG 19 were trained, and tool and ensemble strategy was framed by averaging the predicted probabilities of the individual models. The method was effective in providing a sensitivity of 97.73% and an accuracy of 95.29%. Resnet V1 50 and Resnet V1 152 were utilized in Ref. [9] for classifying breast cancer subtypes. The quantitative analysis illustrated negligible false negative rates and false positive rates.

Convolutional autoencoder was used in Ref. [31] to learn the patterns of abnormalities and normal image patches, and the patterns are presented to SVM and one layer neural network for analysis. The demonstrated results proved the performance of the methodology that can provide better diagnostic solutions. Ref. [32] proposed a method to solve the problem of an imbalanced dataset where the classifier model skews toward the majority class. The methodology is framed using pretrained ResNet50 and kernelized weighted extreme learning model to identify breast cancer.

The method achieved appropriate classification performance from BreakHis and BisQue datasets for both majority and minority classes.

Multiresolution models developed demonstrated noticeable performance in histopathological image analysis. Ref. [33] worked on the multiresolution statistical model in the wavelet domain with higher-order statistical features and gray-level run-length matrix to detect ductal carcinoma. The efficiency of breast cancer performance is improved by combining handcrafted features and deep learning. Ref. [34] developed a hybrid model combining morphological, texture, and statistical features with CNN to detect mitosis in breast cancer pathological images. Model evaluation on the ICPR12 dataset was accurate and provided an F score of 0.7345.

This work considers the integration of the shape and texture descriptors in combination with ensemble classifier methods to achieve acceptable classification or discrimination of IDC breast cancer and normal histopathological images.

3. Methodology formulation

The proposed methodology for classifying breast cancer histopathological images is displayed in Fig. 13.1.

Histopathological image classification is a pattern recognition and classification problem where the patterns are identified and assigned the class label. Patterns are the key attributes that provide information about the variations in cancerous cells/tissues as compared to normal tissues. As indicated, morphological (abnormal shapes, angular edges) and texture (rigid) changes can be observed in the cancerous

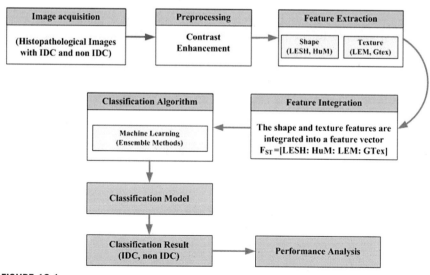

FIGURE 13.1

Proposed methodology to classify breast cancer histopathological images.

cells. These changes are to be recognized, represented, and analyzed to be able to discriminate them from normal cells effectively. The changes are best represented using shape and texture descriptors/features. For the work considered, the shape and texture features are derived from the histopathological images and are integrated to provide complete information. These features are input to the machine learning algorithms to develop the classifier models to identify and categorize the images into normal and cancerous (invasive ductal carcinoma). To indicate the effectiveness of the proposed methodology, the models are assessed based on the quantitative computation of the performance measures: classification accuracy (CA), recall, precision, Mathew's correlation coefficient, area under the curve (AUC), kappa value, and F1 score. The details of the process are provided in the subsequent subsections.

3.1 Materials

The proposed model aims to recognize invasive ductal carcinoma, a subtype of breast cancer from histopathological images. The task of classifying the images into normal and IDC was implemented on the breast cancer histopathological image dataset acquired from Ref. [35]. The dataset has a total of 162 whole-mount images scanned at a magnification level of 40×. For precise identification, the region of interest (RoI) is extracted from the whole-mount images, that is, the region of IDC. So the dataset has 277,524 patches with a dimension of 50 × 50 pixels, of which 198,738 patches are IDC-negative/normal, and the remaining 78,786 are IDC-positive. Sample patches from the dataset are displayed in Fig. 13.2.

3.2 Feature extraction

Features play a significant role in deciding the classification accuracy of pattern recognition systems. For better recognition and classification, the features are to be robust and less redundant. Cancerous cells show variations in shape, edges, and texture compared to normal cells. These variations are to be identified, analyzed, and described for representation. The shape features can describe the changes in morphological characteristics such as abnormal shape and angular edges. Normal

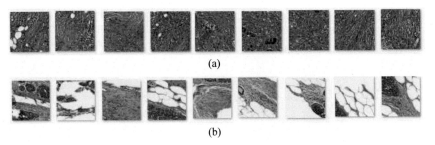

(a)

(b)

FIGURE 13.2

Sample patches (A) with IDC; (B) normal (negative IDC).

breast cells have soft surfaces, whereas cancerous cells are rigid in nature. This variation in surface distinctiveness can be captured by the texture features. The proposed work utilizes LESH, Hu moments, and texture descriptors based on LEM and GLCM for recognizing IDC in histopathological image patches.

3.2.1 Local energy-based shape histogram

LESH [36,37] uses phase congruency for indicating features that have considerable advantages over the gradient-based method. LESH features are invariant to changes in image contrast or brightness. Features are recognized at points in an image with maximum phase congruency (P_C).

The phase congruency function is derived from the Fourier series expansion of a signal:

$$F(x) = \sum_n A_n \, Cos(n\omega x + \phi_n) \tag{13.1}$$

here, ω is a constant and ϕ_n is phase constant

The P_C function is defined as

$$P_C(x) = \max_{\theta\varepsilon \, [0,2\pi]} \frac{\sum_n A_n \, Cos(n\omega x + \phi_n - \theta)}{\sum_n A_n} \tag{13.2}$$

Finding the points of maximum P_C is approximately comparable to the points with a minimum weighted standard deviation of phase angle. It is learned that the maximum P_C can be calculated by searching for peaks in the local energy function. Energy (E) is equal to P_C scaled by the sum of the Fourier amplitudes, that is,

$$E(x) = P_c(x) \sum_n A_n \tag{13.3}$$

From Eq. (13.3), it is understood that E is directly proportional to the P_C function. Thus peaks in E correspond to peaks in P_C. LESH is a local energy-based model for feature observation. The local energy response of an image is obtained by using Gabor filtering, which gives a relatively stable response in terms of high energy on edges, valleys, contours, and corners. These features have good discrimination ability.

Local energy-based shape histogram works by transforming an image into a combination of local energies across different orientations. The features derived at the points of maximum P_C can be helpful in analyzing images.

The framework for computing the P_C in two-dimensional (2D) images while using a high pass filter to obtain image features/descriptors at varied scales is given by

$$P_C(Z) = \frac{\sum_n W(Z)/A_n(Z)\Delta\phi_n(Z) - T)}{\sum_n A_n + \varepsilon} \tag{13.4}$$

where $\Delta\phi_n(Z) = Cos\left(\phi_n(Z) - \overline{\phi_n(Z)}\right) \left|Sin\left(\phi_n(Z) - \overline{\phi_n(Z)}\right)\right|$

here T = noise cancellation factor

$W(Z)$ = weighting of frequency spread

A_n and $\phi(n)$ represent the amplitude and phase angle, respectively, of local complex value

Fourier components at location $Z = (x, y)$ in the image of size n

ε is the constant value included to avoid division by zero.

A_n and $\phi(n)$ are computed by using the logarithmic Gabor wavelets filters. It determines low-level features that are invariant to changes in image contrast, illumination, and magnification. These one-dimensional (1D) symmetric/antisymmetric filters are converted into a 2D form by the application of the Gaussian spreading function along the direction that is perpendicular to its orientation.

The image is further convolved with a bank of Gabor kernels at each of the 16 subregions along eight different orientations

$$G_{u,v}(Z) = I(Z) * \psi_{u,v}(Z) \tag{13.5}$$

where $Z = (x, y)$ represents position of the image.

$G_{u,v}(Z)$ is the convolutional result of the Gabor kernel with an orientation of u and scale v. The Gabor wavelet kernel $\psi u, v(Z)$ can be calculated as depicted in Eq. (13.6)

$$\psi_{u,v}(Z) = \frac{\left\| (K_{u,v}) \right\|^2}{\sigma^2} e^{-\|K_{u,v}\|^2 \|K_{u,v}\|^2 / 2\sigma^2} \left[e^{i\, K_{u,v}\, Z} - \frac{e^{-\sigma^2}}{2} \right] \tag{13.6}$$

$\|.\|$ is the L_2 norm operator.

The wave vector $K_{u,v} - K_v\, e^{i\phi_0}$ with $K_v = K_{max}|$ f v

$\phi_v = \pi\mu/8$, K_{max} is the maximum frequency while f is the space factor between the Kernels in the frequency domain and $\sigma = 2\pi$.

These filters are intended to identify features in all orientations because they uniformly cover the frequency plane.

The energy, $P_C(Z)$, at every orientation is calculated, and then it is aggregated as a whole. The sum of the energies for each orientation is normalized by the overall sum and scales of amplitude of individual wavelet responses at a specific location Z. Hence the calculated 2D P_C is given as

$$P_{C2}(Z) = \frac{\sum_o \sum_n W_0(Z)/A_{n,0}(Z)\Delta\phi_{n,0}(Z) - T_0)}{\sum_o \sum_n A_{n,o} + \varepsilon} \tag{13.7}$$

where o indicates the index of orientation as expressed in Ref. [35].

3.2.2 Hu moments
Hu moments, a set of seven moments, are calculated using central moments that are invariant to transformations in image. Of seven, the first six moments have been proven to be invariant to scale, translation, rotation, and reflection. Whereas, the seventh moment's sign changes for image reflection.

Hu moments are independent to geometric transformation and parallel projection [2,38]. The basic principle of moment is to represent the object by means of measurable quantities called invariants which have sufficient power of discrimination to differentiate between objects belonging to different classes.

For a two-dimensional continuous function $f(x, y)$, the moment of order $(p + q)$ is defined as

$$m_{pq} = \int_{-\infty}^{\infty} \int_{-\infty}^{\infty} x^p y^q f(x, y)dx\, dy \tag{13.8}$$

where $p, q = 0,1,2, \ldots$, and $f(x, y)$ is a piecewise continuous function that has non-zero values only in a fixed portion of the xy-plane; $f(x, y)$ uniquely defines the moments and moment series(m_{pq}) of all orders. On the other hand, m_{pq} uniquely determines $f(x, y)$.

The central moments are defined as represented in Eq. (13.9):

$$\mu_{pq} = \int_{-\infty}^{\infty} \int_{-\infty}^{\infty} (x - \bar{x})^p (y - \bar{y})^q f(x, y)dx\, dy \tag{13.9}$$

where $\bar{x} = \frac{m_{10}}{m_{00}}$ $\bar{y} = \frac{m_{01}}{m_{00}}$ represent the centroids of the image, and m_{00} is the mass of the image

The central moments μ_{pq} computed using the centroid $(\bar{x}, \bar{y})$ of the image $f(x, y)$ is equivalent to the moment m_{pq} whose center has been shifted to centroid of the image. The central moments are invariant to image translation. Scale invariance in moments can be obtained by the normalization process. The central moments after normalization are defined as

$$\eta_{pq} = \frac{\mu_{pq}}{\mu_{00}^{\gamma}} \tag{13.10}$$

where $= \frac{(p+q+2)}{2}$, $p + q = 2,3,4, \ldots$

Based on the centralized central moments, Hu introduced seven-moment invariants,

$$M1 = \eta_{20} + \eta_{02} \tag{13.11}$$

$$M2 = (\eta_{20} - \eta_{02})^2 + 4\eta_{11} \tag{13.12}$$

$$M3 = (\eta_{30} - 3\eta_{12})^2 + (3\eta_{21} - \eta_{03})^2 \tag{13.13}$$

$$M4 = (\eta_{30} + \eta_{12})^2 + (\eta_{21} + \eta_{03})^2 \tag{13.14}$$

$$M5 = (\eta_{30} - 3\eta_{12})(\eta_{30} + 3\eta_{12})[(\eta_{30} + 3\eta_{12})^2 - 3(\eta_{21} + \eta_{03})^2]$$
$$+ (3\eta_{21} - \eta_{03})(\eta_{21} + \eta_{03})[3(\eta_{30} + \eta_{12})^2 - (\eta_{21} + \eta_{03})^2] \tag{13.15}$$

$$M6 = (\eta_{20} - \eta_{02})[(\eta_{30} + \eta_{12})^2 - (\eta_{21} + \eta_{03})^2] + 4\eta_{11}(\eta_{30} + \eta_{12})(\eta_{21} + \eta_{03}) \tag{13.16}$$

$$M7 = (3\eta_{21} - \eta_{03})(\eta_{30} + \eta_{12})[(\eta_{30} + \eta_{12})^2 - 3(\eta_{21} + \eta_{03})^2]$$
$$+ (3\eta_{12} - \eta_{30})(\eta_{21} + \eta_{03})[3(\eta_{30} + \eta_{12})^2 - (\eta_{21} + \eta_{03})^2] \tag{13.17}$$

The combination of all the seven moments forms the Hu moment descriptor $HuM = [M1, M2, M3, M4, M5, M6, M7]$.

3.2.3 Law's texture features

The texture features are significant and aid in contributing toward identification and recognition of IDC from the histopathological image patches. Kenneth ILaws [39] presented texture energy measures (LEM) for analyzing and classifying textures. These features are proved to be invariant to changes in rotation, luminance, and contrast. The Law's texture features are demonstrated to be more precise and perform well in comparison to other methods [40]. The methodology of extracting Law's texture features is presented below.

(i) Initially, one-dimensional (1D) center-weighted and symmetrical/antisymmetrical convolution masks (CM) with varying dimensions of 1×3, 1×5, and 1×7 are used. The CMs proposed by laws explore different characteristics of the image such as level, edge, spot, and ripple. Here the convolution mask with dimension 1×5 has proved to provide a better texture description. The details of the 1×5 symmetrical masks are as shown:

Characteristics	1D-CM
Level($L_{1 \times 5}$)	[1 4 6 4 1]
Edge ($E_{1 \times 5}$)	[−1 2 0 2 1]
Spot($S_{1 \times 5}$)	[−1 0 2 0 −1]
Ripple($R_{1 \times 5}$)	[1 −4 6 −4 1]

(ii) One-dimensional masks with dimension 1×5 were combined by computing the outer product of each CM to generate two-dimensional masks as indicated,

	$L_{1 \times 5}$	$E_{1 \times 5}$	$S_{1 \times 5}$	$R_{1 \times 5}$
$L_{1 \times 5}$	[LL]$_{5 \times 5}$	[LE]$_{5 \times 5}$	[LS]$_{5 \times 5}$	[LR]$_{5 \times 5}$
$E_{1 \times 5}$	[EL]$_{5 \times 5}$	[EE]$_{5 \times 5}$	[ES]$_{5 \times 5}$	[ER]$_{5 \times 5}$
$S_{1 \times 5}$	[SL]$_{5 \times 5}$	[SE]$_{5 \times 5}$	[SS]$_{5 \times 5}$	[SR]$_{5 \times 5}$
$R_{1 \times 5}$	[RL]$_{5 \times 5}$	[RE]$_{5 \times 5}$	[RS]$_{5 \times 5}$	[RR]$_{5 \times 5}$

This outer product computation provided 16 CMs with dimension 5×5. From the set of 16 CMs, a few of the masks were dropped and combined to form a reduced set with nine CMs.

The two-dimensional mask [LL]$_{5 \times 5}$ was dropped as it is sensitive to intensity variations, and the aggregation of all the entries of the mask results in a nonzero value. Further, the CMs that present analogous texture information are combined. The reduced set with 9 CMs is

	$\mathbf{L_{1x5}}$	$\mathbf{E_{1x5}}$	$\mathbf{S_{1x5}}$	$\mathbf{R_{1x5}}$
$\mathbf{L_{1x5}}$		$[LE]_{5x5}$	$[LS]_{5x5}$	$[LR]_{5x5}$
$\mathbf{E_{1x5}}$	$[EL]_{5x5}$	$[EE]_{5x5}$	$[ES]_{5x5}$	$[ER]_{5x5}$
$\mathbf{S_{1x5}}$	$[SL]_{5x5}$	$[SE]_{5x5}$	$[SS]_{5x5}$	$[SR]_{5x5}$
$\mathbf{R_{1x5}}$	$[RL]_{5x5}$	$[RE]_{5x5}$	$[RS]_{5x5}$	$[RR]_{5x5}$

Here, the cells marked with the same color are combined to obtain nine CMs, and the cell with black color is dropped.

(iii) Next, each histopathological image patch H(x, y) is convolved with nine CMs to produce nine Law's texture images that in point of fact represent nine texture attributes of a pixel of a histopathological image patch. Finally, from the nine texture images, LEM mean and standard deviation are derived that is one LEM per texture image and concatenated to form the texture descriptor with length 18.

$$LEM = [Mean_i, Standard\ deviation_i]_{i\ =\ 1:9}$$

3.2.4 Gray level cooccurrence matrix features

GLCM [2,41] is framed to perform texture analysis based on second-order statistics using the spatial relationship between the pixels of an image. This method works on pairs of pixels with a specific spatial relationship between them. GLCM G(x, y) is formulated from an image, wherein every entry of the matrix represents the occurrence of a combination of pixels in an image in given distance "d" and direction "ϕ."

In the proposed work, for every histopathological image patch four GLCMs $G(x,y)$ are formed for different values of d, $d = \{[0\ 1], [-1\ 1], [-1\ 0], [-1\ -1]\}$ and ϕ, $\phi = \{0°, 45°, 90°, 135°\}$. From each $G(x, y)$, Haralick texture features are derived using Eq. (13.18) through (13.21) (Fig. 13.3).

n_{xy} is the occurrence of a combination of pixels x and y.

$$Contrast = \sum_{x=1}^{N} \sum_{y=1}^{N} |x - y|^2 G(x, y) \tag{13.18}$$

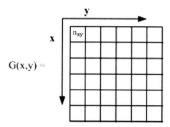

FIGURE 13.3

GLCM of an image.

$$\text{Correlation} = \sum_{x=1}^{N} \sum_{y=1}^{N} \frac{(x - \mu_x)(y - \mu_y)G(x,y)}{\sigma_x \sigma_y} \tag{13.19}$$

where $\mu_x \; \mu_y \; \sigma_x \; \sigma_y$ are mean and standard deviation, respectively

$$\text{Energy} = \sum_{x=1}^{N} \sum_{y=1}^{N} G(x,y)^2 \tag{13.20}$$

$$\text{Homogeneity} = \sum_{x=1}^{N} \sum_{y=1}^{N} \frac{G(x,y)}{1 + [x - y]} \tag{13.21}$$

The Haralick features extracted from all the GLCMs with the given d and ϕ are combined to form the GLCM texture features GTex = [Contrast$_i$, Correlation$_i$, Energy$_i$, Homogeneity$_i$]$_{i\,=\,1:4}$ with dimension of 1×16.

The extracted shape and texture features are integrated to provide complete information about the characteristics of the breast cells. The integrated features F_{ST} = [LESH: HuM: LEM: GTex] are provided to machine learning algorithms with ensemble methods to build classifier models to learn discrimination between IDC and non-IDC histopathological images.

4. Results and discussion

The proposed work focuses on classifying the breast cancer histopathological images. The data set considered for evaluation has image patches belonging to both the categories: IDC and non-IDC (IDC negative). As a preprocessing step, the contrast of all the image patches H(x,y) were enhanced using adaptive histogram equalization to improve the quality for identifying significant variations in the shape and texture characteristics. Subsequently, the histopathological image patches H(x,y) were randomly split for training the ensemble classifiers and testing.

The patches of training and testing set were provided to the feature extraction module to derive the LESH, HuM, LEM, and GTex features and later were integrated or concatenated to form a feature vector F_{ST} of dimension 1×578. The F_{ST} s of training set were labeled with "0" and "1," where 1 represents the presence of IDC while 0 indicates non-IDC. The labeled data set is presented to ensemble classifiers for model creation. The ensemble classifiers [42,43] improve the predictive performance of the classifiers. Here multiple classifier models are built with a provided machine learning algorithm, and the predictions of all the models are combined to improve the classification ability as compared to a single model. Ensemble classifier methods are broadly categorized into averaging and boosting methods. The proposed work is implemented using both methods, and their performance is compared.

(i) Averaging methods: Here, several classifier models are built independently considering subsets randomly from the original labeled dataset, and their predictions are aggregated to find the final predictive decision. These methods improve the performance by reducing the variance. Here in this work (1) bagging methods with base classifiers: decision tree, kNN, and SVM with rbf kernel, (2) random forest classifier, and (3) extra tree classifier are trained to create ensemble classifier models for the classification task.

In case of bagging classifier (BC) with three base classifiers: decision tree (DT), kNN and SVM with rbf kernel, the parameters such as number of estimators (NoE) and number of neighbors (NoN) for kNN are varied during training to find the best model. The model is further tested, and its performance is analyzed quantitatively with the measures [44]: classification accuracy (CA), recall, precision, area under the curve (AUC), Mathew's correlation coefficient (MCC), kappa value, and F1 score. These measures are computed using the elements of the confusion matrix framed by testing the classifier model using Eqs. from (13.22) to (13.27).

$$\text{Classification accuracy}(CA) = (TP + TN)/(TP + TN + FP + FN) \tag{13.22}$$

here, TP = True Positive, TN = True negative,
FP = False Positive and FN = False Negative

$$\text{Precision} = TP/(TP + FP) \tag{13.23}$$

$$\text{Recall} = TP/(TP + FN) \tag{13.24}$$

$$MCC = (TP \times TN - FP \times FN)/Y$$
$$\text{where } Y = \sqrt{((TP + FP) \times (TP + FN) \times (TN + FP) \times (TN + FN))} \tag{13.25}$$

$$\text{F1 score} = (2 \times \text{Precision x Recall})/(\text{Precision} + \text{Recall}) \tag{13.26}$$

$$\text{Kappa} = (P_o - P_e)/(1 - P_e),$$
$$\text{with } P_o = \text{Accuracy}(ACC)$$
$$P_e = ((TP + FN) \times (TP + FP) + (FP + TN) \times (FN + TN))/(TP + TN + FP + FN) \tag{13.27}$$

The results obtained from the bagging classifier methods are presented in Fig. 13.4. The figure indicates the better performance of BC with base models DT and SVM as compared to kNN for higher number of estimators or classifier models. Thus time for training the model was more. A precision of 94.19% specifies the ability of the model in distinguishing the image classes. The kappa and MCC values for all BC were greater than 72%, indicating higher association between the actual and predicted class labels.

Subsequently, the task was repeated with other averaging methods, random forest (RF) and extra tree (ET) classifiers. With these methods, NoE and the quality of split (Gini index and Entropy) are varied to find a good fit for classifications. The performance metrics are presented in Fig. 13.5, which signifies the best performance with criterion entropy.

(ii) Boosting methods: In boosting method, weak classifiers are combined to construct a strong classifier following a sequential process reducing the bias.

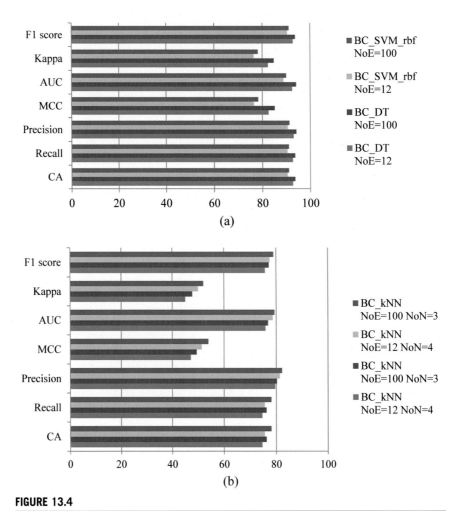

FIGURE 13.4

Performance metrics provided by bagging classifier models with base estimator:
(A) decision tree and SVM; (B) kNN.

In the sequential process, the errors produced by the previous classifier model are corrected by the next model thus improving the performance (Fig. 13.6).

In the proposed task, boosting methods: Adaboost, gradient boosting, and histogram based gradient boosting classifiers are implemented. In this method, the parameters such as NoE, learning rate (LR), and boosting algorithms are varied to design the most favorable classifier model. The simulation was done initially using Adaboost method varying the NoE and keeping the LR constant to "1" and selecting SAMME.R boosting algorithm. The performance metrics computed from the simulation results are displayed in Fig. 13.7A.

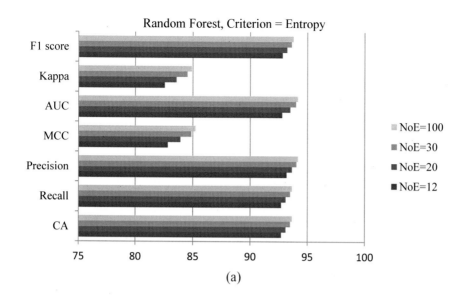

(a)

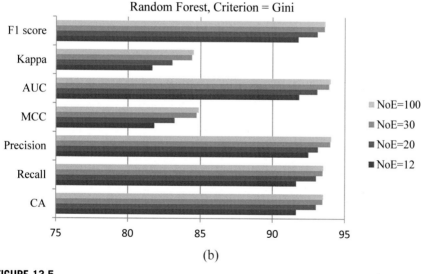

(b)

FIGURE 13.5

Performance metrics provided by random forest ensemble classifier with criterion:
(A) entropy and (B) Gini index.

The result indicates the best performance for 100 estimators. Thus number of estimators was fixed to 100, and LR was varied to investigate the influence of it on
classification rate. The results for the same are presented in Fig. 13.7B.

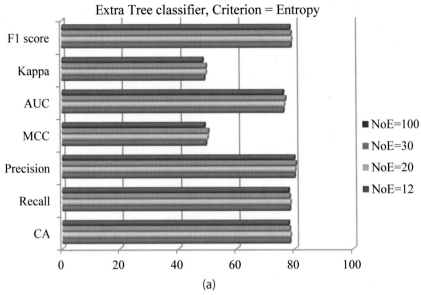

(a)

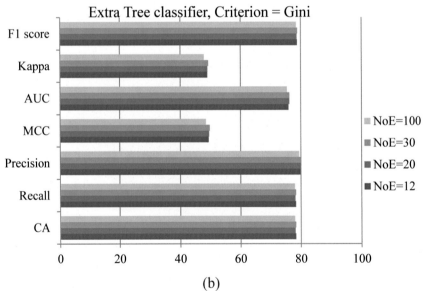

(b)

FIGURE 13.6

Performance metrics provided by extra tree ensemble classifier with criterion: (A) entropy and (B) Gini index.

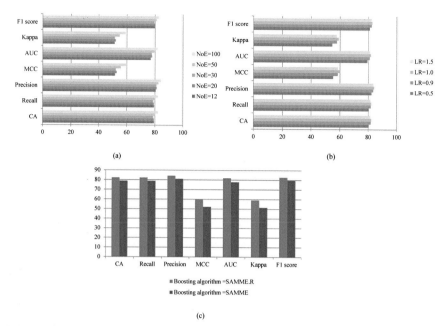

FIGURE 13.7

Performance metrics provided by Adaboost ensemble method with variation in (A) NoE-constant LR, (B) LR-constant NoE, and (C) boosting algorithm-constant NoE and LR.

From Fig. 13.7B, it is observed that the best performance in terms of all metrics is provided by LR = 1.0. Later NoE was fixed to 100 and LR to 1.0 and the boosting algorithm was changed to SAMME, then a drop in the performance was noticed as illustrated in Fig. 13.7C. This demonstrated the best performance of boosting algorithm SAMME.

The simulation was continued further with gradient boosting varying NoE and constant LR and histogram-based gradient boosting with maximum iterations. The results for the same are shown in Figs. 13.8 and 13.9.

The results displayed in Fig. 13.9 were obtained for maximum iterations of 100. The results for variation in maximum iterations are not indicated as the performance was constant irrespective of variations in iterations. Thus the results (exhibited in Fig. 13.9) were acceptable. As compared to averaging methods and other boosting methods, histogram-based gradient boosting converged faster and improved the performance in terms of all the parameters. It can also be noticed that the model had a greater association of 85%.

Finally, the voting ensemble method is implemented combining heterogeneous classifier models to predict the class label. The final decision is based on the voting mode: hard or soft. In the case of hard voting, the predicted class label is based on the majority voting of the predicted class labels from the individual classifier models; whereas, in soft voting, the probabilities of the predictions of all the classifier models

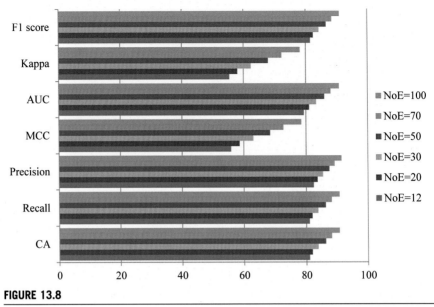

FIGURE 13.8

Performance metrics provided by gradient boosting ensemble method with variation in NoE and constant LR.

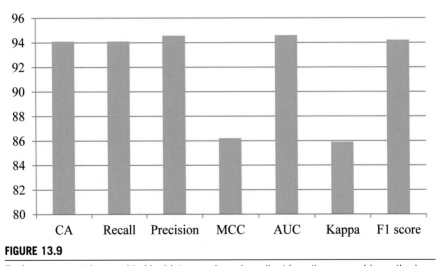

FIGURE 13.9

Performance metrics provided by histogram-based gradient boosting ensemble method.

are combined and the decision is made by selecting the prediction with peak total probability. In the presented work, the classifier models built training the classifier algorithms: decision tree, k nearest neighbor, Gaussian Naive Baye's, support vector

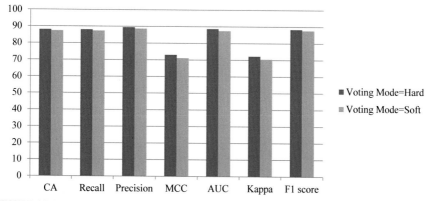

FIGURE 13.10

Performance metrics provided by voting ensemble method.

machine, and random forest are implemented with both hard and soft voting mode, and the performance metrics are computed for analysis. Fig. 13.10 shows the computed parameters. Hard voting mode turned out to be more effective than soft mode.

5. Conclusion

This chapter formulated a pattern recognition-based methodology to classify breast cancer histopathological images into IDC and non-IDC categories. The methodology utilized integration of shape and texture features LESH, Hu moments, Law's energy, and GLCM-based measures to aptly represent the variations in the breast cells. These descriptors significantly capture the key characteristics or patterns from the histopathological images acquired from histopathological dataset. These patterns with class labels trained the homogeneous and heterogeneous ensemble classifier methods. The simulations were completed by varying all the required parameters, and the testing results were analyzed computing the metrics CA, precision, recall, AUC, kappa, MCC and F1 score. The analysis indicated the better performance of boosting methods than the averaging and voting methods in particular histogram-based gradient boosting classifier. This model predicted the class labels of the histopathological images with a CA and recall of 94.11%, precision of 94.57%, and AUC and F1 score of 94.6% and 94.21%, respectively. The model was also significant in delivering a better association (85%) between actual and predicted categories of images. The performance metrics demonstrate the ability of the shape and texture descriptors combined with ensemble classifier models in identifying and discriminating breast cancer and normal images.

References

[1] Zhuang Z, Li N, Joseph Raj AN, Mahesh VG, Qiu S. An RDAU-NET model for lesion segmentation in breast ultrasound images. PLoS One 2019;14(8):e0221535.

[2] Raj ANJ, Nersisson R, Mahesh VG, Zhuang Z. Nipple localization in automated whole breast ultrasound coronal scans using ensemble learning. Ultrasonic Imaging 2021; 43(1):29–45.

[3] Abdelrahman L, Al Ghamdi M, Collado-Mesa F, Abdel-Mottaleb M. Convolutional neural networks for breast cancer detection in mammography: a survey. Computers in Biology and Medicine 2021;131:104248.

[4] Mann RM, Kuhl CK, Moy L. Contrast-enhanced MRI for breast cancer screening. Journal of Magnetic Resonance Imaging 2019;50(2):377–90.

[5] Mahesh VG, Kumar MM. An ensemble classification based approach for breast cancer prediction. InIOP Conference Series: Materials *Science and Engineering* February 2021;1065(1):012049.

[6] Loke SY, Lee ASG. The future of blood-based biomarkers for the early detection of breast cancer. European Journal of Cancer 2018;92:54–68.

[7] Gurcan MN, Boucheron LE, Can A, Madabhushi A, Rajpoot NM, Yener B. Histopathological image analysis: a review. IEEE Transactions on Biomedical Engineering 2009; 2:147–71.

[8] Veta M, Pluim JPW, van Diest PJ, Viergever MA. Breast cancer histopathology image analysis: a review. IEEE Transactions on Biomedical Engineering 2014;61:1400–11.

[9] Motlagh MH, Jannesari M, Aboulkheyr H, Khosravi P, Elemento O, Totonchi M, et al. Breast cancer histopathological image classification: a deep learning approach. BioRxiv; 2018, 242818.

[10] Carvalho ED, Antonio Filho OC, Silva RR, Araujo FH, Diniz JO, Silva AC, Gattass M. Breast cancer diagnosis from histopathological images using textural features and CBIR. Artificial Intelligence in Medicine 2020;105:101845.

[11] Belsare AD, Mushrif MM. Histopathological image analysis using image processing techniques: an overview. Signal & Image Processing 2012;3(4):23.

[12] Aswathy MA, Jagannath M. Detection of breast cancer on digital histopathology images: present status and future possibilities. Informatics in Medicine Unlocked 2017; 8:74–9.

[13] Spanhol FA, Oliveira LS, Petitjean C, Heutte L. A dataset for breast cancer histopathological image classification. IEEE Transactions on Biomedical Engineering 2015; 63(7):1455–62.

[14] Yan R, Ren F, Wang Z, Wang L, Zhang T, Liu Y, Zhang F. Breast cancer histopathological image classification using a hybrid deep neural network. Methods 2020;173:52–60.

[15] Belsare AD, Mushrif MM, Pangarkar MA, Meshram N. Classification of breast cancer histopathology images using texture feature analysis. In: Tencon 2015–2015 IEEE region 10 conference. IEEE; November 2015. p. 1–5.

[16] Gupta V, Bhavsar A. An integrated multi-scale model for breast cancer histopathological image classification with joint colour-texture features. In: In international conference on computer analysis of images and patterns. Cham: Springer; August 2017. p. 354–66.

[17] Alinsaif S, Lang J. Texture features in the Shearlet domain for histopathological image classification. BMC Medical Informatics and Decision Making 2020;20(14):1–19.

[18] Rezk E, Awan Z, Islam F, Jaoua A, Al Maadeed S, Zhang N, Rajpoot N. Conceptual data sampling for breast cancer histology image classification. Computers in Biology and Medicine 2017;89:59—67.

[19] Zhang Y, Zhang B, Lu W. Breast cancer histological image classification with multiple features and random subspace classifier ensemble. In: Knowledge-based systems in biomedicine and computational life science. Berlin, Heidelberg: Springer; 2013. p. 27—42.

[20] Hu H, Qiao S, Hao Y, Bai Y, Cheng R, Zhang W, et al. Breast cancer histopathological images recognition based on two-stage nuclei segmentation strategy. PLoS One 2022; 17(4):e0266973.

[21] Wang P, Hu X, Li Y, Liu Q, Zhu X. Automatic cell nuclei segmentation and classification of breast cancer histopathology images. Signal Processing 2016;122:1—13.

[22] Afify HM, Mohammed KK, Hassanien AE. Multi-images recognition of breast cancer histopathological via probabilistic neural network approach. Journal of System and Management Sciences 2020;1(2):53—68.

[23] Bayramoglu N, Kannala J, Heikkilä J. Deep learning for magnification independent breast cancer histopathology image classification. In: 2016 23rd international conference on pattern recognition (ICPR). IEEE; December 2016. p. 2440—5.

[24] Adeshina SA, Adedigba AP, Adeniyi AA, Aibinu AM. Breast cancer histopathology image classification with deep convolutional neural networks. In: 2018 14th international conference on electronics computer and computation (ICECCO). IEEE; November 2018. p. 206—12.

[25] Zhu C, Song F, Wang Y, Dong H, Guo Y, Liu J. Breast cancer histopathology image classification through assembling multiple compact CNNs. BMC Medical Informatics and Decision Making 2019;19(1):1—17.

[26] Vo DM, Nguyen NQ, Lee SW. Classification of breast cancer histology images using incremental boosting convolution networks. Information Sciences 2019;482:123—38.

[27] Deniz E, Şengür A, Kadiroğlu Z, Guo Y, Bajaj V, Budak Ü. Transfer learning based histopathologic image classification for breast cancer detection. Health Information Science and Systems 2018;6(1):1—7.

[28] Spanhol FA, Oliveira LS, Cavalin PR, Petitjean C, Heutte L. Deep features for breast cancer histopathological image classification. In: 2017 IEEE international conference on systems, man, and cybernetics (SMC). IEEE; October 2017. p. 1868—73.

[29] De Matos J, Britto ADS, Oliveira LE, Koerich AL. Double transfer learning for breast cancer histopathologic image classification. In: 2019 international joint conference on neural networks (IJCNN). IEEE; July 2019. p. 1—8.

[30] Hameed Z, Zahia S, Garcia-Zapirain B, Javier Aguirre J, María Vanegas A. Breast cancer histopathology image classification using an ensemble of deep learning models. Sensors 2020;20(16):4373.

[31] Li X, Radulovic M, Kanjer K, Plataniotis KN. Discriminative pattern mining for breast cancer histopathology image classification via fully convolutional autoencoder. IEEE Access 2019;7:36433—45.

[32] Saxena S, Shukla S, Gyanchandani M. Breast cancer histopathology image classification using kernelized weighted extreme learning machine. International Journal of Imaging Systems and Technology 2021;31(1):168—79.

[33] Vaishali D, Priya PV, Govind N, Prabha K. Higher order statistical analysis in multiresolution domain-application to breast cancer histopathology. In: International conference on service-oriented computing. Cham: Springer; 2021. p. 495—508.

[34] Wang H, Roa AC, Basavanhally AN, Gilmore HL, Shih N, Feldman M, Madabhushi A. Mitosis detection in breast cancer pathology images by combining handcrafted and convolutional neural network features. Journal of Medical Imaging 2014;1(3):034003.

[35] http://gleason.case.edu/webdata/jpi-dl-tutorial/IDC_regular_ps50_idx5.zip. 26 December 2022.

[36] Dominic D, Balachandran K. Lung cancer diagnosis from CT images based on local energy based shape histogram (LESH) feature extration and pre-processing. In: ICCCE 2020: Proceedings of the 3rd international conference on communications and cyber physical engineering. Springer Singapore; 2021. p. 713—20.

[37] Kovesi P. Phase congruency: a low-level image invariant. Psychological Research 2000; 64(2):136—48.

[38] Hu MK. Visual pattern recognition by moment invariants. IEEE Transactions on Information Theory 1962;8(2):179—87.

[39] Laws KI. Textured image segmentation. University of Southern California Los Angeles Image Processing INST; 1980.

[40] Mahesh VG, Chen C, Rajangam V, Raj ANJ, Krishnan PT. Shape and texture aware facial expression recognition using spatial pyramid Zernike moments and law's textures feature set. IEEE Access 2021;9:52509—22.

[41] Faust O, Acharya UR, Meiburger KM, Molinari F, Koh JE, Yeong CH, Ng KH. Comparative assessment of texture features for the identification of cancer in ultrasound images: a review. Biocybernetics and Biomedical Engineering 2018;38(2):275—96.

[42] Sagi O, Rokach L. Ensemble learning: a survey. Wiley Interdisciplinary Reviews: Data Mining and Knowledge Discovery 2018;8(4):e1249.

[43] Kazemi Y, Mirroshandel SA. A novel method for predicting kidney stone type using ensemble learning. Artificial Intelligence in Medicine 2018;84:117—26.

[44] Mahesh VG, Raj ANJ, Nersisson R. Implementation of machine learning-aided speech analysis for speaker accent identification applied to audio forensics. In: Aiding forensic investigation through deep learning and machine learning frameworks. IGI Global; 2022. p. 174—94.

An automatic level set segmentation of breast tumor from mammogram images using optimized fuzzy *c*-means clustering

Kama Ramudu, S.P. Girija and Ganta Raghotham Reddy
Department of Electronics and Communication Engineering, Kakatiya Institute of Technology and
Science, Warangal, Telangana, India

1. Introduction to mammogram image segmentation

One of the most important causes of death for women worldwide is breast cancer. Early recognition and detection, therefore, improve the chance of revitalization and reduce the death rate. According to data from the World Health Organization (WHO), breast cancer is the illness that is most frequently discovered worldwide. Every year, more than 626,700 women die from cancer-related illnesses, and more than 2 million new cases were reported in 2018 [1,2]. Although a biopsy is not necessary, it can significantly lower mortality and treatment costs if it is found in the early stages, making it more bearable for patients. Additionally, studies have shown that due to the high volume of ultrasound images generated daily and the dearth of radiologists presented to examine these medical images, radiologists may misdiagnose breast cancer. The abundance of ultrasound images generated as a result of the rise in breast cancer diagnoses may cause radiologists to feel overworked [3].

One of the dreadful diseases that severely affect women today is breast cancer. One of the various cancers that can develop in breast cells is breast cancer. When these cells start to behave abnormally and form a dense tissue, it happens. External dimpling, breast lumps, fluid oozing from the nipple, nipple inversion or new nipple growth, and inflamed skin are a few signs of breast cancer. Breast cancer can be found and diagnosed in a variety of ways. However, a mammogram is a reliable and effective way to find breast cancer early on. Basically, mammogram image scan is a low-power X-ray image of the human breast taken to examine the breast tissue. Masses and calcifications are the abnormalities that are noted on a mammogram. Calcifications are calcium deposits. The cancer can be classified as benign or malignant based on the shape of the mass. Benign tumors are typically noncancerous and have a round or oval shape, whereas malignant tumors are cancerous and have a

Computational Intelligence and Modelling Techniques for Disease Detection in Mammogram Images
https://doi.org/10.1016/B978-0-443-13999-4.00007-9

partly smoothed mass with a spike or an irregular outline. These mammograms are used by radiologists to detect cancer. However, there may be differences of opinion among radiologists and errors in human perception. As a result, visual examination of mammograms frequently results in incorrect diagnosis. To avoid this, radiologists employ CAD systems to find suspicious areas. Preprocessing, segmentation, feature extraction, and classification make up the general CAD system. Feature extraction and classification are combined into one step in deep learning-based methods. Semiautomatic and fully automatic CAD systems are the two categories into which they are analyzed in terms of evaluated images. In a semiautomatic system, cancerous regions were first marked on the images after a radiologist had examined them. In contrast to fully automatic systems, which do not require radiologist preanalysis of images, those marked images are then classified by the CAD system as cancerous, noncancerous, benign, or malignant. In actuality, the CAD system automatically performs preprocessing and classification after receiving the entire image as input. Segmentation is one of the most crucial phases in a fully automatic system because the classification depends on it, and this stage involves identifying mammogram suspicious areas [4,5].

2. Literature review on mammogram image segmentation

2.1 Introduction

In this section, we present numerous breast cancer segmentation methods are using based on machine learning approaches, and deep learning methods for robust and superior segmentation results. Refs. [6,7] describe the traditional segmentation approaches, which rely on digital image processing and its mathematics to segment the image into three categories, such as edge segmentation (ES) techniques, threshold segmentation (TS) techniques, and finally region segmentation (RS) methods. The first ES methods include contour, active contour, Sobel, energy reduction, and canny edge detection. The second method belongs to TBS; some of the methods, as examples, under this segmentation method are manual thresholding, adaptive thresholding, Otsu thresholding, and global and local thresholding. Finally, the region growing and splitting, rough set theory, and watershed methods are examples under region-based segmentation methods.

The following machine learning segmentation techniques have been proposed in Refs. [8–10].

i. Unsupervised machine learning techniques (USML) include hierarchical k-clustering, fuzzy c-clustering, and k-means clustering.
ii. Support vector machines (SVM) and supervised and unsupervised clustering techniques including extreme learning machines

The following deep learning segmentation techniques have been proposed in Ref. [11].

i. Fully convolutional neural networks (FCNNs), SegNet, UNet, and deep learning (DL) segmentation (FCN).

To detect malignancies in the breast ROI, medical personnel usually use manual segmentation (conventional). However, the number of breast cancer patients is increasing every year, and coming to real-time practice, work involving manual diagnosis takes time and the expertise of medical professionals [12]. As a result, medical professionals are overworked, which might lead to incorrect breast tumor categorization. By physically comparing it to the other sections, medical professionals can identify the aberrant spot. Additionally, this method's disadvantage is that it takes a lot of time for medical professionals to precisely pinpoint aberrant locations [13], so that it is flexible enough to accommodate changes and be used in real time. Additionally, opinions from medical experts are required to assess the performance of segmentation accuracy, even for automatic segmentation [14].

In this chapter, machine learning algorithms are implemented with the help of unsupervised clustering methods for tumor segmentation in breasts is explained in detail. The two main conventional unsupervised clustering methods, which are focused in this chapter, are k-means which is a hard clustering method and fuzzy c-means clustering which comes under the soft clustering method. There are many methods for mammogram segmentation, yet every method has its own limitations. Mammogram image segmentation methods are classified by using the latest algorithms depicted in Fig. 14.1.

2.2 Breast cancer segmentation using *k*-means algorithm

k-means is mainly used for segmentation purpose, and this algorithm comes under unsupervised algorithm where data are not labeled. Given image will be grayscale and if it is not a grayscale then the color (RGB) image will be converted to gray. When clustering is applied for a grayscale image, it generally classifies into three areas: normal breast tissue, the background part, and the tumor area. For any given image with tumor, tumor area will always have high-intensity pixels compared to other areas, so we can easily detect that high-intensity cluster as a tumor part.

In an image, there is only some part of useful information, so instead of processing the entire image, partitioning the image and processing only the required

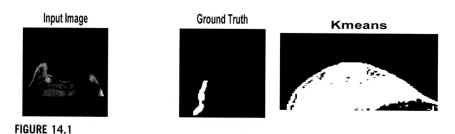

FIGURE 14.1

Mammogram breast cancer segmentation using *k*-means clustering.

informational part seem to be better. Hence, clustering is done based on similarity. Number of cluster centroids will be given prior, and based on that, clusters will be made accordingly.

2.2.1 Introduction to k-means clustering algorithm

In this algorithm, firstly, cluster centers are chosen randomly from the data points present. During the initialization of cluster centers, they should be less in number than training data. Distance is calculated between cluster centers and every data point's recursively as it is an iterative algorithm. New clusters are formed with the minimum distance criteria. In Fig. 14.2A, the given data are made into two clusters. To form optimized centers, centroids need to be recursive to a new location. In this way, the entire cluster centers will be changed until they stop moving, that is, they do not change their position to attain a static point. This indicated that the k-means algorithm converged. Distinct clusters are formed at the end in Fig. 14.2C. Results might change depending on the first initialization stage and number of clusters we take. This is an algorithm that recursively partitions the given pixels of the image in to given number of clusters (k). This forms nonoverlapping clusters where each pixel belongs to only one group. Pixels in each cluster will be similar to the other pixels in that cluster and pixels in different clusters will be very dissimilar with each other.

2.2.2 k-Means clustering' algorithm

k-Means algorithm is an iterative unsupervised clustering algorithm; the objective function of the k-means algorithm is defined in Eq. (14.1).

$$J(V) = \sum_{i=1}^{c} \sum_{j=1}^{c_i} \left(\|x_i - v_j\| \right)^2 \tag{14.1}$$

At which, the Euclidian distance between the cluster centers and data points is $\| x_i - v_j \|$, i-th cluster number of data points is assumed as c_i, and cluster centers are assumed as c.

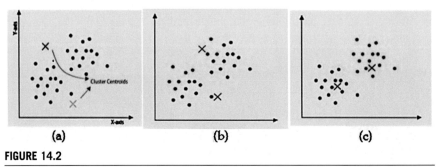

(a)   (b)   (c)

FIGURE 14.2

Process of k-means clustering: (A) random initialization of cluster centers; (B) assigning data points to nearest cluster centers; (C) clusters are grouped based on similarity.

Consider $X = \{x1, x2, x3, ..., xn\}$ which is a set of data points and $V = \{v1, v2, ..., vc\}$ which are the group of cluster centers. The updated cluster centers are calculated using the formula in Eq. (14.2)

$$v_i = (1 / c_i) \sum_{j=1}^{c_i} x_i \tag{14.2}$$

where "c_i" represents the number of data points in the i-th cluster.

2.3 Breast tumor segmentation using fuzzy *c*-means algorithm

The fuzzy c-means is the example for soft clustering algorithm, which is raised due to the problems of classifying outliers, and it is flexible enough to deal with these points. This is also called soft k means because its design mainly centers on the k means design. Generally, hard clustering will allocate pixels or data points or objects to exactly one cluster, and a strong boundary between them is present, whereas soft clustering that is fuzzy clustering allows the pixel or object to belong to one or many clusters. This is the only difference between k-means and fuzzy c-means.

There is also a rule that the sum of the degree membership value of all the objects w.r.t to clusters is equal to 1. Fuzzy c-means (FCM) algorithm is sensitive to additive noise because it does not take local information of the image into account, which degrades the image pixel features (Fig. 14.3).

2.3.1 Fuzzy *c*-means clustering algorithm

A fundamental thought of presenting the fuzzy idea in the FCM algorithm is that a question can have a place at the same time beyond one class and does as such by differing degrees called participation. The FCM algorithm is an iterative technique, which tries to isolate the arrangement of information into various conservative groups. Each cluster is represented by its center. A fuzzy c-means algorithm's primary goal is to limit:

$$J(U, V) = \sum_{i=1}^{n} \sum_{j=1}^{c} (\mu_{ij})^m \|x_i - v_i\|^2 \tag{14.3}$$

where "$\|x_i - v_j\|$" is a Euclidean distance between i-th data as well as j-th cluster center.

FIGURE 14.3

Mammogram breast cancer segmentation using fuzzy c-means clustering.

Algorithm for fuzzy c-means clustering:

In the FCM clustering algorithm, the calculation of fuzzy membership "μ_{ij}" by using Eq. (14.4) is necessary by considering a group of data points $X = \{x_1, x_2, x_3, ..., x_n\}$ and a group of cluster centers $V = \{v_1, v_2, ..., v_c\}$, respectively. In FCM, cluster center "c" is assumed randomly and processed onto the datasets.

$$\mu_{ij} = 1 / \sum_{k=1}^{c} (d_{ij}/d_{ik})^{(2/m-1)} \tag{14.4}$$

Now compute fuzzy centers "v_j" by using

$$v_j = \left(\sum_{i=1}^{n} (\mu_{ij})^m x_i \right) / \left(\sum_{k=1}^{n} (\mu_{ij})^m \right), \forall j = 1, 2, ...c \tag{14.5}$$

Calculate the fuzzy membership and fuzzy centers again until you reach the minimum required "J" values; otherwise, $\|U(k+1) - U(k)\| < \beta$. Here "$\beta$" represents a termination criterion between [0, 1], and "J" represents the objective function of a matrix.

2.4 Challenges faced in breast tumor detection

Existing methods of image segmentation and their algorithm are studied in detail. However, the k-means clustering method is quick, reliable, simple to use, and effective. It gives the best results when the given datasets are distinct in nature. Tight clusters are produced and have better computational costs. Besides, there are limitations for this method, like number of cluster centers needs to be specified in the beginning and two overlapped data cannot be distinguished. Handling noisy data and outliers becomes a problem. This algorithm does not work for nonlinear data. However, in fuzzy c-means, it works best for overlapped data. In both methods, random initialization of centroids is a primary issue that causes the output to fluctuate for every execution. Results do not remain constant for the same data. Hence to get optimal cluster centers, cuckoo search algorithm performs well when applied in the preprocessing stage, which can segment tumors far better than existing methods, and also, output remains the same for the same data taken. In Chapter 3, cuckoo search optimization techniques are explained in detail (Fig. 14.4).

3. Proposed optimized fuzzy c-means clustering using level set method for breast tumor segmentation

In breast tumor detection based on conventional machine learning algorithms, which are called k-means and fuzzy c-means, the performance of those methods is not satisfying because of the random initialization of centroids. Cuckoo search optimization has the capacity of getting optimum pixel values. One of the efficient optimization techniques is applied in the preprocessing stage before applying the clustering

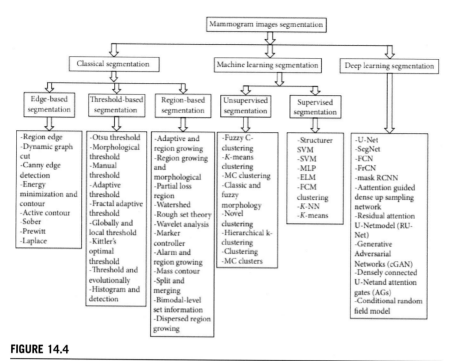

FIGURE 14.4

Mammogram image segmentation methods and their classifications.

methods to improve clustering efficiency. This cuckoo search optimization and its postprocessing stage using the level set method are explained in detail in the following sections. Finally, an automatic level set method for segmentation of breast tumor from mammogram images using optimized fuzzy *c*-mean (OFCM) clustering method is proposed in this chapter for robust and accurate segmentation.

3.1 Cuckoo search algorithm for calculation of optimum cluster centers

The maximum and minimum values of the current position will be calculated by the cuckoo search optimization (CSO). This is the primary objective of this optimization algorithm. In addition to computation time, accuracy and convergence rate are needed to determine the lowest or highest values. Cuckoo search optimization is a strategy based on the parasitic lifestyle of the cuckoo bird. A new solution is to be assumed as a cuckoo egg in this case; each egg in the nest represents a different solution. This method is straightforward because we only take into account two variables: the total number of birds in the nest and the possibility of discovering a cuckoo egg. For the majority of optimization issues, these are adequate. The primary goal is to replace existing, ineffective solutions with new, better ones (Fig. 14.5).

FIGURE 14.5

Initialization of nests.

1. At first, total nests are formed.
2. Probability of finding a cuckoo egg with pa $= 0.25$
3. Initializing the Max. numbers of iterations
4. A new cuckoo position is obtained at random via Levy's Flight. The Levy's Flight will at random select a cuckoo ($i = 1$).
   The formula for carrying out Levy's Flight is

$$x_i^{t+1} = x_i^t + \alpha \oplus Levy\,(\lambda) \tag{14.6}$$

Here α_i is the step size,

$\oplus$ is the entry-wise multiplication,

x_i^{t+1} is assumed as the new solution,

$Levy\,(\lambda)$ is Levy's exponent, and

x^t represents the current location of the egg

Based on the present location, the next position of the egg is determined by using Levy's Flight. The cuckoo's hunt for new solutions is the main aspect of this optimization method based on Levy's Flight (Fig. 14.6).

Levy's distribution may be used to create random steps, which are a succession of the smallest distributions. Step size is defined in Eq. (14.7)

$$S = \frac{\sigma_u * u}{|v|^{1/\beta}} \tag{14.7}$$

Here if s is too large, v can be assumed as a normal variable and also if "s" is small means the new solution is far from the previous solution, then the position will only slightly change. Levy's Flight-based Random walk strategy is defined in Eq. (14.8).

$$x_i^{t+1} = x_i^t + \alpha * S \oplus \left(X_i^t - X_{gbest}^t\right) \tag{14.8}$$

Z_i change of position

$$Z_i = \alpha * S \oplus \left(X_i^t - X_{gbest}^t\right) \tag{14.9}$$

$$X_{gbest}^t = 0$$

FIGURE 14.6

Randomly chosen nest.

5. Pick a random nest n say j

$$f(X_i) \geq f(X_j)$$

The egg with the lowest ranking will be destroyed if the test to determine whether the cuckoo egg is comparable to the host fails, and a new egg will develop close to the older one (Fig. 14.7).

6. If the criterion is true, a cuckoo egg that resembles the host egg bird will be present. As a result, the bottom egg nest is destroyed and the randomly chosen nest is replaced with the new solution.

7. The counter is incremented by 1 until the best solution is to be retained; otherwise, repeat until the condition is satisfied.

8. A new solution is constructed using Eq. (3.4).

The main advantage of using this cuckoo search algorithm is the usage of fewer numbers of parameters. Convergence is achieved faster. It can hybridize with other swarm-based algorithms, and it can also be used in multicriteria optimization problems; these characteristics make it better than other metaheuristic algorithms.

3.1.1 Cuckoo search algorithm steps

1. Objective function $f(x)$, $x = (x_1, \ldots, x_d)T$

2. Initial population generation of n host nests x_i ($i = 1, 2, \ldots, n$)

3. while (stop criterion)

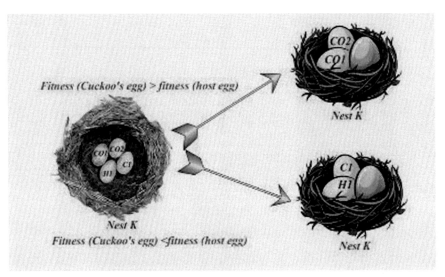

FIGURE 14.7

Replacement of cuckoo egg using fitness value.

    4. By Levy flights get a cuckoo randomly

    5. Evaluate its fitness value F_i

    6. Choosing nest randomly out of n nest(j)

    7. if ($F_i > F_j$),

        a. Put back j by the new solution;

    8. end

    9. A fraction (pa) of bad nests are rejected and the latest ones are put up

  10. Keep the best solutions

  11. Best solutions are found from the ranking of the solutions

    end while

    end.

3.1.2 The flow chart of cuckoo search optimization algorithm

See Fig. 14.8.

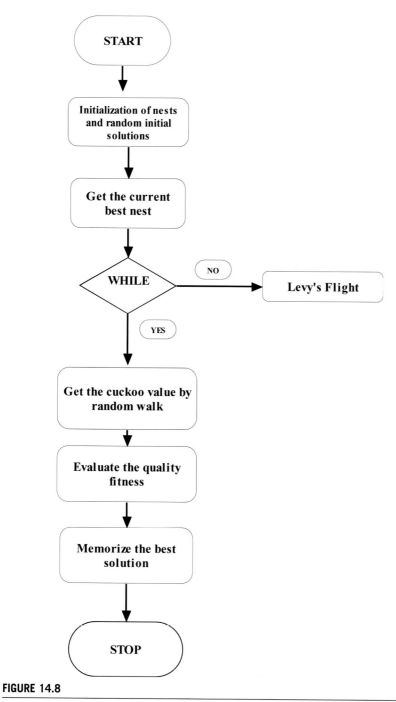

FIGURE 14.8

The flow chart of cuckoo search optimization.

3.2 Optimized *k*-means clustering algorithm for breast tumor segmentation

The hybrid approach of cuckoo search optimization and k-means clustering method is termed as "optimized k-means" clustering. In this, there are two stages involved. Firstly, after reading the input image, the cuckoo search algorithm is applied to get optimal centroids, which are not possible by k-means alone. Now in the second stage, these optimal centroids act as cluster centers for the k-means clustering methods, which provide the best results for the diagnosis of breast tumor segmentation.

3.2.1 Algorithm steps of optimized k-means clustering

1. Reading the input image (mammogram breast scan)
2. For the random selection of the cuckoo, the new position of that cuckoo is calculated using Levy's Flight Eq. (3.1).
3. Using the global best position, the new solution is calculated by Eq. (3.3).
4. Iterating the steps until the condition gets satisfied at the end all the best positions are obtained.
5. Depending on the number of clusters (k) required and as it is a global minimum condition, k optimal centroids are treated as cluster centers for k-means.
6. All the data points are allocated to the cluster centers.
7. Reassigning the pixels depending on the minimum distance criteria.
8. Recalculating the cluster centers using Eq. (2.1).
9. Repeating the above two steps until cluster centers do not change.

3.2.2 Flowchart

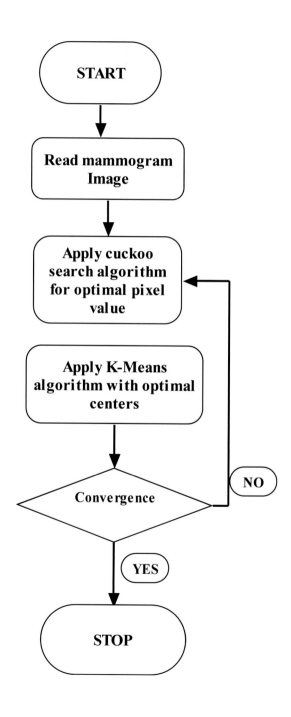

3.3 Optimized fuzzy *c*-means clustering algorithm

The hybrid approach of cuckoo search optimization and FCM clustering method is termed as "optimized fuzzy *c*-means." Though fuzzy *c*-means is good at overlapped dataset, its cluster centers are initialized randomly, which is a primary disadvantage that is removed by applying cuckoo search. As discussed in the optimized fuzzy *c*-means, there are two stages involved. Firstly, after reading, the input image cuckoo search algorithm is applied to get optimal centroids. Now, in the second stage, these optimal centroids act as cluster centers for the FCM clustering methods that help in the accurate segmentation of breast tumors.

3.3.1 Algorithm steps of OFCM clustering

1. Reading the input image (mammogram breast scan)
2. For the random selection of the cuckoo, the new position of that cuckoo is calculated using Levy's Flight Eq. (3.1).
3. Using the global best position, the new solution is calculated by Eq. (3.3).
4. Iterating the steps until the condition gets satisfied at the end all the best positions are obtained.
5. Depending on the number of clusters (k) required and as it is a global minimum condition, k optimal centroids are treated as cluster centers for k-means.
6. All the data points are allocated to the cluster centers.
7. Fuzzy membership is calculated for each data in the cluster using Eq. (2.4).
8. Centroids v_j are calculated for each cluster from Eq. (3.3).
9. Repeating the above two steps until the minimum j value is achieved.

3.3.2 Flowchart of OFCM clustering algorithm

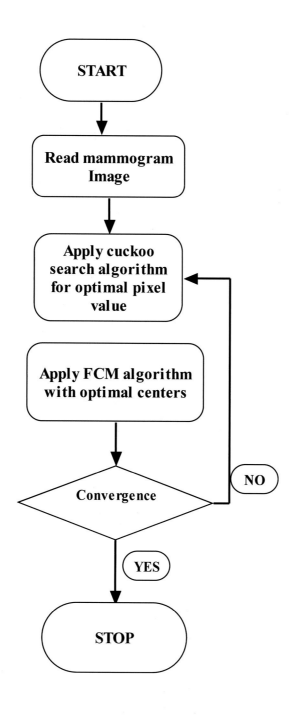

3.4 Postprocessing on optimized FCM clustering using level set method

The level set approach is a postprocessing technique. For image segmentation and analysis, level set approaches are effective numerical algorithms. The definition of a speed function that determines curve evolution is required for this procedure. The fundamental step of the level set is to explain the curves as zero-level set of the higher-dimensional hypersurface.

The suggested method handles noise and intensity in homogeneity while segmenting mammogram images. The novel hybrid model is having two stages: they are preprocessing and postprocessing of medical images. In the preprocessing stage, by considering the different regions of the input image, optimized fuzzy clusters are formed based on image intensities; based on the area of interest, one of the images of the cluster is selected and followed by postprocessing. Basically, at the postprocessing, level set curve evolution on the selected clustered image to lock the region of interest effectively by creating contour on it. The implementation of the level set method on optimized fuzzy c-means clustering and its energy minimization is explained in the following section to perfectly lock the tumor region in mammogram RIDER image scans.

3.4.1 Energy minimization

The segmented image, bias-corrected image, and bias field are obtained by energy minimization about level set function ϕ, the optimal weighting coefficients $\mathbf{w}$, and also cluster center $\mathbf{c}$. It is accomplished by a series of iterations [15−18].

The energy minimization of $F(\phi, w, c)$ concerning ϕ is derived by using the standard gradient flow method.

$$\frac{\partial \phi}{\partial t} = -\frac{\partial F}{\partial \phi} = -\delta(\phi)(\lambda_1 e_1 - \lambda_2 e_2) + \mu \text{div} d_p(|\nabla\phi|\nabla\phi) + \lambda\delta_e(\phi)\text{div}\left(\frac{\nabla\phi}{|\nabla\phi|}\right) \qquad (14.10)$$

where $\delta(\phi)$ is the Dirac function, which is the derivative of the Heaviside term and $d_p(S) \triangleq p'(s)/s$. The following is a representation of e_i's numerical calculation.

$$e_i(x) = I_{fcm}^2 1_k - 2c_i I_{fcm}(x)\left(w^T G * K\right) + c_i^2\left(\left(w^T G\right)^2 * K\right), i = 1, 2 \qquad (14.11)$$

by solving $\frac{\partial \mathscr{F}}{\partial w}$, we will get the energy minimization function, which is given as

$$\frac{\partial \mathscr{F}}{\partial w} = -2v + 2Aw \qquad (14.12)$$

where the equation $\frac{\partial \mathscr{F}}{\partial w} = 0$, $\mathbf{w}$ is given by

$$W = A^{(-1)}v \qquad (14.13)$$

where $\mathbf{v}$ denotes the column vector of M dimension represented by

$$v = \int\left(K * \left(I_{ofcm}(x)\sum_{i=1}^{2}\lambda_i c_i M_i(\varnothing)\right)\right)(y)G(y)dy \qquad (14.14)$$

where A is a $M \times M$ dimension matrix that can be written as

$$A = \int \left(K * \left(\sum_{i=1}^{2} \lambda_i c_i^2 M_i(\varnothing) \right) \right)(y) G(y) G^T(y) dy \qquad (14.15)$$

By fixing the level set method and weight coefficient **w**, the energy function of $\mathscr{F}$ is minimized by the most desirable c with

$$c_i = \frac{\int I_{fcm}(x) M_i(\phi(x)) \left(K * \left(w^T G \right) \right) dx}{\int M_i(\phi(x)) \left(K * \left(w^T G \right)^2 \right) dx}, i = 1, 2 \qquad (14.16)$$

4. Simulation results and discussions

To evaluate the performance of the tumor segmentation using clustering and optimized methods, we considered the RIDER dataset of mammogram scans. Using this, the tumor is segmented accurately which helps in further diagnosis. Hence segmentation results provide the area covered by the tumor accurately to some extent.

Fig. 14.9 depicts the simulation results of the conventional and proposed methods, respectively. The first row images show the original image, which is a mammogram scan taken from the RIDER dataset, and its ground truth images from left to right; conventional clustering methods such as k-means and FCM and their results in each image are depicted in the second row from left to right, respectively. Similarly, our proposed optimized k-means and optimized FCM using cuckoo search algorithm simulation results are depicted in the third row in Fig. 14.9 (from image 1 to image 10). In the above two optimized clustering techniques, optimized fuzzy c-means clustering depicted in the third row in Fig. 14.9 (from image 1 to image 10) has segmented breast tumors accurately over existing segmentation methods.

Fig. 14.10 shows the proposed optimized fuzzy c-means (OFCM) using level set method simulation results for breast tumor segmentation. This section assesses the accuracy, reliability, and robustness of the breast cancer segmentation results obtained using the proposed approach, the OFCM clustering algorithm with the level set method. The simulation results were tested on MATLAB-2022 software on Reference Image Database to Evaluate Therapy Response (RIDER), and this integrating approach is called OFCM clustering using the level set method to accurately segment tumors from mammogram MRI scans and simultaneously correct the inhomogeneous intensity and noise. The experimental results tested on RIDER database demonstrate the superiority of the proposed method over the most existing level set models in terms of accuracy and robustness to noise. By observing the qualitative and quantitative metric values, we conclude that optimized fuzzy c-means using level set method provides predominant results than existing level set segmentation methods such as Image and Vision Computing (IVC 2010) [15], Image and Vision Computing (IVC 2013) [16], and Expert System with application (ESA2021) [17], respectively.

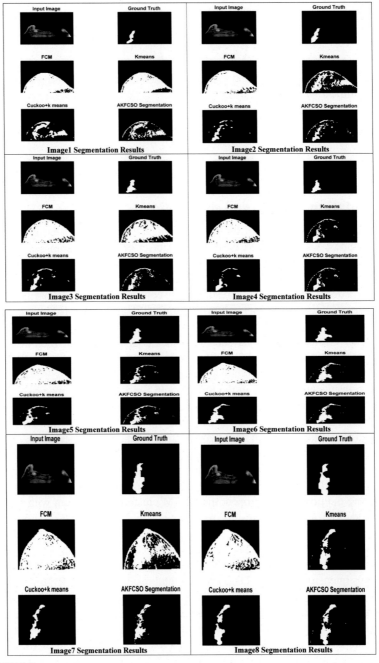

FIGURE 14.9

Segmentation results of the various mammogram image scans using conventional *k*-means, FCM, optimized *k*-means (OKMs), and optimized FCM (OFCM) clustering algorithms, respectively.

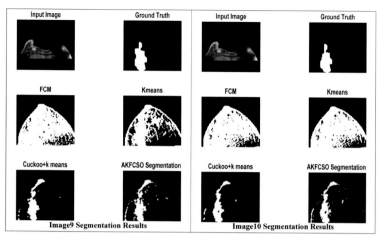

FIGURE 14.9 cont'd.

4.1 Performance analysis of *k*-means, FCM, optimized *k*-means, and optimized FCM

4.1.1 Metric analysis of k-means algorithm
See Table 14.1.

4.1.2 Metric analysis of fuzzy c-means algorithm
See Table 14.2.

4.1.3 Metric analysis of cuckoo with k-means algorithm
See Table 14.3.

4.1.4 Metric analysis of cuckoo with fuzzy c-means algorithm
See Table 14.4.

To understand the performance of the existing methods like k-means clustering and fuzzy c-means clustering and the proposed methods like optimized k-means (cuckoo + k-means) and optimized fuzzy c-means (cuckoo + fuzzy c means), in this work, we conducted many experiments to obtain various sets of mammogram scans from the RIDER dataset, out of which 10 of the results are depicted in Tables 14.1–14.4. By observing the above results for all the seven input scans, we found that, in the mammogram scans, the output of k-means and fuzzy c-means clustering can be compared; however, none of them gave accurate results of tumor detection. Compared with the ground truth values, the segmentation and detection of tumor failed with k-means and fuzzy c-means alone. Now, the same can be proved with the metric analysis for the above two existing methods.

Hence an optimization technique is taken to get optimal centroids for the above existing clustering methods. Now, combining this optimization technique with the clustering methods like k-means and fuzzy c-means gave better results than the

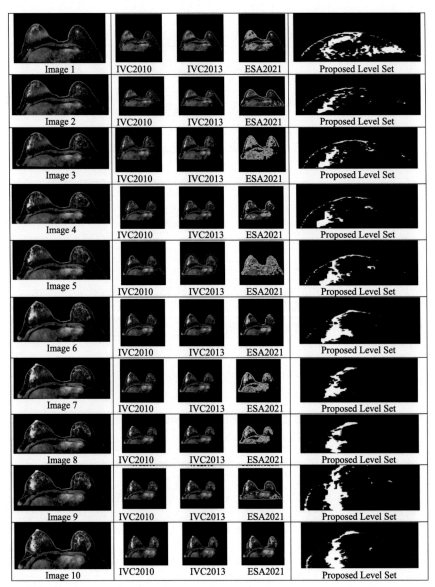

FIGURE 14.10

Simulation results of mammogram RIDER image scans based on optimized fuzzy
c-means (OFCMs) using level set method. The first column depicted in the figure shows
the RIDER input images and the second, third, and fourth column shows the simulation
results of existing level set methods and finally the last column shows the proposed
OFCMs using level set method simulation results for breast tumor segmentation.

Table 14.1 Performance of k-means clustering methods.

Input	nGT	nResult	q factor	Dice
Image 1	747	3397	0.5988	0.6780
Image 2	726	3367	0.5067	0.6656
Image 3	624	3547	0.5673	0.6543
Image 4	731	3361	0.6398	0.6412
Image 5	652	3283	0.6855	0.6477
Image 6	640	4184	0.5895	0.6388
Image 7	486	4000	0.5435	0.6399
Image 8	476	3988	0.5355	0.6500
Image 9	403	3790	0.6670	0.6709
Image 10	400	3670	0.6578	0.6904

Table 14.2 Performance of fuzzy c-means clustering methods.

Input	nGT	nResult	q factor	Dice
Image 1	747	700	0.5581	0.8780
Image 2	726	3445	0.4621	0.7656
Image 3	624	2510	0.5146	0.8343
Image 4	731	2289	0.5835	0.7412
Image 5	652	845	0.6077	0.8467
Image 6	640	4059	0.5100	0.7302
Image 7	486	3995	0.4985	0.6899
Image 8	476	3800	0.4355	0.6900
Image 9	403	3689	0.4670	0.7809
Image 10	400	3545	0.4588	0.7940

Table 14.3 Performance of cuckoo + k-means optimized method.

Input	nGT	nResult	q factor	Dice
Image 1	747	664	0.5581	0.9411
Image 2	726	650	0.2156	0.9447
Image 3	624	572	0.2345	0.9565
Image 4	731	685	0.2937	0.9675
Image 5	652	605	0.5877	0.9626
Image 6	640	772	0.4572	0.9065
Image 7	486	735	0.3170	0.7960
Image 8	476	712	0.3089	0.8013
Image 9	403	696	0.2970	0.7333
Image 10	400	677	0.3088	0.7428

Table 14.4 Performance of cuckoo + fuzzy c-means optimized method.

Input	nGT	nResult	q factor	Dice
Image 1	747	702	0.2181	0.9689
Image 2	726	704	0.2107	0.9846
Image 3	624	699	0.1714	0.9433
Image 4	731	759	0.2087	0.9812
Image 5	652	631	0.1986	0.9836
Image 6	640	712	0.1525	0.9467
Image 7	486	694	0.1170	0.8237
Image 8	476	559	0.1989	0.9198
Image 9	403	450	0.1770	0.9449
Image 10	400	432	0.1508	0.9615

two existing methods. Out of the two optimized methods, optimized fuzzy c-means gave accurate results in comparison with the other three methods. From the above metric table, we can conclude that nGT and nResult are almost comparable in the case of optimized fuzzy c, where nGT is the number of pixels covered by the ground truth and nResult is the number of pixels covered by the particular method; q factor indicates the degree of loss, which is very less, and the dice value represents the similarity between the ground truth and the result of the particular method, which is high in case of the optimized fuzzy c means. So, we can clearly conclude that optimized fuzzy c-means performs accurately rather than the other three methods mentioned.

5. Conclusion

This research work proposed a novel segmentation method for breast tumor detection by assimilating the cuckoo search optimization technique in addition to the clustering methods called k-means and fuzzy c-means. The centroids obtained from this cuckoo search are optimal; hence, the random initialization in the k-means and fuzzy c-means is avoided by this optimization technique included in the preprocessing stage. Inculcating cuckoo with the k-means is termed as "optimized k-means" and similarly with the fuzzy c-means is termed as "optimized fuzzy c-means." After the qualitative and quantitative analysis of both optimized methods, we concluded that optimized fuzzy c-means outperforms the breast tumor segmentation accurately from the RIDER mammogram datasets, but few boundary leakages occur even though optimization is used. To overcome this problem, postprocessing is used to refine and lock the region of interest in the mammogram images by using the level set method, which is effectively a control the boundary leakages from the optimized FCM clustering results. This integrating approach is called OFCM clustering, using the level set method to accurately segment tumors from mammogram MRI scans and simultaneously correct the inhomogeneous intensity and noise. The experimental results tested on the RIDER database demonstrate the superiority of the proposed method over the most existing level set models in terms of accuracy and robustness to noise. By observing the qualitative and quantitative metric values, we conclude that optimized fuzzy c-means using the level set method provides the predominant results than existing segmentation methods.

References

[1] American Cancer Society. Global cancer facts and figures. Atlanta: American Cancer Society; 2018. p. 1−76. no. 4, http://www.cancer.org/content/dam/cancer-org/research/cancer-facts-and-statistics/global-cancer-facts-and-figures/global-cancer-facts-and-figures-4th-edition.pdf.

[2] Singh L, Jaffery ZA, Zaheeruddin Z, Singh R. Segmentation and characterization of breast tumor in mammograms. In: 2010 international conference on advances in recent technologies in communication and computing; 2010. p. 213−6. Kottayam, India.

[3] Bray F, Ferlay J, Soerjomataram I, Siegel RL, Torre LA, Jemal A. Global cancer statistics 2018: globocan estimates of incidence and mortality worldwide for 36 cancers in 185 countries. CA: A Cancer Journal for Clinicians 2018;68(6):394−424.

[4] Makandar A, Halalli B. Threshold based segmentation technique for mass detection in mammography. Journal of Computers 2016;11(6):472−8.

[5] Michael E, He M, Hong L, Frank K, Jing L. Breast cancer segmentation methods: current status and future potentials. BioMed Research International 2021;2021. https://doi.org/10.1155/2021/9962109. 29 pages.

[6] Punitha S, Ravi S, Anousouya Devi M. Breast cancer detection in digital mammograms using segmentation techniques. International Journal of Control Theory and Applications 2016;9(3):167−82.

[7] Arokiyamary Delphia A, Kamarasan M, Sathiamoorthy S. Image processing for identification of breast cancer: a literature survey. Asian Journal of Electrical Sciences 2018;7(2):28−37.

[8] Lou JY, Yang XL, Cao AZ. A spatial shape constrained clustering method for mammographic mass segmentation. Computational and Mathematical Methods in Medicine 2015;2015. 16 pages.

[9] Hassan Shayma'a A, Sayed MS, Abdalla MI, Rashwan MA. Detection of breast cancer mass using mser detector and features matching. Multimedia Tools and Applications 2019;78(14):20239−62.

[10] de Oliveira HN, de Avelar CS, Machado AMC, de Albuquerque Araujo A, dos Santos JA. Exploring deepbased approaches for semantic segmentation of mammographic images. In: Progress in pattern recognition, image analysis, computer vision, and applications. Cham: Springer; 2018. p. 690−8.

[11] Ball JE, Butler TW, Bruce LM. Towards automated segmentation and classification of masses in digital mammograms. In: The 26th annual international conference of the IEEE engineering in medicine and biology society; 2004. p. 1814−7. San Francisco, CA, USA.

[12] Oliver A, Freixenet J, Marti J, et al. A review of automatic mass detection and segmentation in mammographic images. Medical Image Analysis 2010;14(2):87−110.

[13] Palkar P, Agrawal P. A technique to extract statistical parameters of digital mammogram to detect breast cancer. International Journal of Advanced Research in Science, Engineering and Technology 2016;3(12):3033—8.

[14] Zhang X, Weng G. Level set evolution driven by optimized area energy term for image segmentation. Optik 2018;168:517—32.

[15] Zhang K, Zhang L, Song H, et al. Active contours with selective local or global segmentation: a new formulation and level set method. Image and Vision Computing 2010; 28(4):668—76.

[16] Dong F, Zengsi C, Wang J. A new level set method for inhomogeneous image segmentation. Image and Vision Computing 2013;31:809—22. https://doi.org/10.1016/j.imavis.2013.08.003.

[17] Weng G, Dong B, Lei Y. A level set method based on additive bias correction for image segmentation. Expert Systems with Applications 2021;185:115633. https://doi.org/10.1016/j.eswa.2021.115633. ISSN 0957-4174.

[18] Srinivas A, Prasad VVKDV, Leela Kumari B. Computer-aided diagnosis for early cancer detection using adaptive Kernel based fuzzy cuckoo search optimization clustering from mammogram images. Computers & Electrical Engineering 2022;103:108343. ISSN 0045-7906.

Further reading

[1] Mustra M, Grgic M, Rangayyan RM. Review of recent advances in segmentation of the breast boundary and the pectoral muscle in mammograms. Medical, & Biological Engineering & Computing 2016;54:1003—24. no. 7.

Index

323